BINARY MATHEMATICS

PART 1

Using Simple Symbols

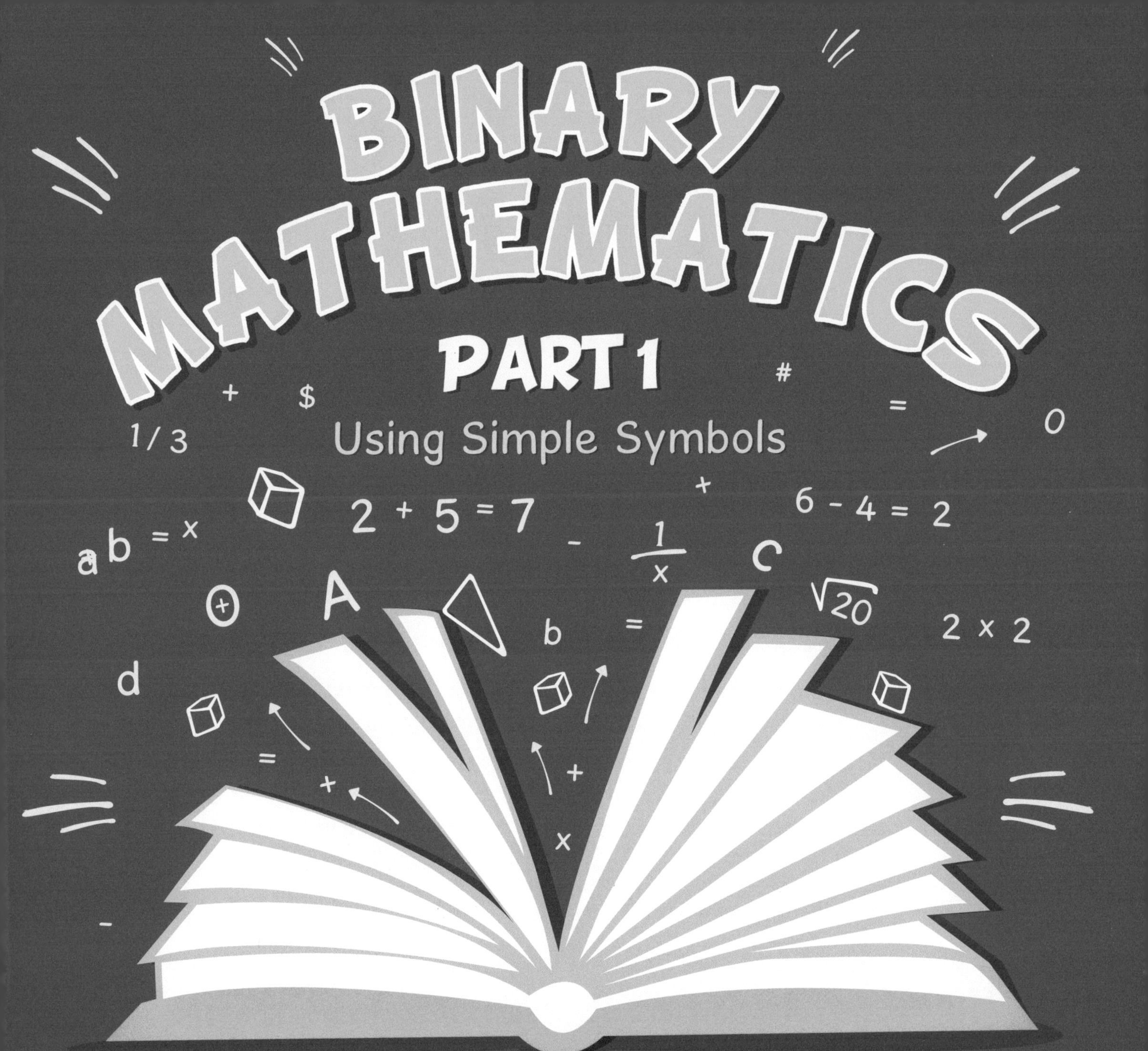

Chester Litvin

Order this book online at www.trafford.com
or email orders@trafford.com

Most Trafford titles are also available at major online book retailers.

www.trafford.com
North America & international
toll-free: 1 888 232 4444 (USA & Canada)
fax: 812 355 4082

Our mission is to efficiently provide the world's finest, most comprehensive book publishing service, enabling every author to experience success. To find out how to publish your book, your way, and have it available worldwide, visit us online at www.trafford.com

ISBN: 978-1-4907-9689-5 (sc)
ISBN: 978-1-4907-9688-8 (e)

Print information available on the last page.

Trafford rev. 12/21/2019

From Author

Welcome to mathematics by using simple symbols. Variety of simple symbols can be visual, audio, kinesthetic, olfactory, tactile, musical and etc. The reason that I developed simple symbols approach was fallowing. In the past, I observed students that have difficulty with decimal calculation and regular grammar and wanted to help them. I saw that they done much better when they used binary mathematics and binary grammar. For example, my dyslexic students that, switched to the simple symbols, were having fewer problems with math and grammar. The average grades students that used binary approach also improved in math and grammar. The autistic and hyperactive students, who used binary system were more focused and were able to do better with their school assignments. After working with binary arithmetic they were more effectively used decimal math and with grammar by using regular alphabet. To have process of learning more effective, I was introducing mixture of symbols of diverse modalities.

Psychoconduction is approach of doing mathematics by using simple symbols. I was able to build a system by using simple symbols in many modes of expressions that included visual, audio, and kinesthetic and etc. In my book I explained how to translate simple symbols to variety modes of expression.

I understand that the increased ability of brain by solving the math problems and using grammar has more complex bases. I patented my discovery and called "Psychocondaction", the neurological process of brain restoration. My US patent is in details explaining my hypothesis. In my patent I have listed the researches in the brain stimulation that were using sensory areas.

To correctly identify areas of the brain affected by psychoconduction I done some research. I just mentioned the approximate areas of the brain. The future steps that would prove my hypothesis of brain restoration can be MRI or many other controlling methods to monitor the changes in brain structure. My subjects did great and I am sure that my hypothesis worked. When I was doing the problems with simple symbols, I was enjoying the new approach of doing mathematics and grammar. Try it and see if somehow it would help you.

With respect,

Chester Litvin, PhD

Psychologist

Dedicated to my nephew, David Gimelfarb,
Lost in Costa Rico in 2009

Chester Litvin, Ph.D., Clinical Psychologist

How to increase learning - the psychological stimulation of the brain

by method of psychoconduction.

If you were called lazy or stupid and have problems with your memory, concentration, decision making, it can be because your brain is not stimulated enough to process complex information. We are using the new approach for non-invasive brain stimulation to wake up your sleeping brain. By using only the simple symbols we are significantly simplifying the identification of the information and capacity to manipulate it. The process of translating the same information to visual, audio, kinesthetic, tactile, olfactory and etc. modes of expression we call psychoconduction. In this book we explore the concept of positions, as part of binary arithmetic, and translation of the symbol. The different forms, which contains the translated information called the Litvin's Code. In future publications we hope to explore the mathematic and grammar by using psychoconduction and Litvin's Code instead of recognition of numbers and letters by contours.

The task to keep information in most simple form is always a struggle. In our discovery we are not changing any information, but somehow we want to change the structure of the brain's cells to process information. It is not a problem anymore to have simple representation of information. The complex information can become simple and acceptable by simple brain cells. For new process of information we are substituting the complex cells by simple brain cells. Our goal is to change the way how the information is carried but is not changing the information itself. We are providing opportunities to use simple brain cells instead of damaged or not effective complex cells. The psychoconduction is the process of educating the simple cell to act as a complex. Our work was done with the brain cells but in future can be done with other body cells.

By translating the complex information to the simple form we are educating the simple brain cells. We can use the simple form to transmit the complex information. The goal is to stimulate the simple cells to increase learning. The stimulated brain can learn how to process information in more efficient way. In our work we enhance and make more effective the previous not sufficient ways of processing the information. We stimulate brain by various combinations of positions, which are empty or filled up with simple symbol.

The wide brain stimulation with simple symbols is an important part of learning. When we are disregarding the stimulation of different part of the brain, we are processing information through the visual stimuli and putting the big emphasis on the visual faculty. The educational contribution of psychoconduction is to widen opportunity of the brain stimulation. When we are using memory of entire brain, we are enhancing the learning, increasing the attention and concentration. By using psychoconduction we are translating visual, audio and kinesthetic stimuli between various modes of expression.

The psychoconduction is bringing the fun to the learning process. Some. have difficulty processing video information, but others are more impaired in processing of audio or kinesthetic. During the translation of information between different modes we are using the empty and full positions, and we are not changing any content of learning. We are using the same logical system for different perceptions. The result is very promising. The students, with problems in learning, from the very young to the old age, were learning to process information with the different levels of complexity. As a result, the youngsters, later in life did not drop out of the high school and all of them continued the education in the college. A student with severe dyslexia, despite all predictions, has not dropped from the school and completed her education in the university. The elderly people with dementia got increased recall of information.

The psychoconduction helped variety of students to increase their academic performance and increase attention, concentration, also ability to focus and to complete tasks. It provides the easy translation to visual, audio, kinesthetic mode, the automatic use of groups, combinations, and permutation and factorial representations. The binary sequence represented by empty and full positions and the content of the position depends on the sequenced number of position, filled up with symbol. The combination part of psychoconduction is the empty and filled up positions joint together and uniquely represents the numbers and the letters.

The content of empty position is always zero and does not change. In binary arithmetic the sequential numbers of position are starting with one with the increment by one for the next position and ext. It could be unlimited number of positions in the sequence. The content of the filled up position is a binary number, which depends on the sequential number of the position. The content of the first positions start with 0, next position will be one, then the next position twice bigger and has content equal 2, then the next position is twice bigger and has content of four, and again the next position has content of eight and etc. To arrive to the needed number or letter we are adding contents only of filled up positions, because the empty position has value of zero.

The little children without any difficulties are memorizing the combination of empty and full position, which could be represented by the audio, video, kinesthetic, olfactory, music and tactile modes of expression. The best way for little children is to use kinesthetic or tactile representation of the positions, because the learning for them also is the fun. The brain is easily processing the patterns of simple symbol that in audio, visual kinesthetic and olfactory modes has the same meaning. By using psychoconduction the complex information which people with learning disabilities had difficult and sometimes were not able to process at all, became very simple.

The psychoconduction allows us:

- understand the notion of ascending direction and the quick recognition of information.

- understand the relationship between empty, and filled position and we are simplifying the process of addition, subtraction, division and multiplication.

- have logical connection between empty and filled up positions and to spell words and sentences.

When the brain is not performing adequately due to physiological limitation, the process of the recognition of the complex spatial information is very complicated and confusing. Intuitively, we can use approximation by contours. The approximation helps us to make a guess. The guess could be right or wrong. We need to recognize the tremendous effort of the brain to bring us to this point.

By using psychoconduction we make the process of recognition less painful and less time consuming. We don't look for matching of spatial image through complicated scanning. We are providing constant numerical reference by checking if the position is filled or empty. We are successfully training the brain to overcome weakness in the processing of information. The psychoconduction protects the brain against unnecessary stress of recognition and allows focusing on the process. We tune areas of the brain against the other areas to achieve the congruency in the responses.

In this book the psychoconduction utilizes the notion of positions as part of binary arithmetic, which is completely different from memorizing contours. When the symbol is in the different position then the meaning is changing. It allows the person to have a stable relationship with the simplest symbol, instead of complicated contours. We also are providing the mathematical references. The psychoconduction allows to process information through audio, visual, kinesthetic and tactile expressions.

CONTENTS

Green Lessons

MATHEMATICS

GREEN LESSONS, LESSON 1, FIGURE 1

In Figure 1, the ***visual*** display has two pictures with boxes. The pictures have different audio representations and different visual displays. The audio representation of a knock represents a filled up position and a double knock represents an empty position. In the first picture we have two empty spaces and the audio representations are **double knock and double knock**. In the second picture, the first position is filled up with a symbol and the second position is empty.

The ***audio*** representation for the second picture consists of a **knock and double knock**.

Figure 1

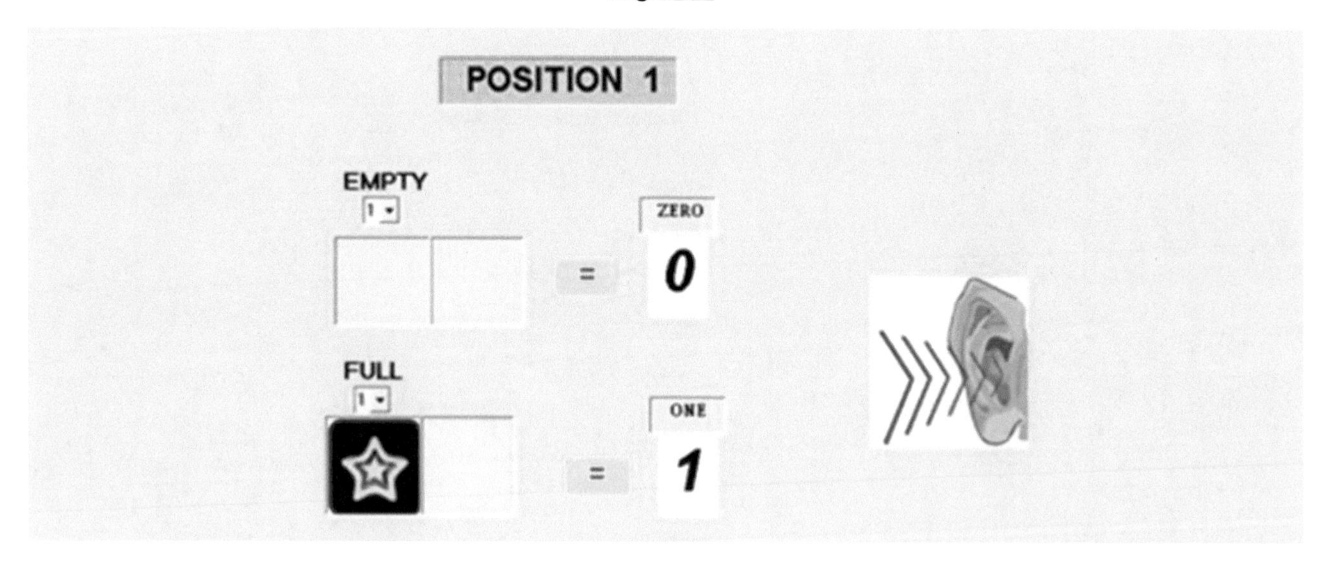

The ***kinesthetic*** representation of a filled up position is the clenching of the right hand. An empty position is represented by the clenching of the left hand. The kinesthetic representation for the first picture is the clenching of the left hand twice which is equal to the value of zero. The kinesthetic representation of the second picture is the clenching of the right hand, then the left hand. This signifies a mathematical value equal to one.

GREEN LESSONS, LESSON 1, FIGURE 2

In Figure 2, we have a ***visual*** display of two pictures with boxes. Both pictures have the same audio representations. The figures have different symbols in the filled positions. In the first picture the symbol is a star and in the second picture, a red ball. The shape of the symbols does not change the mathematical meaning and the pictures correspond to the same audio signals. In both pictures the first position is filled with a symbol and the second position is empty.

The ***audio*** representation for both pictures consists of a **knock and double knock**, where a knock represents a filled up position and a double knock represents an empty position.

Figure 2

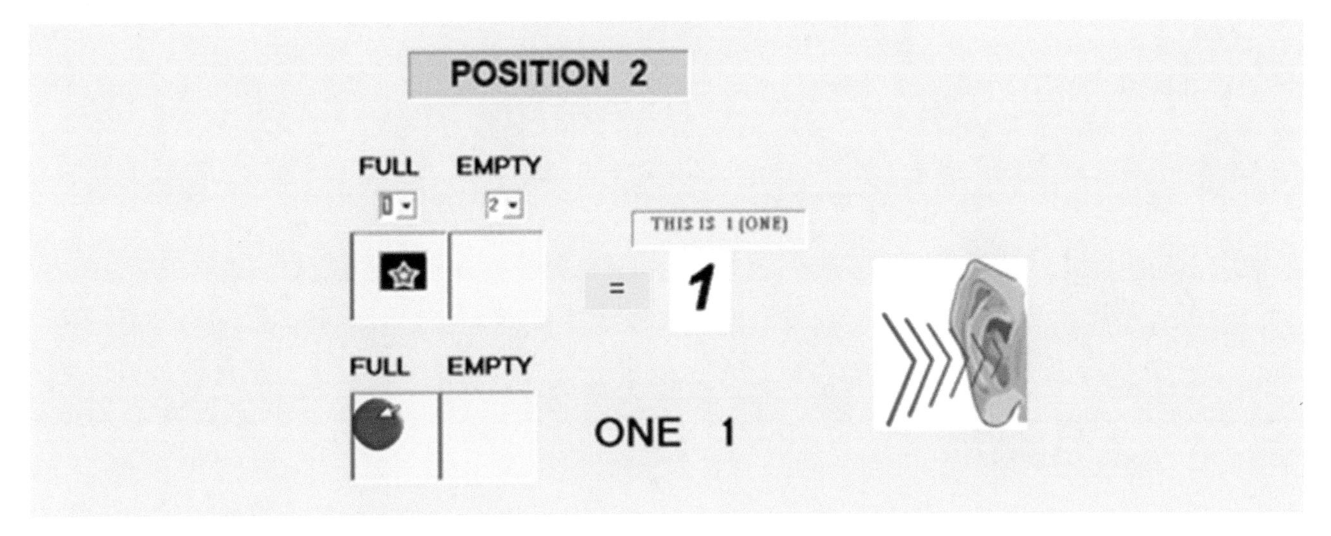

The ***kinesthetic*** representation does not depend on the shapes of the symbols. In our examples a filled up position is announced by the clenching of the right hand. An empty position is represented by the clenching of the left hand. The kinesthetic representation for both pictures is the clenching of the right hand once and the left hand once which is equal to a value of one.

MATHEMATICS

GREEN LESSONS, LESSON 1, FIGURE 3

In Figure 3, we have a ***visual*** display of two pictures with two positions each. Both pictures have the same audio representations. The figures have different symbols in filled positions. In the first picture the second position is filled with two stars, and in second picture, a red ball. The shape and the number of symbols in the second position do not change the mathematical meaning.

The ***audio*** representation for both pictures consists of a **double knock and knock**, where a double knock represents an empty position and a knock represents a filled up position.

Figure 3

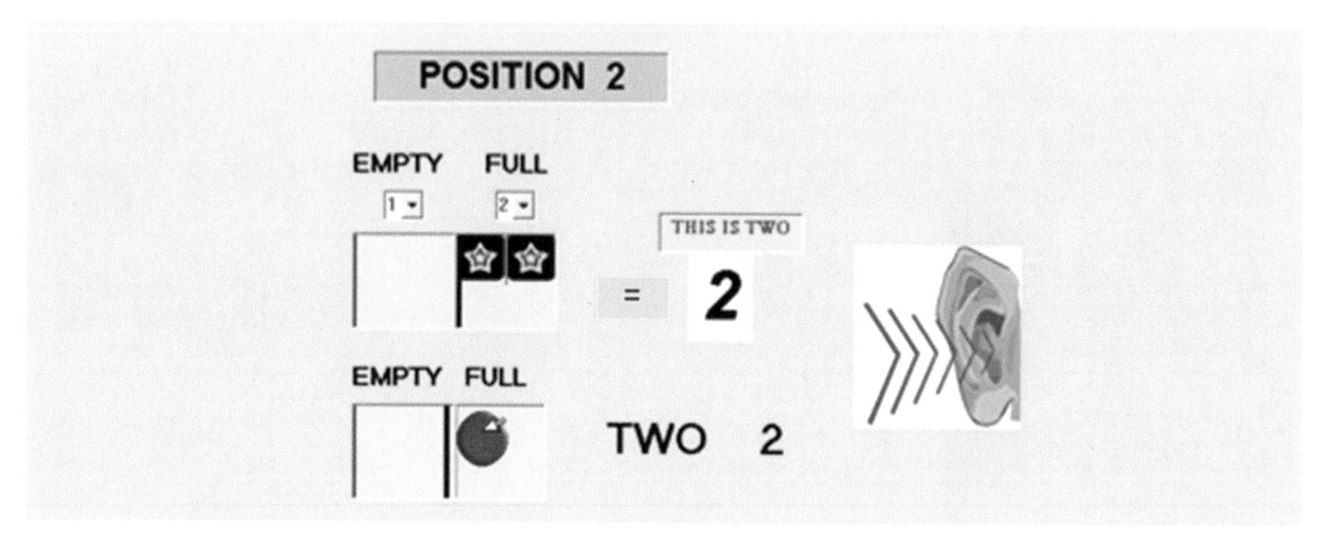

The ***kinesthetic*** representation of both figures does not depend on the shape or the number of the symbols in filled positions. In both pictures the filled position is represented by the clenching of the right hand. The empty position is represented by the clenching of the left hand. The kinesthetic representation for both pictures is the clenching of the left hand once and the right hand once, which is equal to a value of two.

GREEN LESSONS, LESSON 1, FIGURE 4

In Figure 4, we have a ***visual*** display of two pictures with two positions each. Both pictures have the same audio representations. In both pictures the two positions are filled with symbols. The first picture has stars in the filled positions and the second picture has red balls. In the first picture we have two positions filled with three stars. The shape and the number of the symbols in the picture do not change the mathematical meaning and correspond to the same audio signal. The ***audio*** representation for both pictures consists of **knock and knock**, where a knock is represents a filled position.

Figure 4

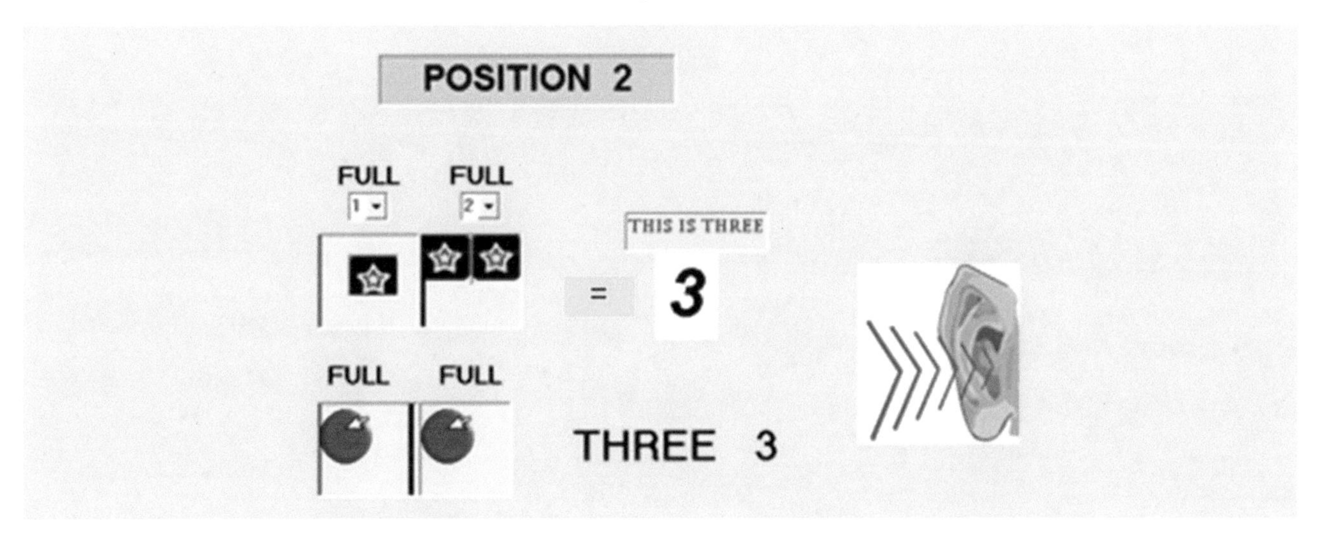

The ***kinesthetic*** representation of both figures does not depend on the shape and the number of the symbols in the filled positions. In both pictures the filled up position is represented by the clenching of the right hand twice, which is equal to a value of three.

MATHEMATICS GREEN LESSONS, LESSON 1, FIGURE 5

In Figure 5, we have a ***visual*** display of two pictures with three positions each. Both pictures have the same audio representations. The figures have different symbols in filled positions. There are four stars in the third position of the first picture and a red ball in the third position of the second picture. The shape and the number of the symbols in the third position of both pictures do not change the mathematical value. In both pictures the first two positions are empty and the third position is filled with a symbol.

The ***audio*** representation for both pictures consists of a **double knock, double knock and knock**, where double knock represents an empty position and knock represents a filled up position.

Figure 5

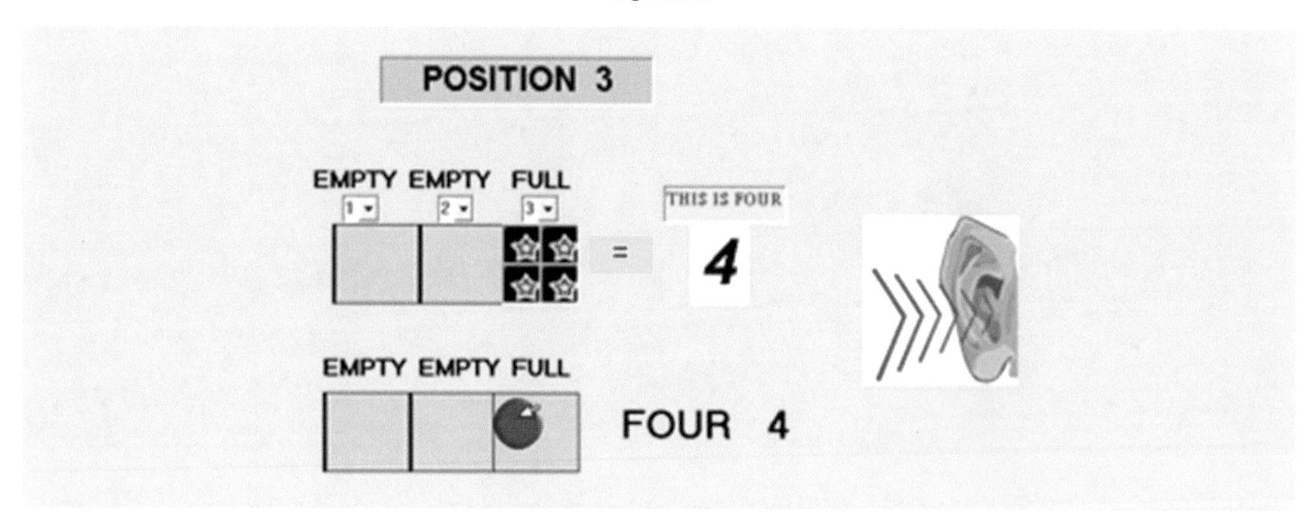

The ***kinesthetic*** representation of both figures does not depend on the shape and the number of the symbols in filled position. In both pictures the filled position three is represented by the clenching of the right hand. An empty position is represented by the clenching of the left hand. The **kinesthetic** representation for both pictures is the clenching of the left hand twice and the right hand once, which is equal to a value of four.

GREEN LESSONS, LESSON 1, FIGURE 6

In Figure 6, we have a ***visual*** display of two pictures with three positions each. Both pictures have the same audio representations. Positions one and three are filled with symbols. The first and third positions in the first picture are filled with stars and in the second picture, with red balls. In the first picture, the number of filled stars is equal to five. The shape and the number of symbols in the first and third positions of both pictures do not change the mathematical value.

The ***audio*** representation for both pictures consists of a **knock, double knock and knock**, where a knock represents a filled up position and a double knock represents an empty position.

Figure 6

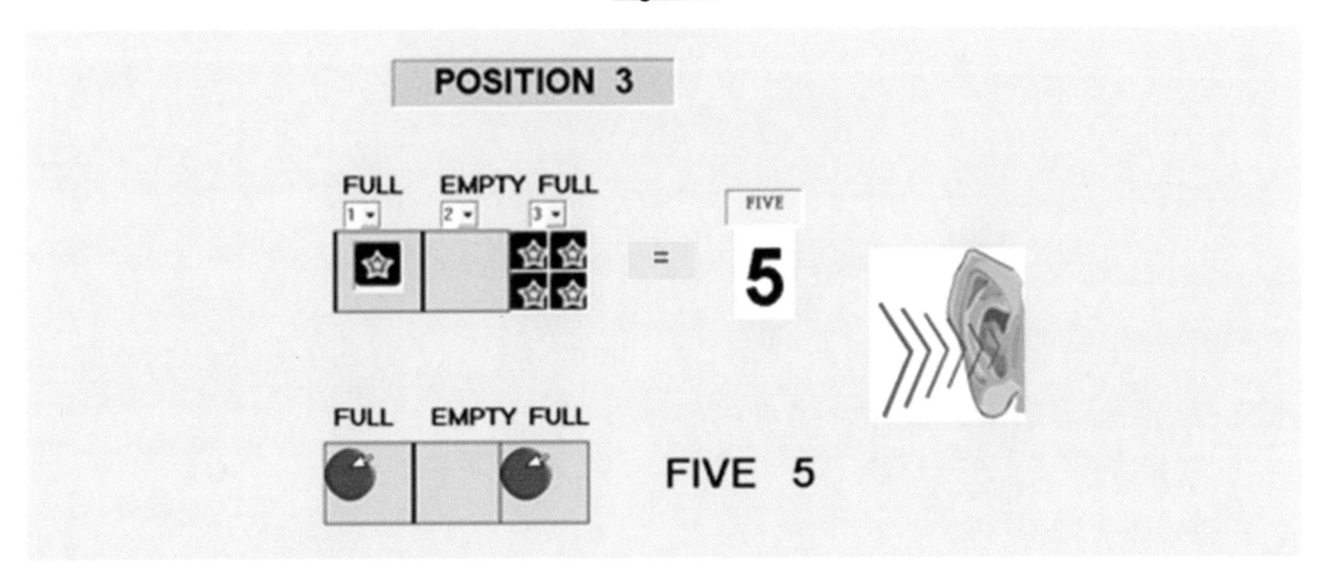

The ***kinesthetic*** representation of both figures does not depend on the shape or the number of symbols in the filled positions. In both pictures the first and third filled positions are represented by the clenching of the right hand. The empty position is represented by the clenching of the left hand. The kinesthetic representation for both pictures is the clenching of the right hand, left hand and right hand, which is equal to a value of four.

MATHEMATICS

GREEN LESSONS, LESSON 1, FIGURE 7

In Figure 7, we have a ***visual*** display of two pictures with three positions each. Both pictures have the same audio representations. The second and third positions in the first pictures are filled with stars and the second and third positions in the second picture are filled with red balls. The shape and the number of symbols in second and third positions of both pictures do not change the mathematical meaning and have the same audio signals. In both pictures the first position is empty and the second and third positions are filled with a symbol.

The ***audio*** representation for both pictures consists of a **double knock, knocks and knock**, where a double knock represents an empty position and a knock represents a filled up position.

Figure 7

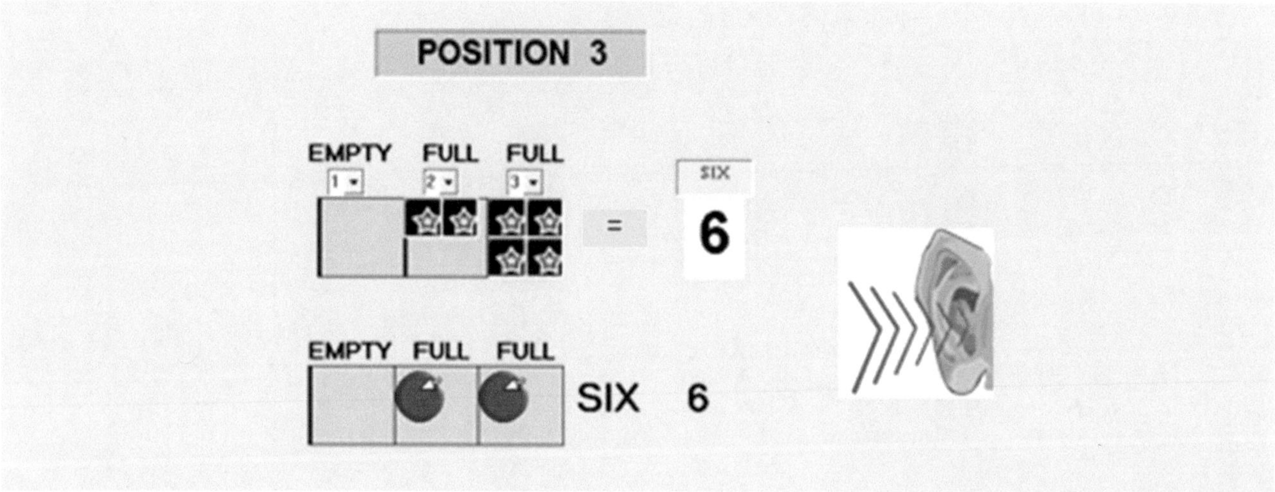

POSITION 3

EMPTY FULL FULL

1 2 3

= SIX 6

EMPTY FULL FULL

SIX 6

The ***kinesthetic*** representation of both figures does not depend on the shape or the number of the symbols. In both pictures, the second and third filled positions are represented by the clenching of the right hand twice. The empty position is represented by the clenching of the left hand. The kinesthetic representation for both pictures is the clenching of the left hand once and the right hand twice, which is equal to a value of six.

GREEN LESSONS, LESSON 1, FIGURE 8

In Figure 8, we have a ***visual*** display of two pictures with three positions each. Both pictures have the same audio representations. In both pictures all three positions are filled with symbols. The first picture is filled with seven stars and the second picture is filled with red balls. When we count the amount of the stars in the first picture, the number equals the mathematical value of the image. The shape and the number of the symbols in both pictures do not change the mathematical value. The ***audio*** representation for both pictures consists of **knock, knock and knocks**, where a knock represents a filled up position.

Figure 8

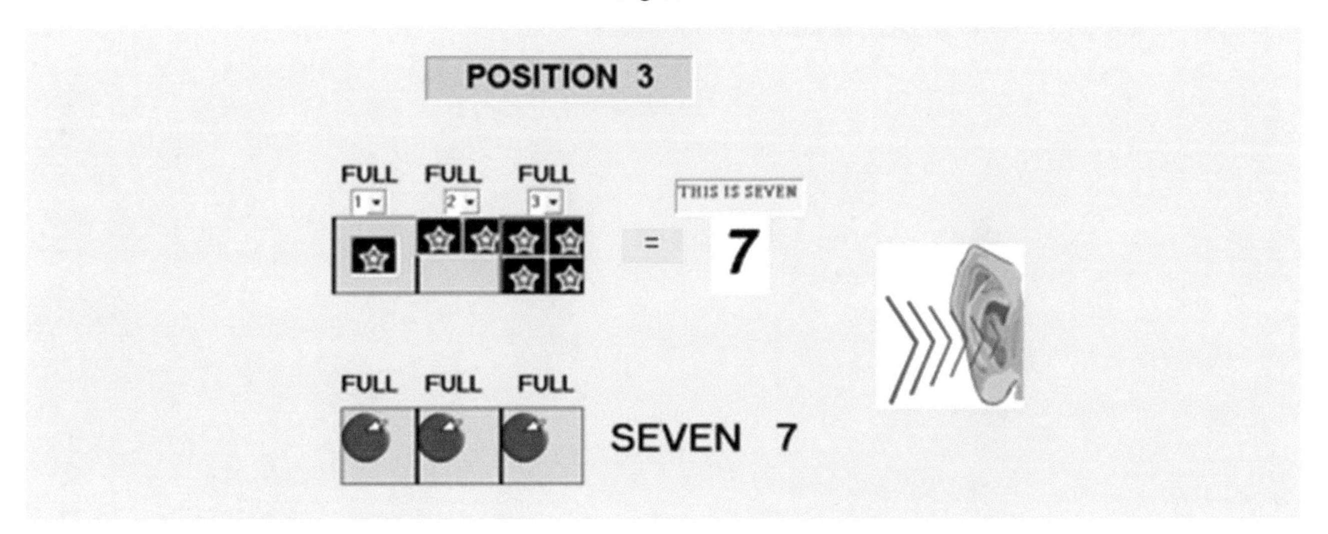

The ***kinesthetic*** representation of both figures does not depend on the shape or the number of the symbols in the filled positions. In both pictures the filled positions are represented by the clenching of the right hand three times, which is equal to a value of seven.

MATHEMATICS GREEN LESSONS, LESSON 1, FIGURE 9

In Figure 9, we have a ***visual*** display of two pictures with four positions each. The pictures have two different audio representations. The figures correspond to the visual representation of different numbers which are the combination of filled up and empty positions.

The ***audio*** representation of filled up positions is a knock. A double knock represents an empty position. In the first picture, we have four empty spaces and the audio illustration is **double knock, double knock, double knock and double knock**. In the second picture, only the fourth position is filled with a symbol and the positions one, two and three are empty. The audio representation for the second picture consists of a **double knock, double knock, double knock and knock**.

Figure 9

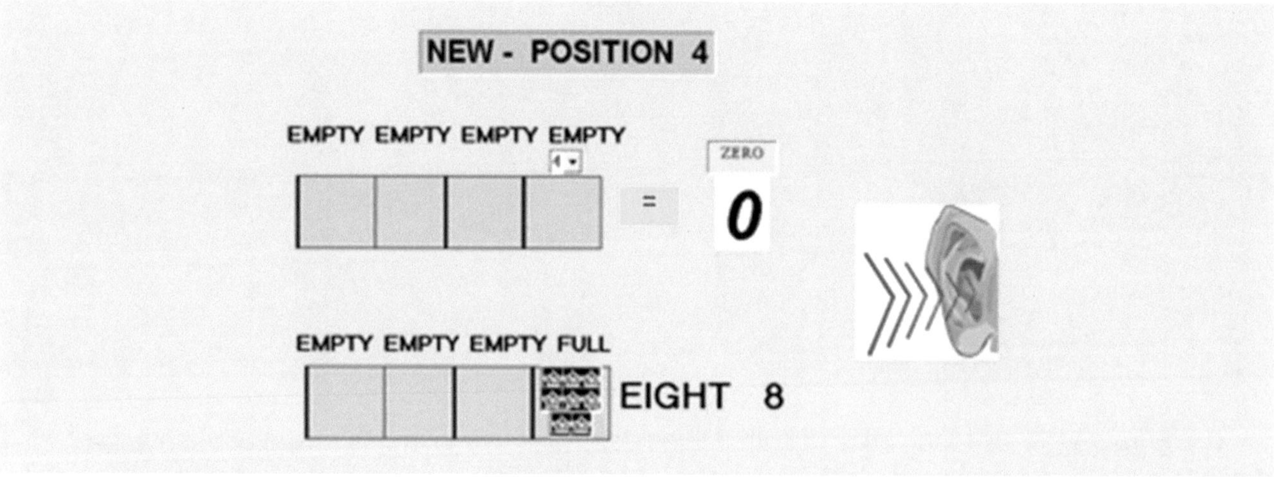

NEW - POSITION 4

EMPTY EMPTY EMPTY EMPTY

ZERO

= 0

EMPTY EMPTY EMPTY FULL

EIGHT 8

The ***kinesthetic*** representation of a filled up position is represented by the clenching of the right hand. An empty position is represented by the clenching of the left hand. The kinesthetic representation for the first picture is the clenching of the left hand four times, which is equal to a value of zero. The kinesthetic representation of the second picture is the clenching of the left hand three times, then the right hand once, which signifies a mathematical value equal to eight.

GREEN LESSONS, LESSON 1, FIGURE 10

In Figure 10, we have a ***visual*** display of two pictures with four positions each. The fourth position of the first picture is filled with eight balls and the fourth position of the second picture is filled with a red ball. The shape and the number of the symbols in fourth position do not change the mathematical meaning. In both pictures the first three positions are empty and the fourth position is filled with a symbol.

The ***audio*** representation for both pictures consists of a **double knock, double knock, double knock, and knock**, where a double knock represents an empty position and a knock represents a filled position.

Figure 10

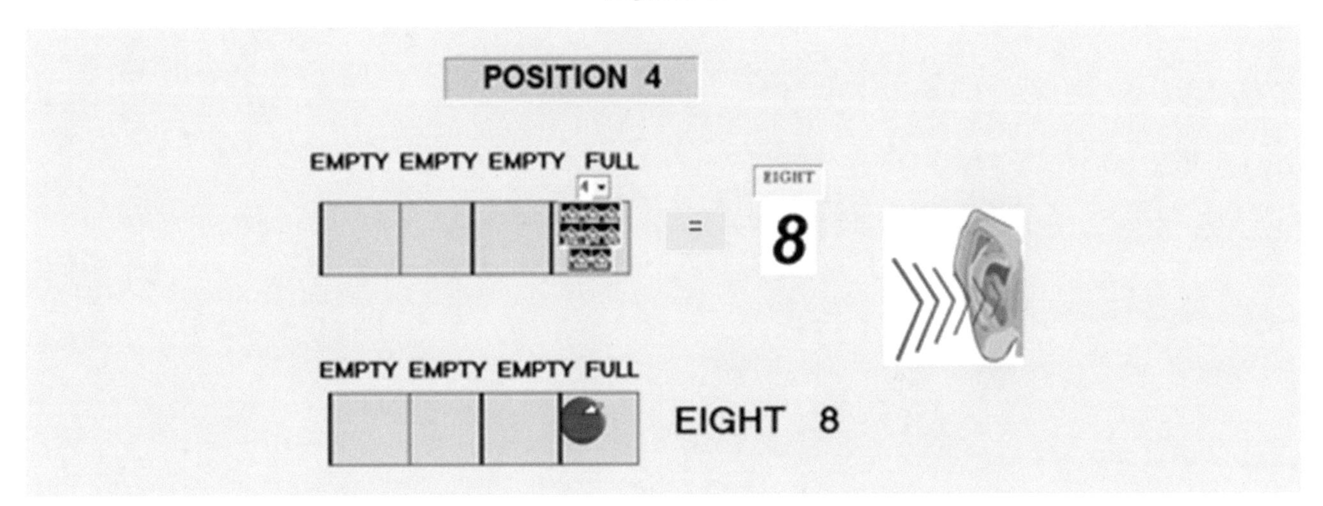

The ***kinesthetic*** representation of both figures does not depend on the shape or the number of the symbols in the filled position. In both pictures the filled position is represented by the clenching of the right hand. An empty position is represented by the clenching of the left hand. The kinesthetic representation for both pictures is the clenching of the left hand three times and the right hand once, which is equal to a value of eight.

MATHEMATICS GREEN LESSONS, LESSON 1, FIGURE 11

In Figure 11, we have a ***visual*** display of two pictures with four positions each. Both pictures have the same audio representations. In both pictures, positions one and four are filled with symbols. The first picture has one star in the first and eight stars in the fourth positions. The second picture has red balls in both first and fourth positions. The shape and the number of symbols in the first and forth positions of both pictures do not change the mathematical meaning. The ***audio*** representation for both pictures consists of a **knock, double knock, double knock and knock**, where a knock represents a filled up position and a double knock represents an empty position

Figure 11

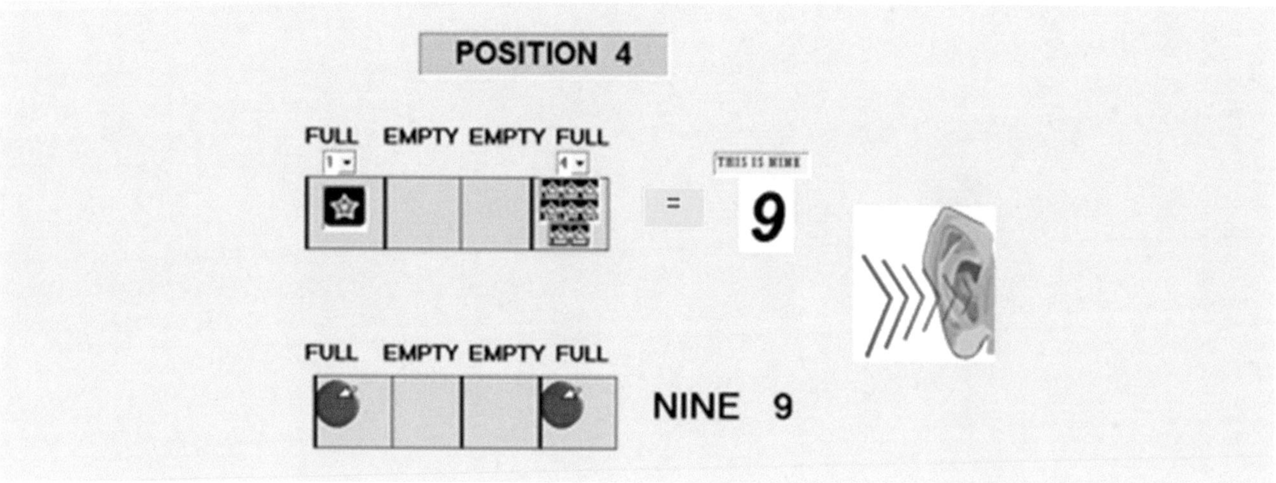

POSITION 4

FULL EMPTY EMPTY FULL

= THIS IS NINE 9

FULL EMPTY EMPTY FULL

NINE 9

The ***kinesthetic*** representation of both figures does not depend on the shape or the number of the symbols in filled positions. In both pictures the filled first and fourth positions are represented by the clenching of the right hand. An empty position is represented by the clenching of the left hand. The kinesthetic representation for both pictures is the clenching of the right hand, then left hand twice and the right hand once, which is equal to a value of nine.

GREEN LESSONS, LESSON 1, FIGURE 12

In Figure 12, we have a ***visual*** display of two pictures with four positions each. Both pictures have the same audio representations. The second and fourth positions in the first picture are filled with two and eight stars, which is equal to a mathematical value of ten. The second and fourth positions of the second picture are filled with red balls. The shape and the number of the symbols in the second and forth positions of both pictures do not change the mathematical meaning. The ***audio*** representation for both pictures consists of a **double knock, knock, double knock and knock**, where a knock represents a filled up position and a double knock represents an empty position.

Figure 12

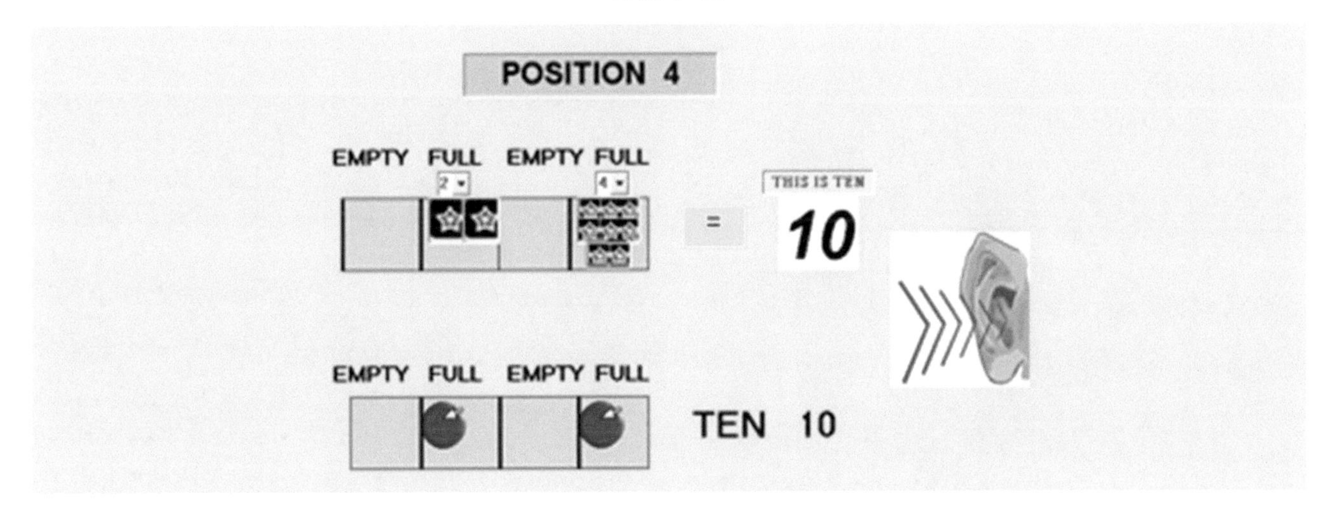

The ***kinesthetic*** representation of both figures does not depend on the shape or the number of the symbols in the filled positions. In both pictures the second and fourth filled positions are represented by the clenching of the right hand. An empty position is represented by the clenching of the left hand. The kinesthetic representation for both pictures is the clenching of the left hand, right, left and right, which is equal to a value of ten.

GREEN LESSONS, LESSON 2, FIGURE 1

In Figure 1, we have a ***visual*** display of two pictures with four positions each. Both pictures have the same audio representations. Positions one, two and four are filled with symbols. The first, second, and fourth positions in the first picture are filled with one, two, and eight stars. The number of stars corresponds to the mathematical value of the picture. The first, second, and fourth positions of the second picture are filled with red balls. The shape and the number of the symbols in the first, second and fourth positions of both pictures do not change the mathematical value.

The ***audio*** representation for both pictures consists of a **knock, knock, double knock and knock**, where a knock represents a filled up position and a double knock represents an empty position.

Figure 1

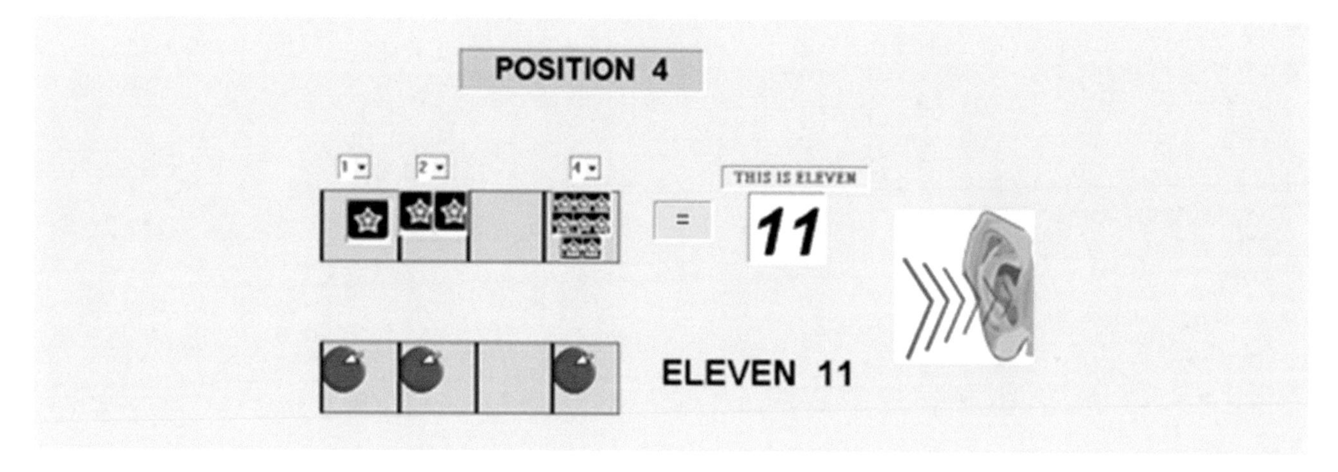

The ***kinesthetic*** representation of both figures does not depend on the shape or the number of the symbols in filled positions. In both pictures the first, second, and fourth filled positions are represented by the clenching of the right hand. An empty position is represented by the clenching of the left hand. The kinesthetic representation for both pictures is the clenching of the right hand twice, left once and again the right hand once, which is equal to a value of eleven.

GREEN LESSONS, LESSON 2, FIGURE 2

In Figure 2, we have a ***visual*** display of two pictures with four positions each. Both pictures have the same audio representations. In the first picture, in the filled positions we see twelve stars, and in the second picture we see two red balls. In both pictures the first two positions are empty and positions three and four are filled with symbols.

The ***audio*** representation for both pictures consists of a **double knock, double knock, knock and knock**, where a double knock represents an empty position and a knock represents a filled up position.

Figure 2

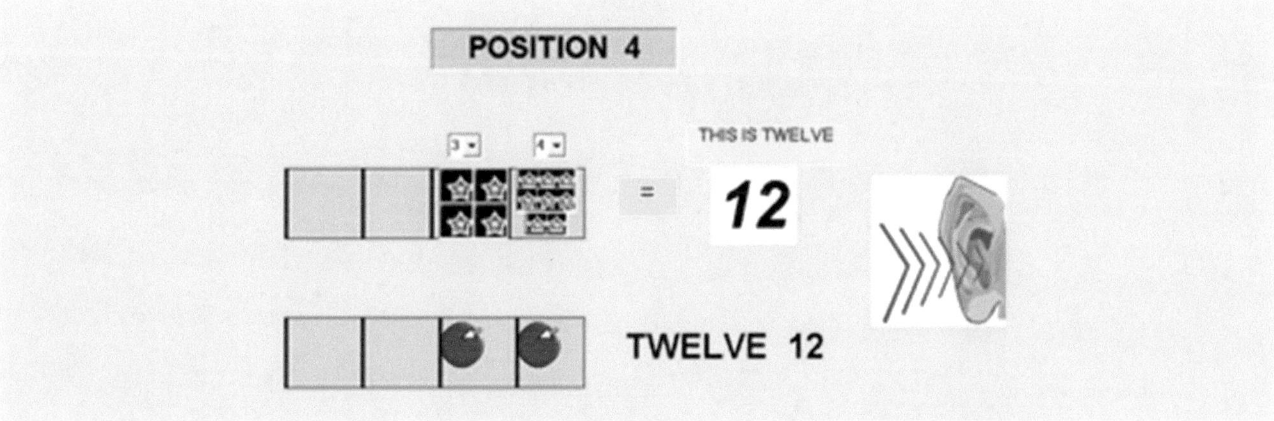

The ***kinesthetic*** representation of both figures does not depend on the shape or the number of the symbols in the filled positions. In both pictures the filled up position is represented by the clenching of the right hand. An empty position is represented by the clenching of the left hand. The kinesthetic representation for both pictures is the clenching of the left hand twice and the right hand twice, which is equal to a value of twelve.

MATHEMATICS

GREEN LESSONS, LESSON 2, FIGURE 3

In Figure 3, we have a ***visual*** display of two pictures with four positions each. Both pictures have the same audio representations. In both pictures the first, third and fourth positions are filled with symbols. We have one star in the first position, four stars in the third position and eight stars in the fourth position. In the second picture in the first, third and forth positions are filled with a red ball. In the first picture the amount of stars is equal to thirteen. The shape and the number of the symbols in the first, third and forth positions of in pictures does not change the mathematical value. The ***audio*** representation for both pictures consists of a **knock, double knock, knock and knock**, where a knock represents a filled up position and a double knock represents an empty position.

Figure 3

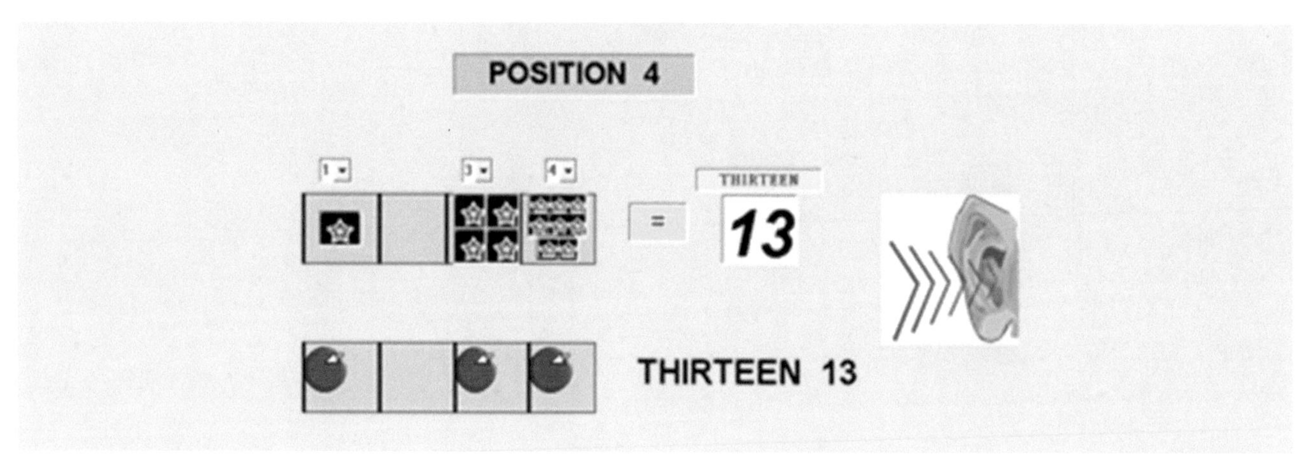

The ***kinesthetic*** representation of both figures does not depend on the shape or the number of the symbols in filled positions. In both pictures the first, third, and fourth filled positions are represented by the clenching of the right hand. An empty position is represented by the clenching of the left hand. The kinesthetic representation for both pictures is the clenching of the right hand, left and the right hand twice, which is equal to a value of thirteen.

GREEN LESSONS, LESSON 2, FIGURE 4

In Figure 4, we have a ***visual*** display of two pictures with four positions each. Both pictures have the same audio representations. In the first picture, in filled up positions we see fourteen stars and in the second picture we see three red balls. The shape and the number of the symbols in second, third and fourth positions do not change the mathematical value.

The ***audio*** representation for both pictures consists of a **double knock, knock, knock and knock**, where a double knock represents an empty position and a knock represents a filled up position.

Figure 4

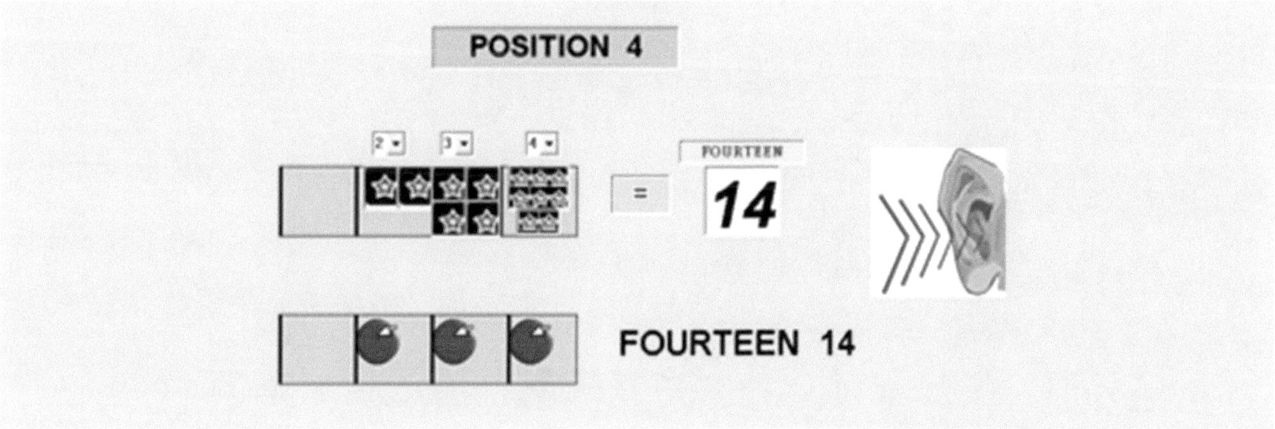

POSITION 4

FOURTEEN

14

FOURTEEN 14

The ***kinesthetic*** representation of both figures does not depend on the shape or the number of symbols in filled positions. In both pictures the filled position is represented by the clenching of the right hand. An empty position is represented by the clenching of the left hand. The kinesthetic representation for both pictures is the clenching of the left hand once then the right hand three times, which is equal to a value of fourteen.

MATHEMATICS GREEN LESSONS, LESSON 2, FIGURE 5

In Figure 5, we have a ***visual*** display of two pictures with four positions each. Both pictures have the same audio representations. In both pictures all four positions are filled with symbols, but the figures have different symbols in the filled positions. In the first picture, we see fifteen stars and in the second picture we see four red balls. The shape and the number of the symbols in all positions do not change the mathematical value.

The ***audio*** representation for both pictures consists of a **knock, knock, knock and knock**, where a knock represents a filled up position.

Figure 5

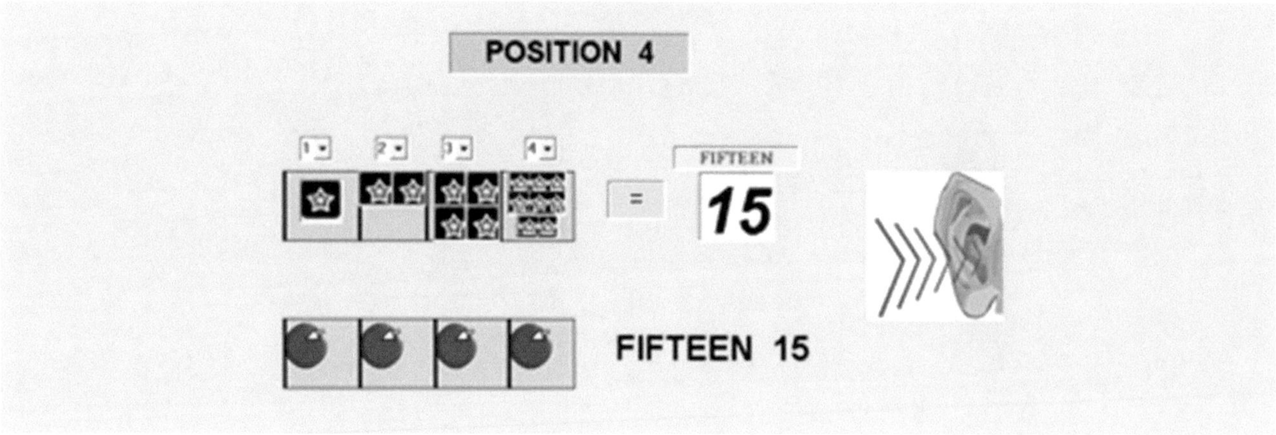

The ***kinesthetic*** representation of both figures does not depend on the shape or the number of symbols in the filled positions. In both pictures the filled position is represented by the clenching of the right hand four times, which is equal to a value of fifteen.

GREEN LESSONS, LESSON 3, FIGURE 1

In Figure 1, we have a ***visual*** display of two pictures with four positions each. The pictures have different audio representations.

The ***audio*** representation of a filled up position is a knock, and a double knock represents an empty position. In the first picture, we have five empty spaces and the audio representation is **double knock, double knock, double knock, double knock and double knock**. In the second picture, the fifth position is filled with a symbol and positions one, two, three, and four are empty. The audio representation for the second picture consists of a **double knock, double knock, double knock, double knock and knock**.

Figure 1

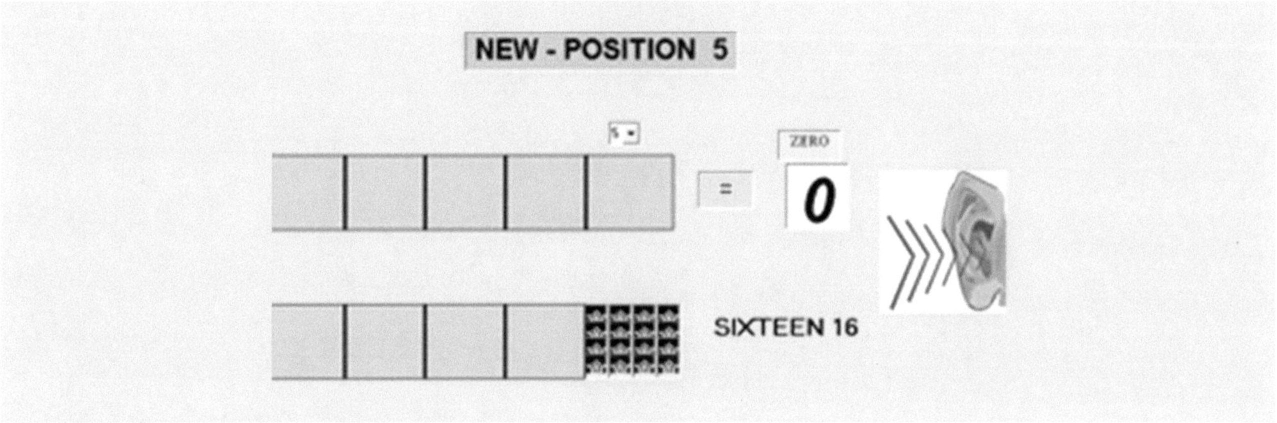

NEW - POSITION 5

= ZERO 0

SIXTEEN 16

The ***kinesthetic*** representation of a filled up position is represented by the clenching of the right hand. An empty position is represented by the clenching of the left hand. The kinesthetic representation for the first picture is the clenching of the left hand five times, which is equal to zero. The kinesthetic representation of the second picture is the clenching of left hand four times, then the right hand once, which signifies a mathematical value equal to sixteen.

MATHEMATICS

GREEN LESSONS, LESSON 3, FIGURE 3

In Figure 3, we have a ***visual*** display of two pictures with four positions each. Both pictures have the same audio representations. In both pictures position one and five are filled up with symbols. The figures have different symbols in filled positions. In the first picture, the first and fifth positions are filled with symbols. We have one star in the first position and sixteen stars in the forth position. In the second picture, there is a red ball in the first and fifth positions. The shape and the number of symbols in the first and fifth positions of both pictures do not change the mathematical value. The ***audio*** representation for both pictures consists of a **knock, double knock, double knock, double knock and knock**, where a knock represents a filled up position and a double knock represents an empty position.

Figure 3

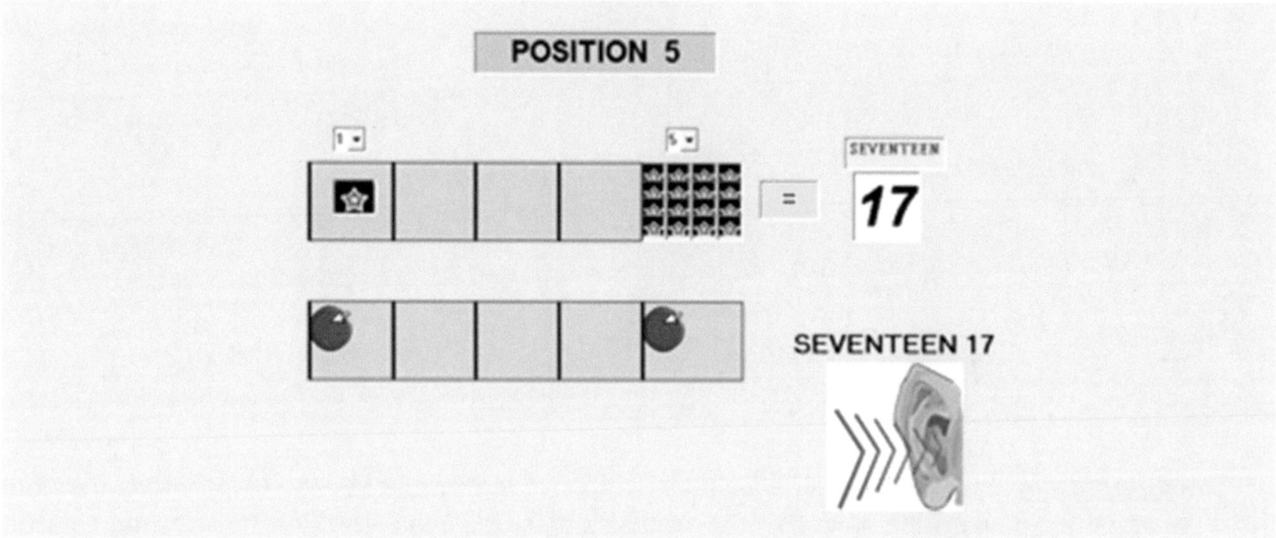

The ***kinesthetic*** representation of both figures does not depend on the shape and the number of symbols in the filled positions. In both pictures the first and fifth filled positions are represented by the clenching of the right hand. An empty position is represented by the clenching of the left hand. The kinesthetic representation for both pictures is the clenching of the right hand, then left hand three times and the right hand once, which is equal to a value of seventeen.

GREEN LESSONS, LESSON 3, FIGURE 4

In Figure 4, we have a ***visual*** display of two pictures with four positions each. Both pictures have the same audio representations. In both pictures position two and five are filled with symbols. The figures have different symbols in filled positions. In the first picture, the second and fifth positions are filled with symbols. We have two stars in the second position and sixteen stars in the fifth position. In the second picture, there is a red ball in the second and fifth positions. The shape and the number of the symbols in the second and fifth positions of both pictures do not change the mathematical value. The ***audio*** representation for both pictures consists of a **double knock, knock, double knock, double knock and knock**, where a knock represents a filled up position and a double knock represents an empty position.

Figure 4

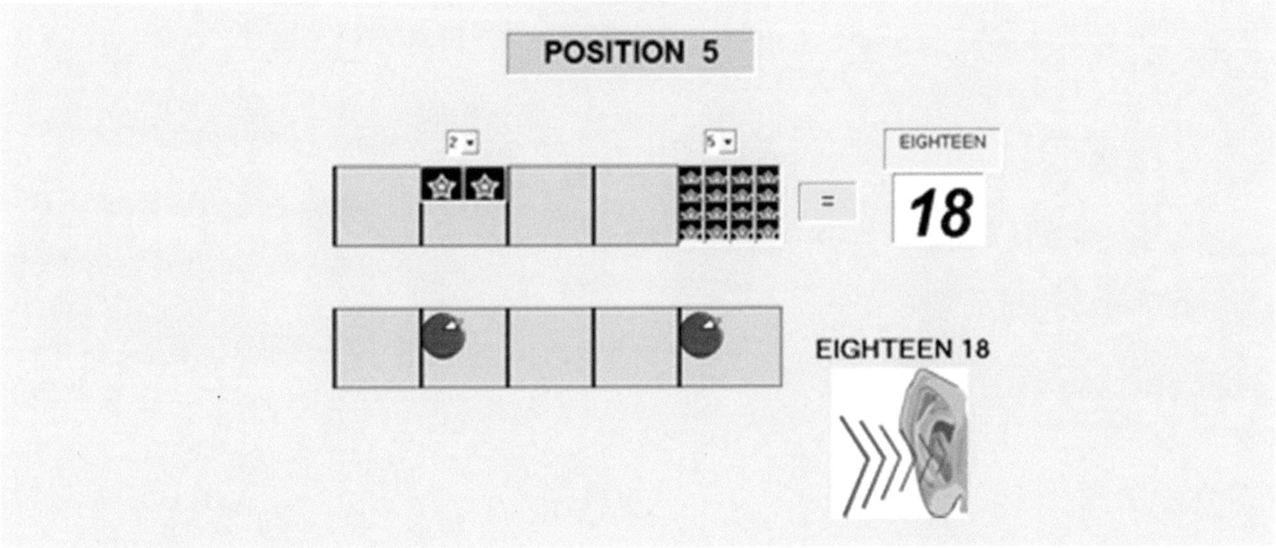

POSITION 5

EIGHTEEN

18

EIGHTEEN 18

The ***kinesthetic*** representation of both figures does not depend on the shape or the number of symbols in the filled positions. In both pictures the second and fifth filled positions are represented by the clenching of the right hand. An empty position is represented by the clenching of the left hand. The kinesthetic representation for both pictures is the clenching of the left hand, right, left twice and right again, which is equal to a value of eighteen.

MATHEMATICS

GREEN LESSONS, LESSON 3, FIGURE 5

In Figure 5, we have a ***visual*** display of two pictures with four positions each. Both pictures have the same audio representations. The figures have different symbols in filled up positions. In the first picture, the first, second and fifth positions are filled with symbols. We have one star in the first position, two stars in the second position and sixteen stars in the fifth position. In the second picture, the first, second and fifth positions are filled with red balls. The shape and the number of symbols in the first, second and fifth positions of both pictures do not change the mathematical value. The ***audio*** representation for both pictures consists of a **knock, knock, double knock, double knock and knock**, where a knock represents a filled up position and a double knock represents an empty position.

Figure 5

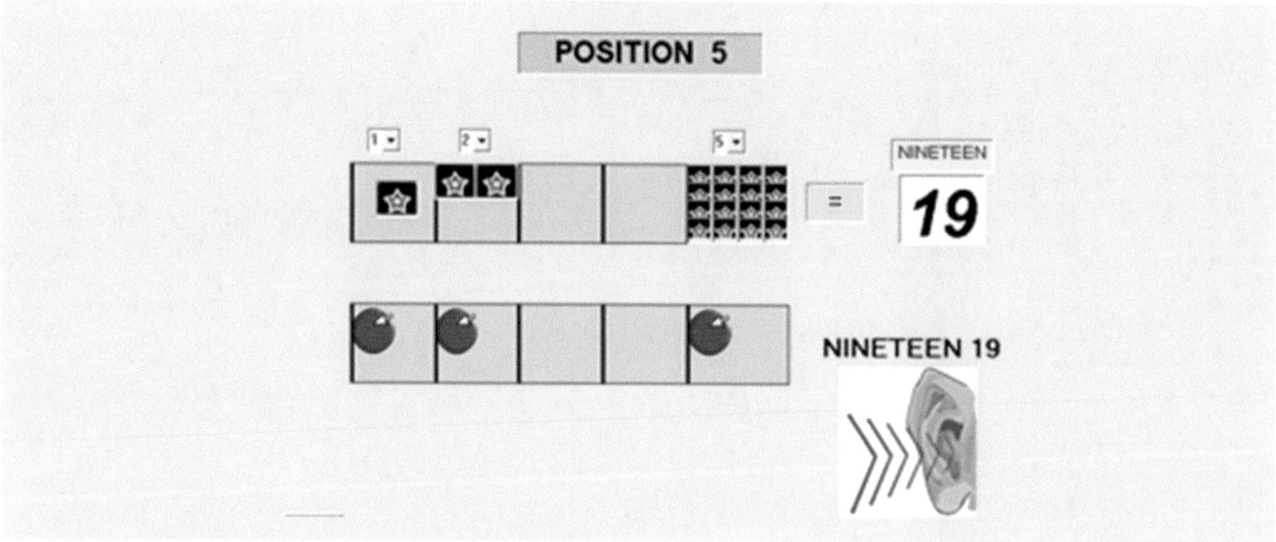

The ***kinesthetic*** representation of both figures does not depend on the shape and the number of symbols in the filled positions. The first, second and fifth filled positions are filled with symbols represented by the clenching of the right hand. An empty position is represented by the clenching of the left hand. The kinesthetic representation for both pictures is the clenching of the right hand twice, left twice and the right hand again, which is equal to a value of nineteen.

GREEN LESSONS, LESSON 3, FIGURE 6

In Figure 6, we have a ***visual*** display of two pictures with four positions each. Both pictures have the same audio representations. The figures have different symbols in filled positions. In the first picture, the third and fifth positions are filled with symbols. We have four stars in the third position and sixteen stars in the fifth position. In the second picture, there are red balls in third and fifth positions. The shape and the number of symbols in the third and fifth positions of both pictures do not change the mathematical value. The ***audio*** representation for both pictures consists of a **double knock, double knock, knock, double knock and knock**, where a knock represents a filled up position and a double knock represents an empty position.

Figure 6

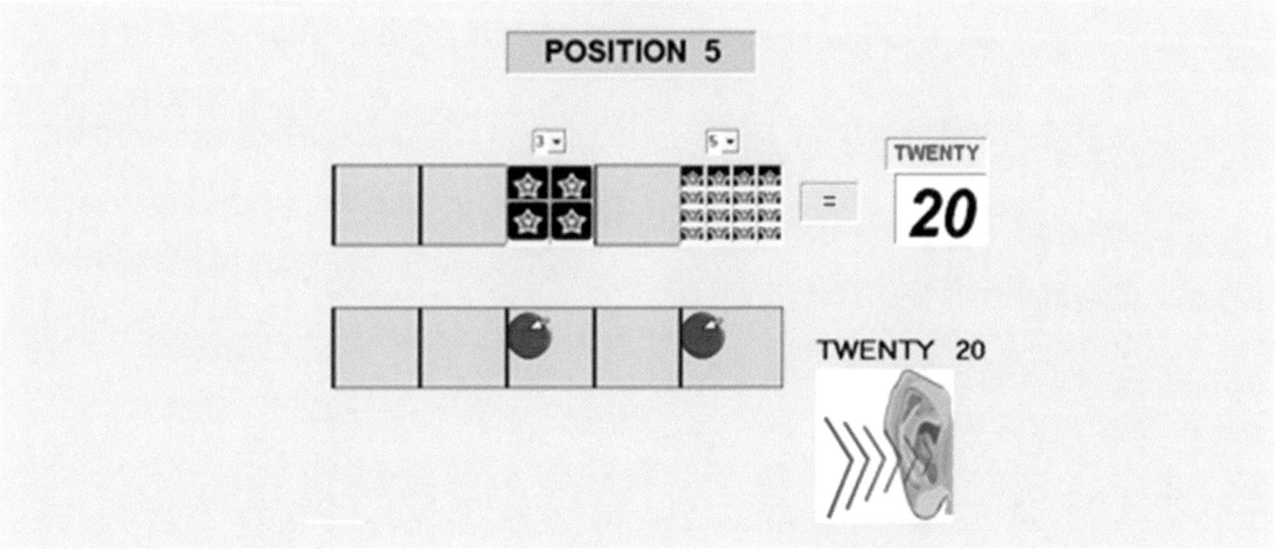

The ***kinesthetic*** representation of both figures does not depend on the shape or the number of the symbols in the filled positions. In both pictures the third and fifth filled positions are represented by the clenching of the right hand. An empty position is represented by the clenching of the left hand. The kinesthetic representation for both pictures is the clenching of the left twice, right, left and right again, which is equal to a value of twenty.

MATHEMATICS

GREEN LESSONS, LESSON 4, FIGURE 1

In Figure 1, the ***visual*** display of both pictures has five positions each. Both pictures have the same audio representation. The figures have different symbols in filled positions. In the first picture the first, third and fifth positions are filled with symbols. We have one star in the first position, four stars in the in third position and sixteen stars in the fifth position. In the second picture, there are red balls in the first, third, and fifth positions. The shape and the number of symbols in the first, third and fifth positions in both pictures do not change the mathematical value. The ***audio*** representation for both pictures consists of a **knock, double knock, knock, double knock and knock** where a knock represents a filled position and a double knock represents an empty position

Figure 1

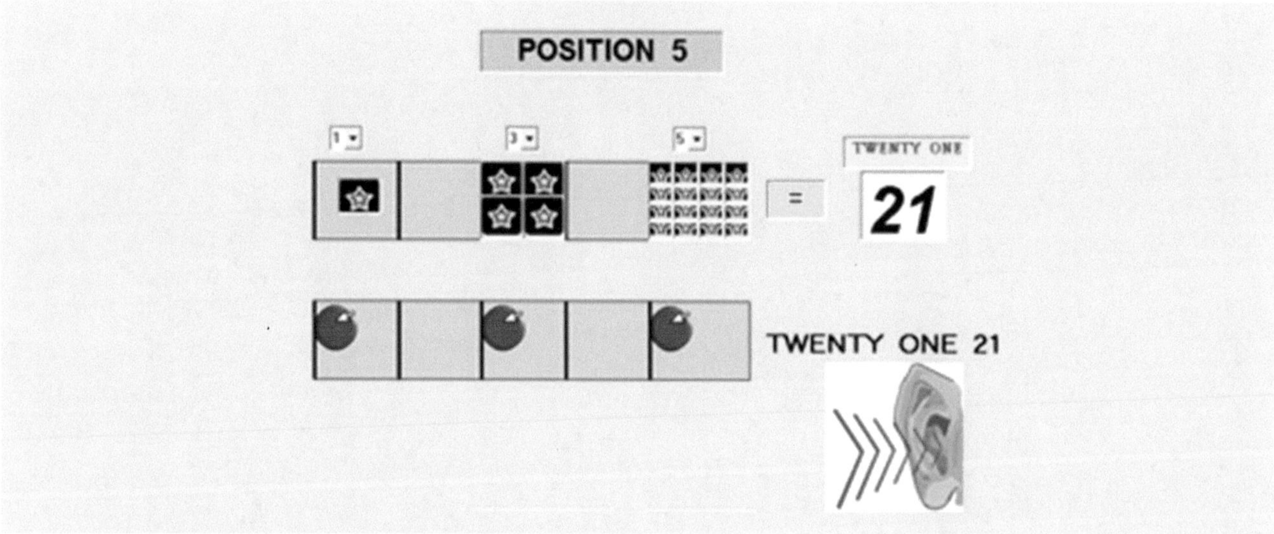

The ***kinesthetic*** representation of both figures does not depend on the shape or the number of symbols in the filled positions. The first, third, and fifth filled positions in the picture are represented by the clenching of the right hand. An empty position is represented by the clenching of the left hand. The kinesthetic representation for both pictures is the clenching of the right hand, left, right, left and right again, which is equal to a value of twenty one.

GREEN LESSONS, LESSON 4, FIGURE 2

In Figure 2, we have a ***visual*** display of two pictures with four positions each. Both pictures have the same audio representations. The figures have different symbols in filled positions. In the first picture the second, third and fifth positions are filled with symbols. We have two stars in the second position, four stars in the in third position and sixteen stars in the fifth position. In the second picture, there are red balls in the second, third, and fifth positions. The shape and the number of symbols in the second, third and fifth positions in both pictures do not change the mathematical value. The ***audio*** representation for both pictures consists of a **double knock, knock, knock, double knock and knock,** where a knock represents a filled up position and a double knock represents an empty position.

Figure 2

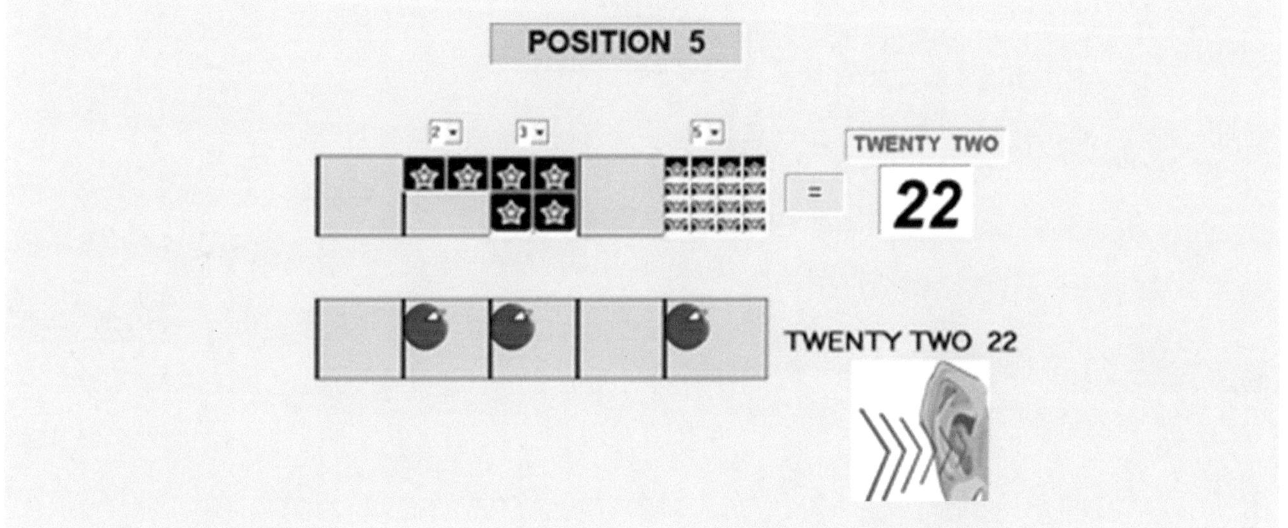

The ***kinesthetic*** representation of both figures does not depend on the shape or the number of filled positions. In both pictures the second, third, and fifth filled positions are represented by the clenching of the right hand. An empty position is represented by the clenching of the left hand. The kinesthetic representation for both pictures is the clenching of the left hand, right twice, left and right again, which is equal to a value of twenty two.

MATHEMATICS

GREEN LESSONS, LESSON 4, FIGURE 3

In Figure 3, we have a ***visual*** display of two pictures with four positions each. Both pictures have the same audio representations. The figures have different symbols in filled positions. The first, second, third, and fifth positions in the first picture are filled with stars, and the number of the stars adds up to twenty three. The first, second, third, and fifth positions in the second picture are filled with red balls. The shape and the number of the symbols in the first, second, third and fifth positions in both pictures do not change the mathematical value. The ***audio*** representation for both pictures consists of a **knock, knock, knock, double knock and knock,** where a knock represents a filled position and a double knock represents an empty position.

Figure 3

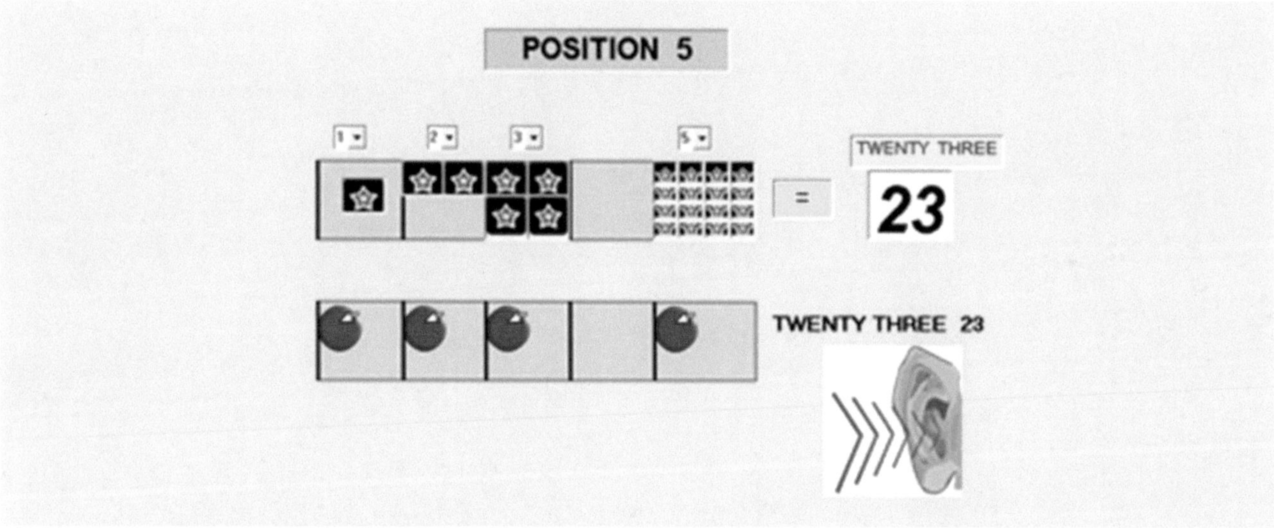

The ***kinesthetic*** representation of both figures does not depend on the shape or the number of symbols in the filled positions. In both pictures, the first, second, third, and fifth filled positions are represented by the clenching of the right hand. An empty position is represented by the clenching of the left hand. The kinesthetic representation for both pictures is the clenching of the right hand three times, left once and right once more, which is equal to a value of twenty three.

GREEN LESSONS, LESSON 4, FIGURE 4

In Figure 4, we have a ***visual*** display of two pictures with four positions each. Both pictures have the same audio representations. The figures have different symbols in filled positions. The fourth and fifth positions in the first picture are filled with twenty four stars, and the fourth and fifth positions in the second picture are filled with red balls. The shape and the number of the symbols in the filled positions do not change the mathematical value. In both pictures the first three positions are empty, and positions four and five are filled with a symbol.

The ***audio*** representation for both pictures consists of a **double knock, double knock, double knock, knock and knock**, where a double knock represents an empty position and a knock represents a filled up position.

Figure 4

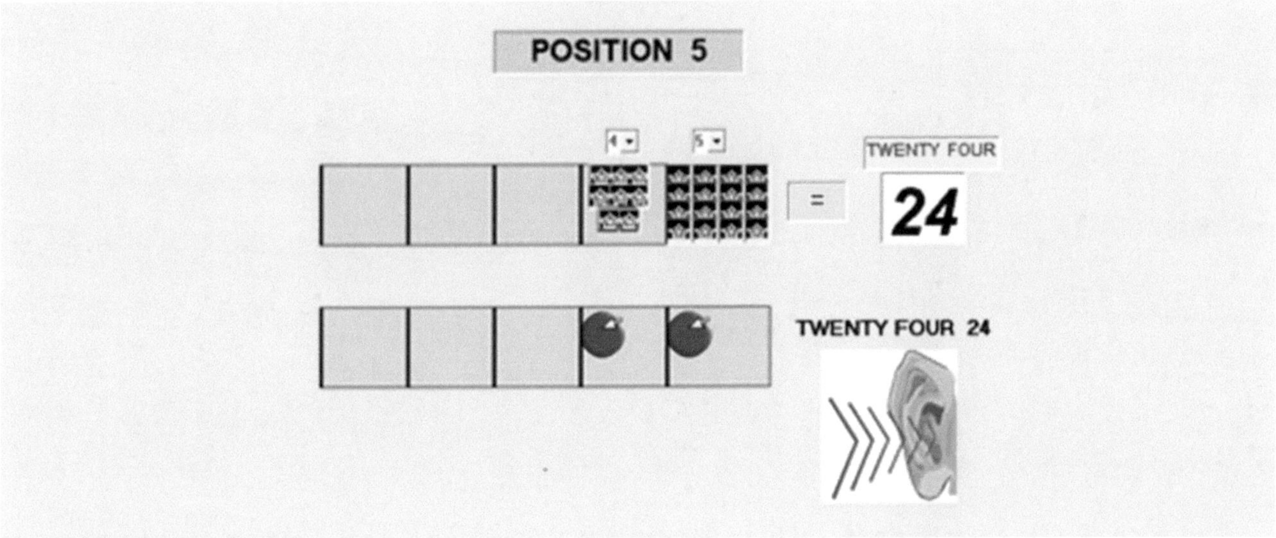

The ***kinesthetic*** representation of both figures does not depend on the shape or the number of the symbols in the filled positions. In both pictures the filled position is represented by the clenching of the right hand. An empty position is represented by the clenching of the left hand. The kinesthetic representation for both pictures is the clenching of the left hand three times and the right hand twice, which is equal to a value of twenty four.

MATHEMATICS GREEN LESSONS, LESSON 4, FIGURE 5

In Figure 5, we a have ***visual*** display of two pictures with four positions each. Both pictures have the same audio representations. The figures have different symbols in filled positions. The first, fourth, and fifth positions in the first picture are filled with twenty five stars, and the same positions in the second picture are filled with red balls. The shape and the number of symbols in the first, fourth and fifth positions in both pictures do not change the mathematical value.

The ***audio*** representation for both pictures consists of a **knock, double knock, double knock, knock and knock**, where a knock represents a filled up position and a double knock represents an empty position.

Figure 5

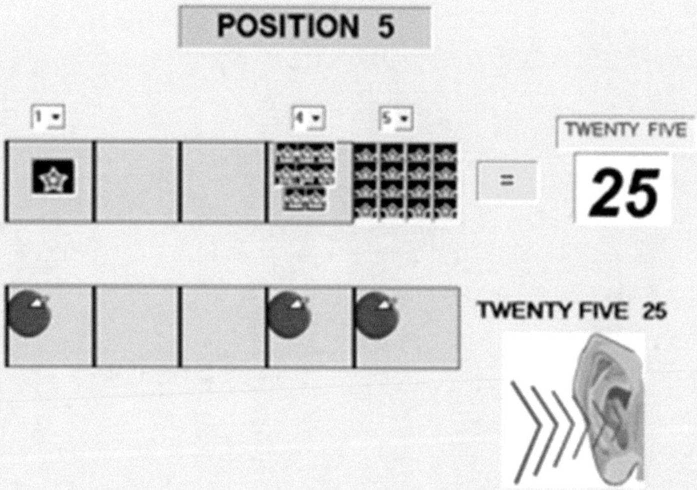

POSITION 5

TWENTY FIVE

25

TWENTY FIVE 25

The ***kinesthetic*** representation of both figures does not depend on the shape or the number of symbols in the filled positions. In both pictures, the first, fourth, and fifth filled positions are represented by the clenching of the right hand. An empty position is represented by the clenching of the left hand. The kinesthetic representation for both pictures is the clenching of the right hand, left twice, and the right hand twice, which is equal to a value of twenty five.

GREEN LESSONS, LESSON 4, FIGURE 1

In Figure 1, we have a ***visual*** display of two pictures with four positions each. Both pictures have the same audio representations. The figures have different symbols in filled positions. The second, fourth, and fifth positions in the first picture are filled with twenty six stars, and the same positions in the second picture are filled with red balls. The shape and the number of symbols in the second, fourth and fifth positions in both pictures do not change the mathematical value. The ***audio*** representation for both pictures consists of a **double knock, knock, double knock, knock and knock,** where a knock represents a filled position and a double knock represents an empty position.

Figure 1

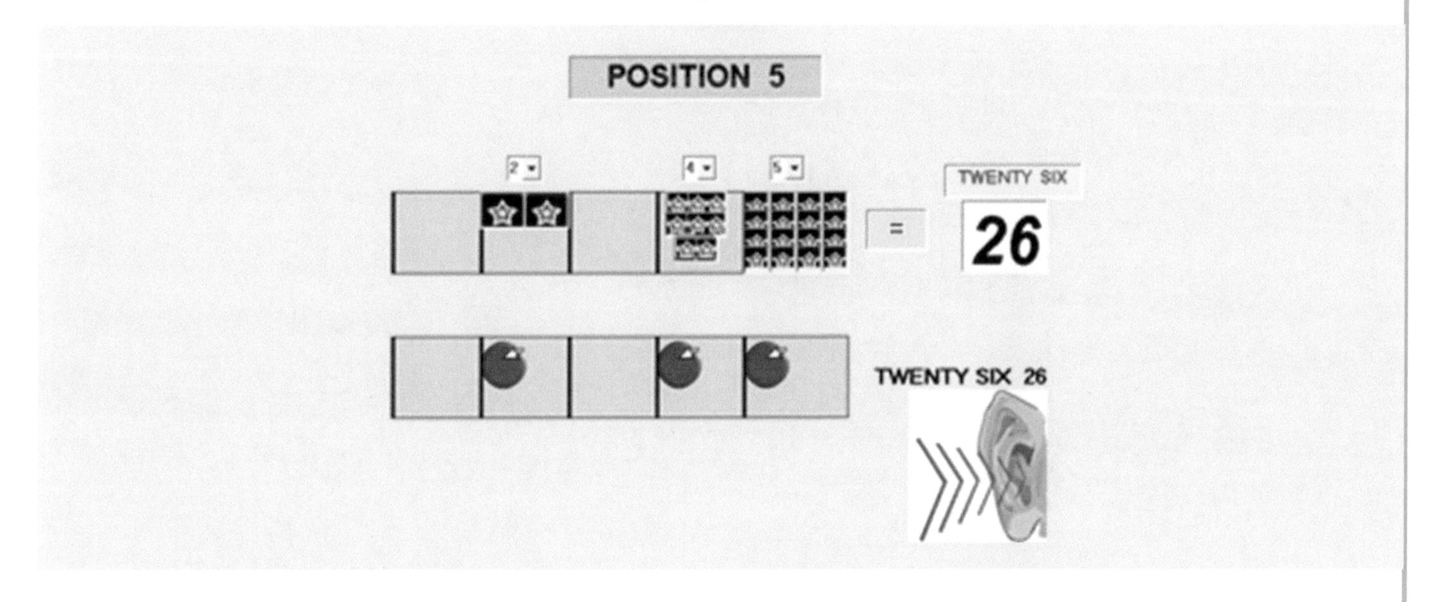

The ***kinesthetic*** representation of both figures does not depend on the shape or the number of symbols in the filled positions. In both pictures, the second, fourth, and fifth filled positions are represented by the clenching of the right hand. An empty position is represented by the clenching of the left hand. The kinesthetic representation for both pictures is the clenching of the left hand, right, left and right hand twice, which is equal to a value of twenty six.

GREEN LESSONS, LESSON 4, FIGURE 2

In Figure 3, we have a ***visual*** display of two pictures with four positions each. Both pictures have the same audio representations. The figures have different symbols in filled positions. The first, second, fourth and fifth positions in the first picture are filled with twenty seven stars, and the same positions in the second picture are filled with red balls. The shape and the number of symbols in the first, second, fourth and fifth positions in both pictures do not change the mathematical value.

The ***audio*** representation for both pictures consists of a **knock, knock, double knock**, **knock, knock,** where a knock represents a filled up position and a double knock represents an empty position.

Figure 2

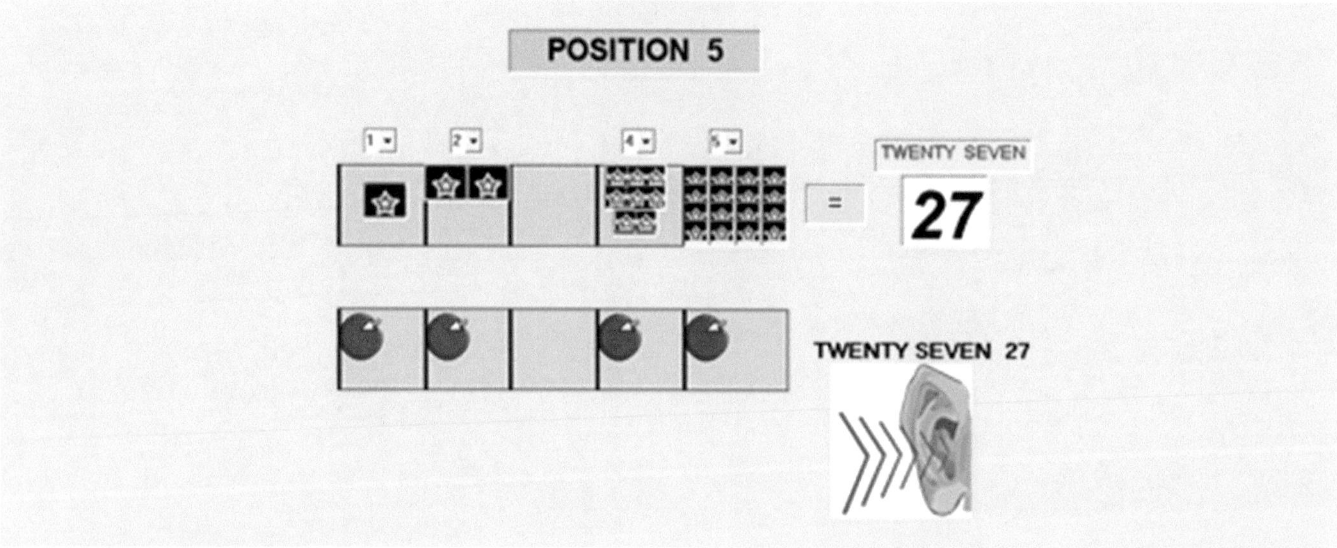

POSITION 5

1 2 4 5

= TWENTY SEVEN 27

TWENTY SEVEN 27

The ***kinesthetic*** representation of both figures does not depend on the shape or the number of symbols in the filled positions. In both pictures, the first, second, fourth, and fifth filled positions are represented by the clenching of the right hand. An empty position is represented by the clenching of the left hand. The kinesthetic representation for both pictures is the clenching of the right hand twice, left once and right twice again, which is equal to a value of twenty seven.

GREEN LESSONS, LESSON 5, FIGURE 3

In Figure 3, we have a ***visual*** display of two pictures with four positions each. Both pictures have the same audio representations. The figures have different symbols in filled positions. The third, fourth, and fifth positions in the first picture are filled with twenty eight stars, and the same positions in the second picture are filled with red balls. The shape and the number of symbols in the filled positions do not change the mathematical value.

The ***audio*** representation for both pictures consists of a **double knock, double knock, knock, knock and knock**, where a double knock represents an empty position and a knock represents a filled up position.

Figure 3

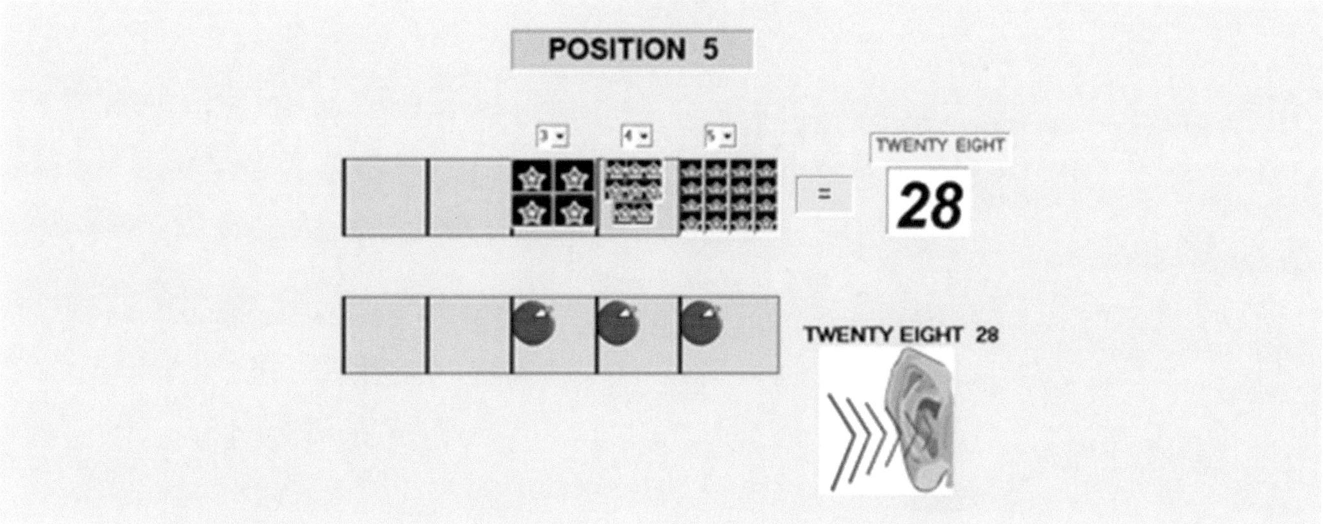

The ***kinesthetic*** representation of both figures does not depend on the shape or the number of symbols in the filled positions. In both pictures, the filled positions are represented by the clenching of the right hand. An empty position is represented by the clenching of the left hand. The kinesthetic representation for both pictures is the clenching of the left hand two times and the right hand three times, which is equal to a value of twenty eight.

GREEN LESSONS, LESSON 5, FIGURE 4

In Figure 3, we have a ***visual*** display of two pictures with four positions each. Both pictures have the same audio representations. The figures have different symbols in the filled positions. The first, third, fourth, and fifth positions in the first picture are filled with twenty nine stars, and the same positions in the second picture are filled with red balls. The shape and the number of symbols in the first, third, fourth and fifth positions of both pictures do not change the mathematical value.

The ***audio*** representation for both pictures consists of a **knock, double knock, knock, knock and knock**, where a knock represents a filled up position and a double knock represents an empty position.

Figure 4

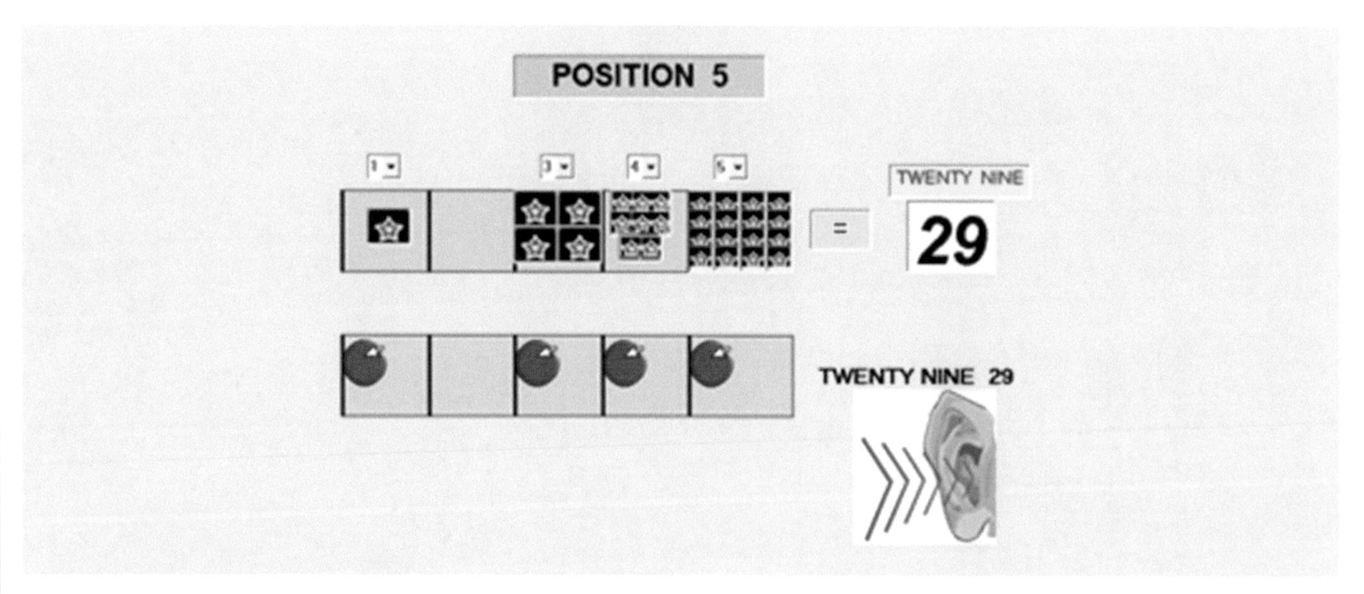

The ***kinesthetic*** representation of both figures does not depend on the shape or the number of symbols in the filled positions. In both pictures, the filled up positions are represented by the clenching of the right hand. An empty position is represented by the clenching of the left hand. The kinesthetic representation for both pictures is the clenching of the right hand, left and the right hand three times, which is equal to a value of twenty nine.

GREEN LESSONS, LESSON 5, FIGURE 5

In Figure 5, we have a ***visual*** display of two pictures with four positions each. Both pictures have the same audio representations. The figures have different symbols in the filled positions. The second, third, fourth and fifth positions in the first picture are filled with thirty stars, and the same positions in the second picture are filled with red balls. The shape or the number of symbols in the filled positions does not change the mathematical value.

The ***audio*** representation for both pictures consists of a **double knock, knock, knock, knock and knock**, where a double knock represents an empty position and a knock represents a filled up position.

Figure 5

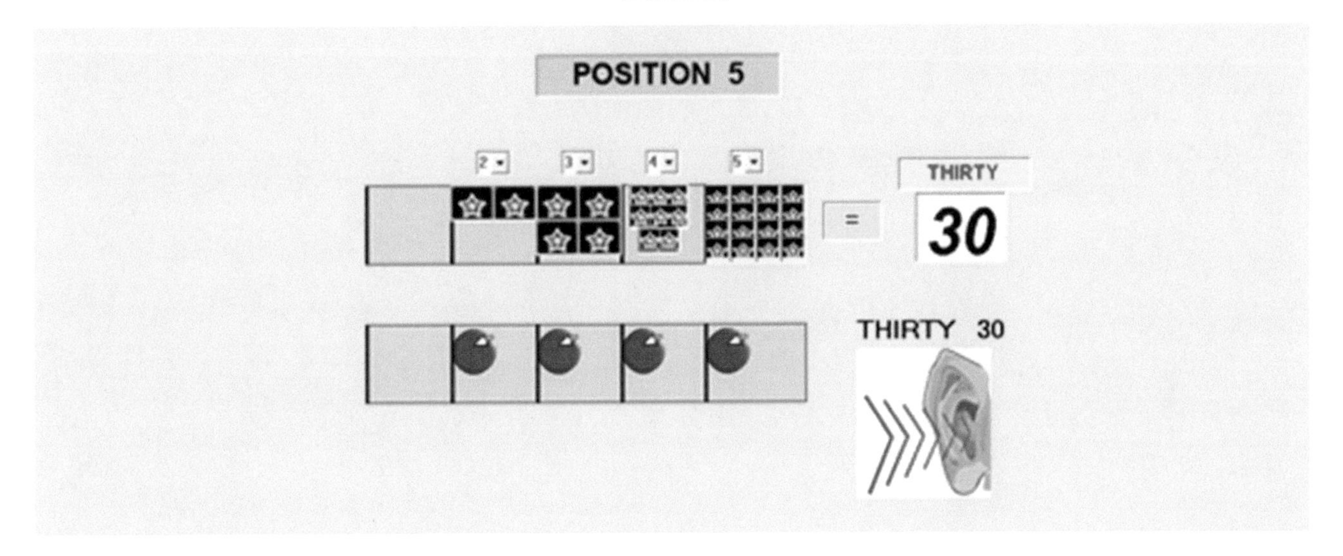

The ***kinesthetic*** representation of both figures does not depend on the shape or the number of symbols in the filled positions. In both pictures, the filled positions are represented by the clenching of the right hand. An empty position is represented by the clenching of the left hand. The kinesthetic representation for both pictures is the clenching of the left hand once and the right hand four times, which is equal to a value of thirty.

MATHEMATICS

GREEN LESSONS, LESSON 5, FIGURE 6

In Figure 6, we have a ***visual*** display of two pictures with four positions each. Both pictures have the same audio representations. In both pictures all five positions are filled with symbols. The first picture is filled with stars and the second is filled with red balls. The shape and the number of symbols in all positions do not change the mathematical value.

The ***audio*** representation for both pictures consists of a **knock, knock, knock, knock and knock**, where a knock represents a filled position.

Figure 6

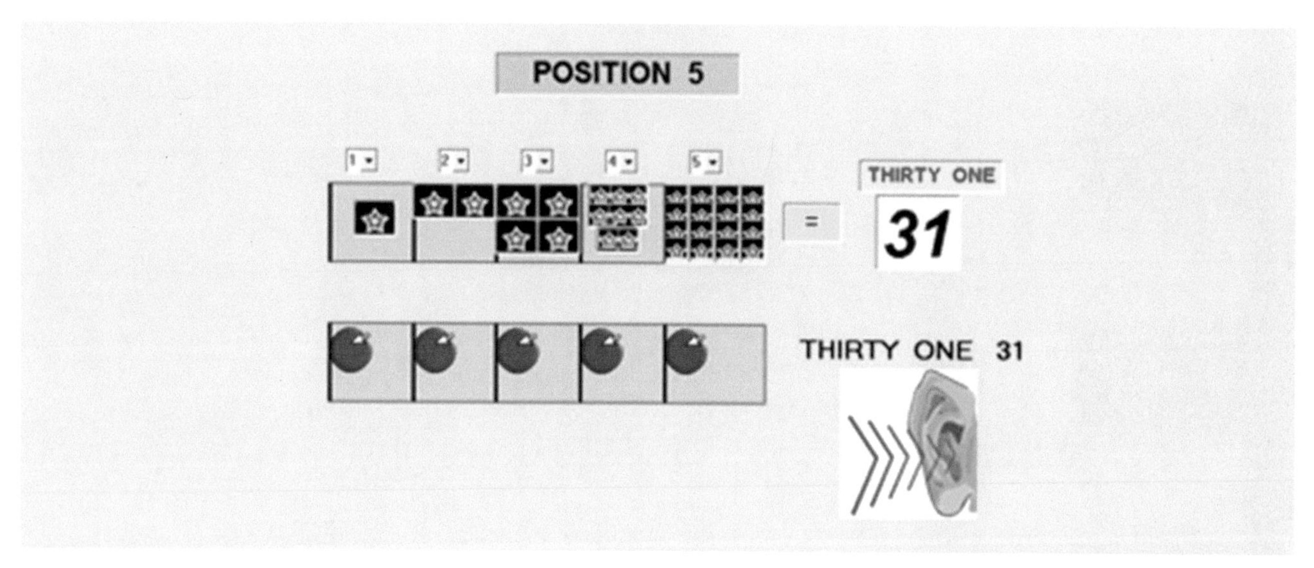

The ***kinesthetic*** representation of both figures does not depend on the shape or the number of symbols in the filled positions. In both pictures the filled positions are represented by the clenching of the right hand five times, which is equal to a value of thirty one.

MATHEMATICS

Count number of objects in Each Picture

MATHEMATICS

BLUE LESSONS, FIGURE 1

In Figure 1, we have a **video** display of two pictures with two positions each, where position one is filled up with symbols and the second position is empty. The **audio** representation of those figures is **knock and double knock**.

Figure 1

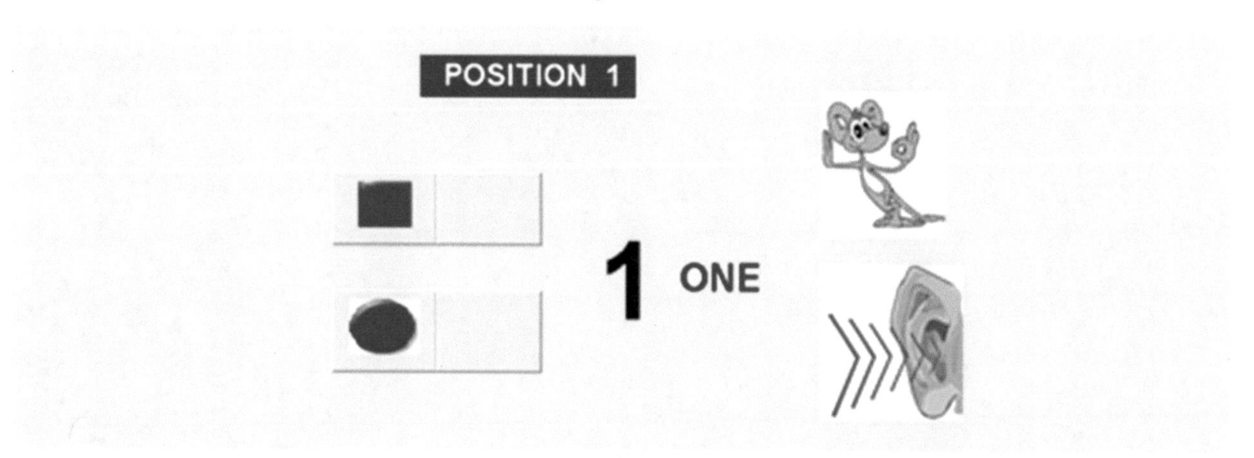

The knock represents the filled up position and double knock represents an empty position. In our example, the **audio** signals knock and double knock correspond to the both pictures. The first picture of boxes has a **visual** display of the square symbol in the first box and empty space in the second box. In the first position of the second picture includes a circle and the second position has an empty space. The different symbols in the first positions of the pictures do not change the mathematical representation of both configurations. The shape of the symbols also does not have any effect on the mathematical meaning. The symbols exist just to show that positions are already filled up. The mathematical value of the both first and second configurations is equal to one. Any shape of the symbols in the first position represents number one.

The **kinesthetic** representation of the knock is the clenching of the right hand and for double knock is the tightening the left hand into a fist. The filled up position is represented by the clenching of the right hand and an empty position represented by the clenching of the left hand. The **kinesthetic** representation for the first and second configurations of boxes is the clenching of the right and then the left hand.

BLUE LESSONS, FIGURE 2

Figure 2 is the **visual** display of two pictures with two positions each, where position one is empty and the second position is filled up with symbols. The **audio** representation of this figure is **double knock and knock**

Figure 2

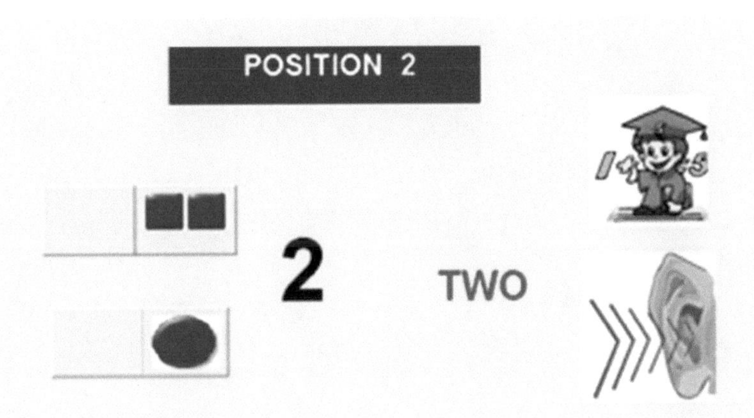

The double knock is the **audio** representation of the empty position and knock represents a filled up position. In our example double knock and knock correspond to both pictures. The first picture has an empty space in the first position and square symbol in the second. In the second picture, we see an empty space in the first positions and a circle in the second. The different **visual** shape of symbols in the second position does not change the mathematical meaning. The presence of the symbol is only to show that the position is already filled up. The mathematical value of the first configuration as well as the second is equal to two. Any shape of symbols in the second position represents number two. The **kinesthetic** representation of the double knock is the tightening of the left hand into a fist and for knock is the clenching of the right hand. An empty position is represented by the clenching of the left hand. The filled up position is represented by the clenching of the right hand. The **kinesthetic** representation for the first and second configurations of boxes is the clenching of the left hand and then the right.

MATHEMATICS

BLUE LESSONS, FIGURE 3

In Figure 3, we have a **visual** presentation of two pictures with two positions each, where both positions are filled up with symbols. The **audio** representation of this figure is **knock and knock**.

Figure 3

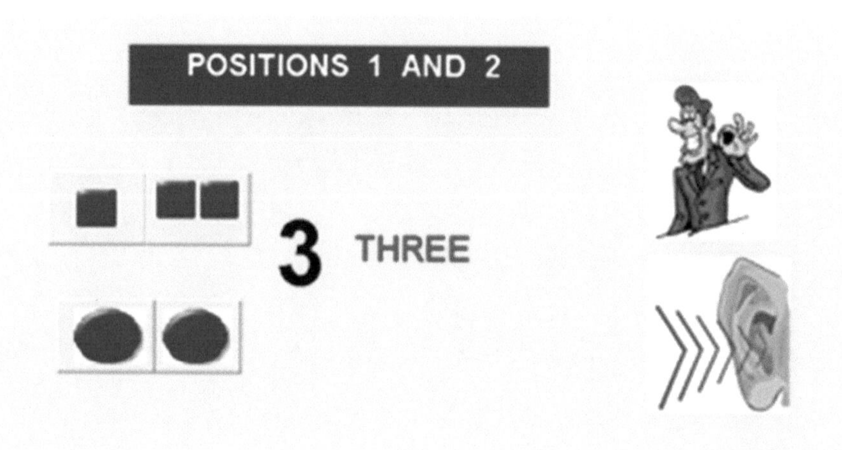

The two pictures correspond to the same mathematical value of number three. In our example, the **audio** signals knock and knock correspond to both pictures. The **visual** display of the first picture has a square symbol in both positions, where in first position is only one square and in the second position we have two squares. In the second picture we have circles in both positions. The different symbols in the first position do not change the mathematical representation of both configurations. The shape of the symbols does not affect the mathematical meaning. The symbols exist just to show that positions are already filled up. The one box in the first position and two boxes in the second position of the first picture are visually illustrating the mathematical quantity.

The **kinesthetic** representation of filled up positions is the clenching of the right hand. The **kinesthetic** representation for both pictures is the clenching of the right hand twice.

The two boxes in the second position of the first picture are **visually** illustrating the quantity of the mathematical representation.

BLUE LESSONS, FIGURE 4

In Figure 4, we have a <u>**visual**</u> representation of two pictures where each one has three positions. In both pictures positions one and two are empty. Only position three is filled up with a symbol. The ***audio*** representation of these figures is **double knock, double knock and knock**.

<u>Figure 4</u>

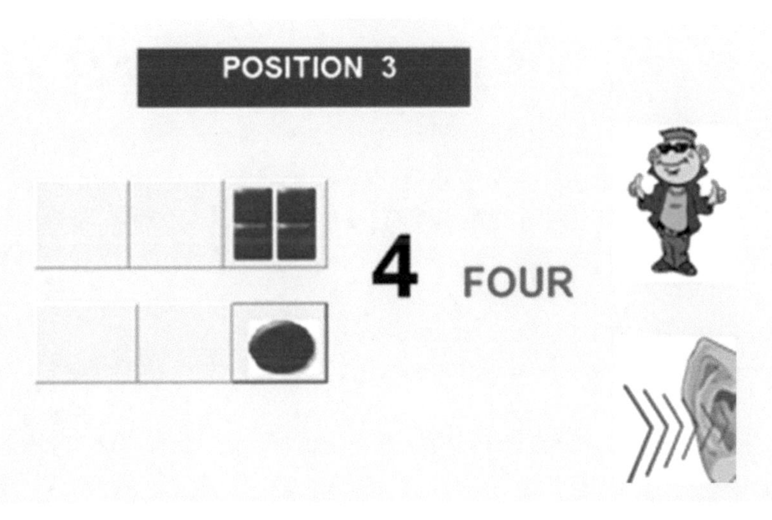

In the first and second positions of both pictures we have an empty space. In the third position we have rectangles or a circle. The different symbols in the third position of the pictures do not change the mathematical representation of both configurations. The shape of the symbols also does not affect the mathematical value. The symbols exist just to show that positions are already filled up. The mathematical value of the first and second configurations is equal to four.

The <u>***kinesthetic***</u> representation of the knock is the clenching of the right hand and for double knock is the tightening the left hand into a fist. The filled up position is represented by the clenching of the left hand two times and then by the clenching of the right hand.

MATHEMATICS BLUE LESSONS, FIGURE 5

In Figure 5 we have a **visual** display of two pictures, which have three positions each. The first and third positions are filled up with symbols. The second position is left empty. The **audio** representation of this figure is **knock, double knock, and knock**.

Figure 5

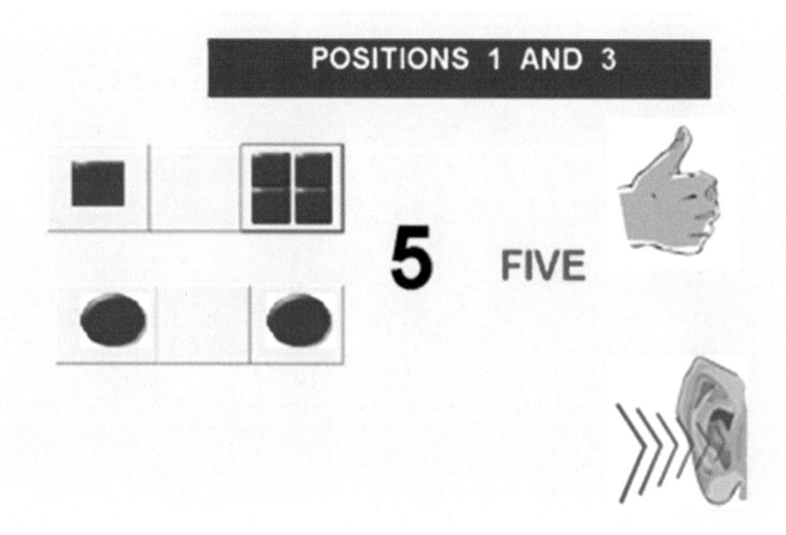

Pictures in boxes correspond to the mathematical value of number five.

The filled up position is represented by the **audio** signal of a knock, and an empty position is represented by a double knock. In our example knock, double knock, and knock correspond to both pictures. The first picture of boxes has a **visual** square symbol in the first and third place, but empty space in the second place. The different symbols in the first position of the pictures do not change the mathematical representation of both configurations. The shape of the symbols also does not affect the mathematical meaning. The symbols exist just to show that positions are already filled up. The **kinesthetic** representation of the knock is the clenching of the right hand and for double knock is the tightening of the left hand into a fist. The **kinesthetic** representation for the first and second configurations of boxes is the clenching of the right, left and right hand again.

BLUE LESSONS, FIGURE 6

In Figure 6, we have a **visual** display of two pictures with three positions each. Position one is empty, but position two and position three are filled up with symbols. The **audio** representation of those figures is **double knock, knock and knock**.

Figure 6

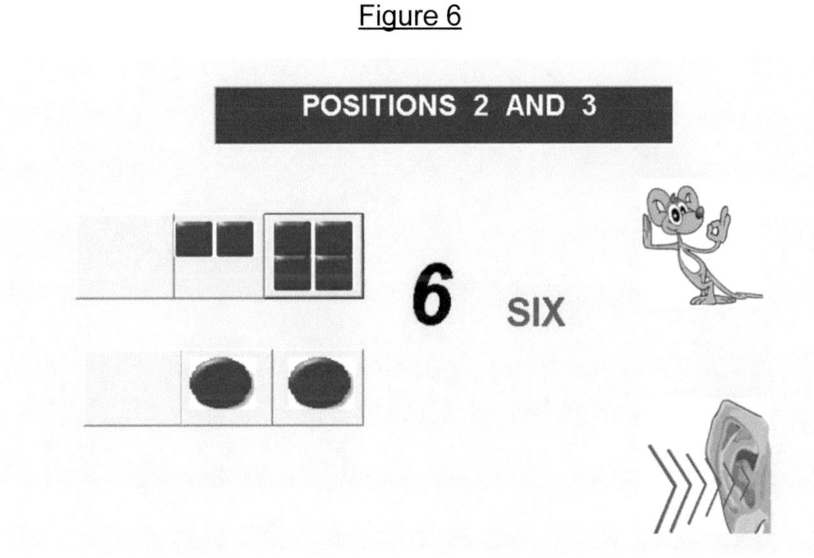

Pictures of boxes correspond to the mathematical value of number six.

The **audio** signal of a knock represents the **video** image of a filled up position and double knock represents an empty position. In our example double knock, knock and knock correspond to both pictures. The first picture of boxes has square symbols in the second and third place, but empty space in the first place. In the first position of the second picture we see an empty space and in the second and third positions we see a circle. The shape of the symbols does not affect the mathematical meaning. The symbols exist just to show that positions are already filled up. The mathematical value of the first configuration as well as the second is equal to six. The kinesthetic representation of the knock is the clenching of the right hand and for double knock is the tightening of the left hand into a fist. The **kinesthetic** representation for the first and second configurations of boxes is the clenching of the left hand once and the clenching the right hand twice.

MATHEMATICS

BLUE LESSONS, FIGURE 7

In Figure 7 we have the **video** display of two pictures, which have three positions each. All positions are filled up with symbols. The **audio** representation of this figure is **knock, knock and knock**.

Figure 7

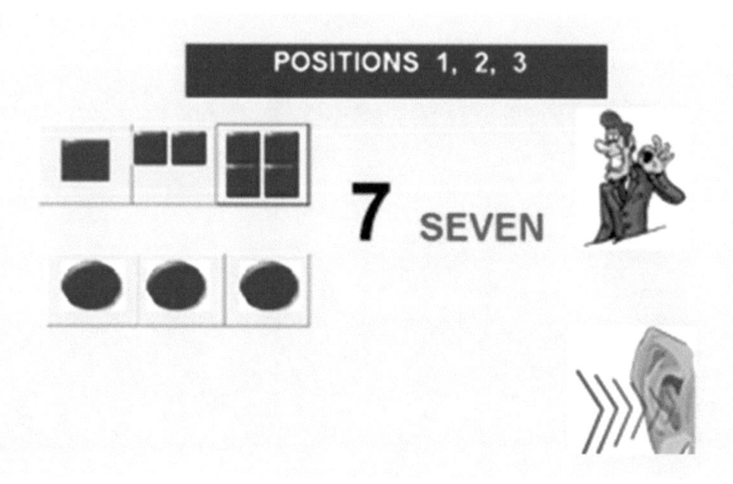

The **audio** signal of knock represents the filled up position and double knock represents an empty position. The **video** display of the first picture of boxes has square symbols in all places. In all positions of the second picture we see a circle. The different symbols in the first position of the pictures do not change the mathematical representation of both configurations. The shape of the symbols also does not affect the mathematical meaning. The symbols exist just to show that positions are already filled up. The mathematical value of the first configuration as well as the second is equal to seven. The **kinesthetic** representation of the knock is the clenching of the right hand and for double knock is the tightening of the left hand into a fist. The filled up position is represented by the clenching of the right hand and an empty position represented by the clenching of the left hand. The **kinesthetic** representation for the first and second configurations of boxes is the clenching of the right hand three times.

BLUE LESSONS, FIGURE 8

In Figure 8 we have a **visual** display of two pictures with four positions each. Positions one through three are empty and position four is filled up with symbols. The **audio** representation of those figures is **double knock, double knock, double knock and knock**.

Figure 8

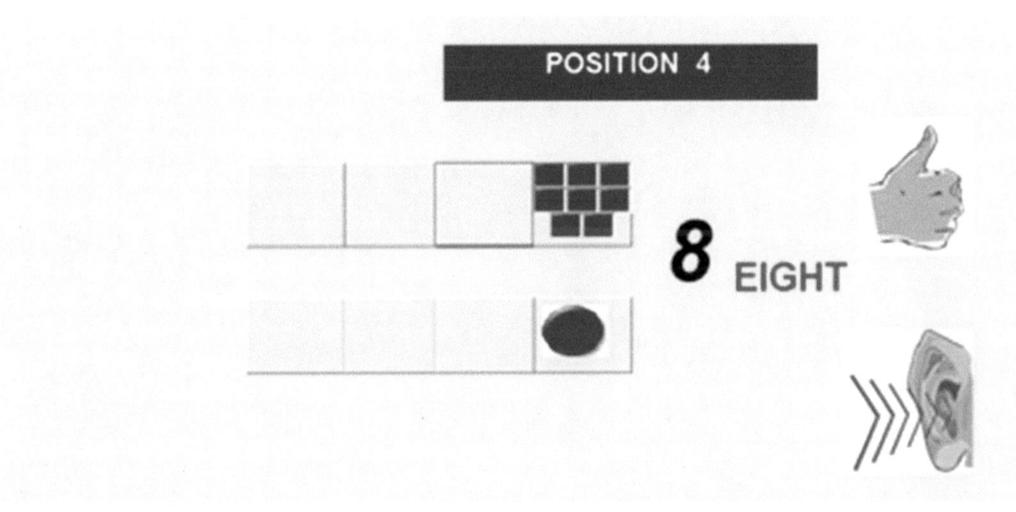

The **audio** signal of knock represents a filled up position and double knock represents an empty position. The **video** display of the first picture of boxes has a square symbol in the fourth place and empty spaces in the first place through third place. In the fourth position of the second picture we see a circle, but in the first, second, and third positions we have empty space. The shape of the symbols does not affect the mathematical meaning. The symbols exist just to show that positions are already filled up. The mathematical value of the first configuration as well as the second is equal to eight.

The kinesthetic representation of the knock is the clenching of the right hand and for double knock is the clenching of the left. The filled up position is represented by the clenching of the right hand and an empty position represented by the clenching of the left hand. The **kinesthetic** representation for the first and second configurations of boxes is the clenching of the left hand three times and then the right hand once.

MATHEMATICS

BLUE LESSONS, EXERCISE, FIGURE 1

There is a combination of positions, filled up with symbols or empty spaces. In Figure 1 the audio representation of one of the pictures below is **knock and double knock**.

Figure 1

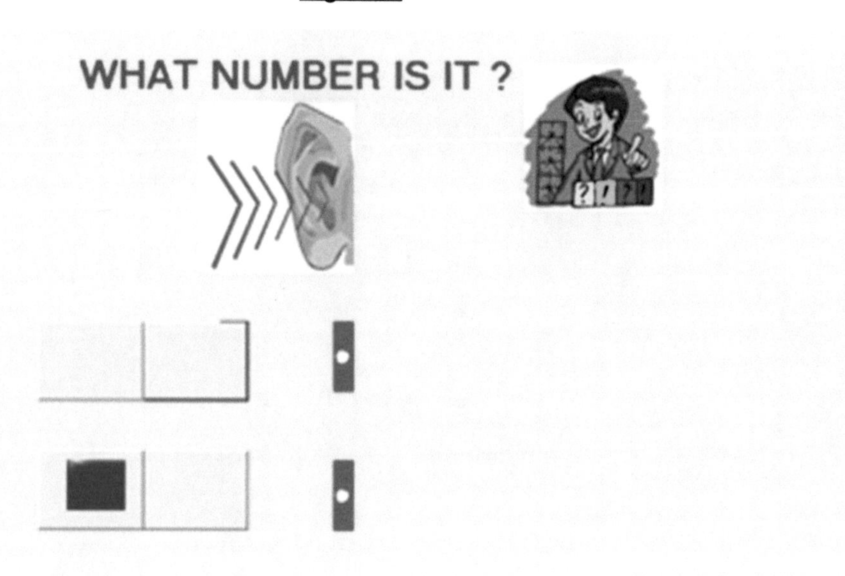

A knock represents a filled up position and a double knock represents an empty position. The clenching of the right hand represents a filled up position and the clenching of the left hand represents an empty position. The ***kinesthetic*** representation for the answer which corresponds to the second picture is the clenching of the right hand and left hand. The kinesthetic representation for the first picture is the clenching of the left hand twice.

BLUE LESSONS, EXERCISE, FIGURE 2

In Figure 2, we have three pictures with ***visual*** combinations of filled up symbols or empty spaces. The ***audio*** representation of one of the pictures below is **double knock and knock**

Figure 2

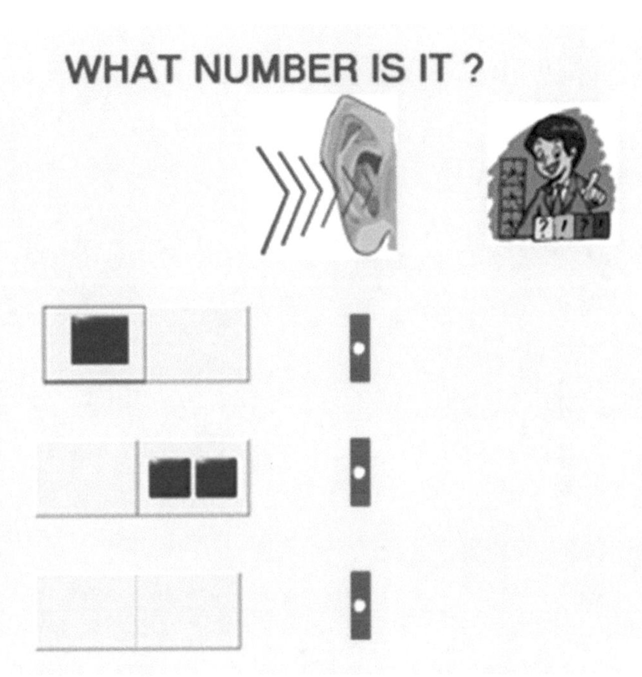

A knock represents a filled up position and a double knock represents an empty position. The clenching of the right hand represents a filled up position and the clenching of the left hand represents an empty position. The ***kinesthetic*** representation for the answer which corresponds to the second picture is the clenching of the left hand and right hand. The kinesthetic representation for the first picture is the clenching of the right hand and left hand. The kinesthetic representation for the third picture is the clenching of the left hand twice.

MATHEMATICS BLUE LESSONS, EXERCISE, FIGURE 3

In Figure 3, we have three different ***visual*** combinations of positions, filled up with symbols or empty positions. The ***audio*** representation of one of the pictures below is represented by **knock, knock.**

Figure 3

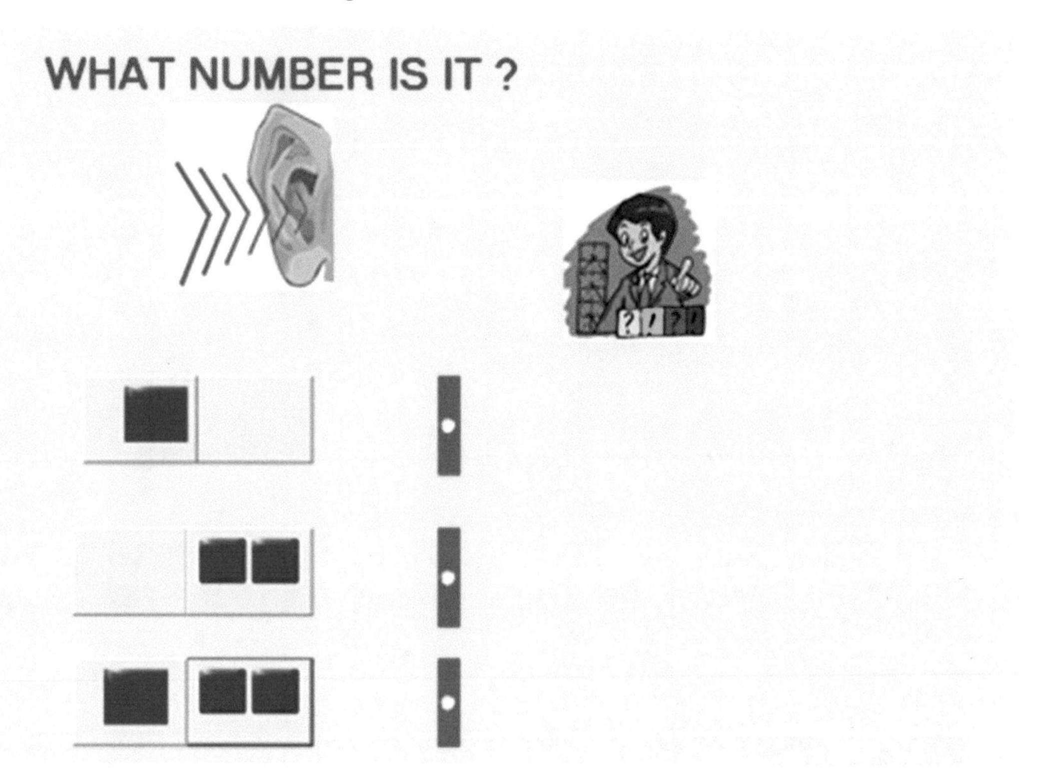

A knock represents a filled up position and a double knock represents an empty position. The clenching of the right hand represents a filled up position and the clenching of the left hand represents an empty position. The ***kinesthetic*** representation for the answer which corresponds to the third picture is the clenching of the right hand twice. The kinesthetic representation for the first picture is the clenching of the right hand and left hand. The kinesthetic representation for the second picture is the clenching of the left hand and right hand.

BLUE LESSONS, EXERCISE, FIGURE 4

In Figure 4, we have three ***visual*** combinations, filled up with symbols or empty positions. The ***audio*** representation of one of the pictures below is **double knock, double knock, and knock**.

Figure 4

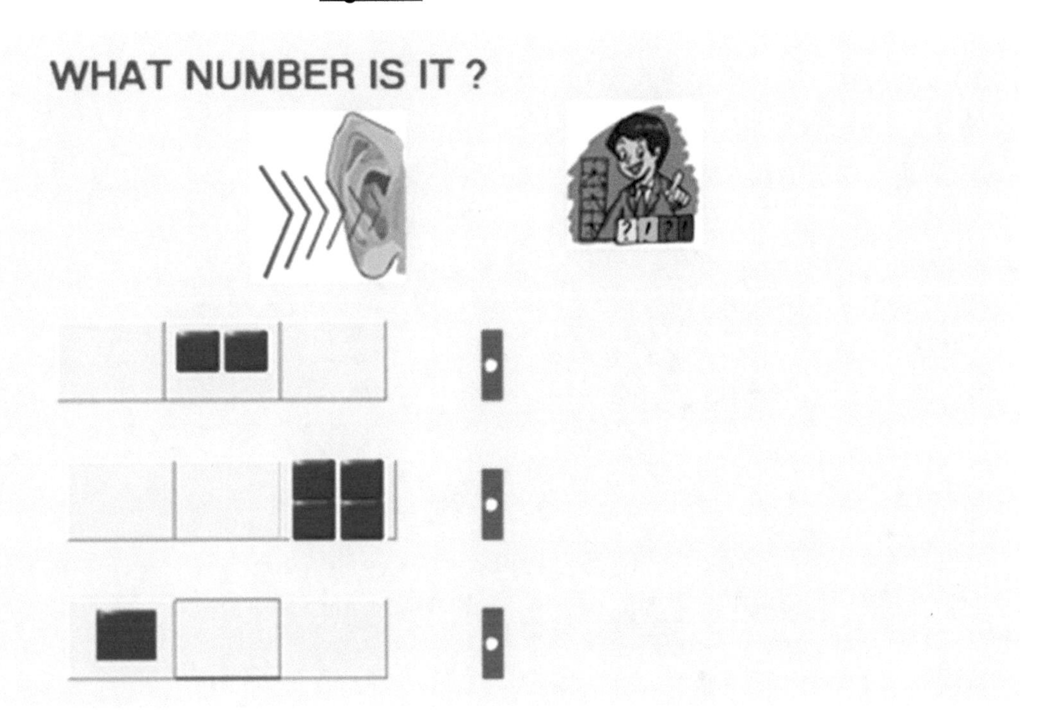

A knock represents a filled up position and a double knock represents an empty position. The clenching of the right hand represents a filled up position and the clenching of the left hand represents an empty position. The ***kinesthetic*** representation for the answer corresponding to the second picture is the clenching of the left hand twice and right hand. The kinesthetic representation for the first picture is the clenching of the left hand and right hand. The kinesthetic representation for the third picture is the clenching of the right hand and left hand twice.

MATHEMATICS

BLUE LESSONS, EXERCISE, FIGURE 5

In Figure 5, we have three ***visual*** combinations, filled up with symbols or empty positions. The ***audio*** representation of one of the pictures below is **knock, double knock and knock**.

Figure 5

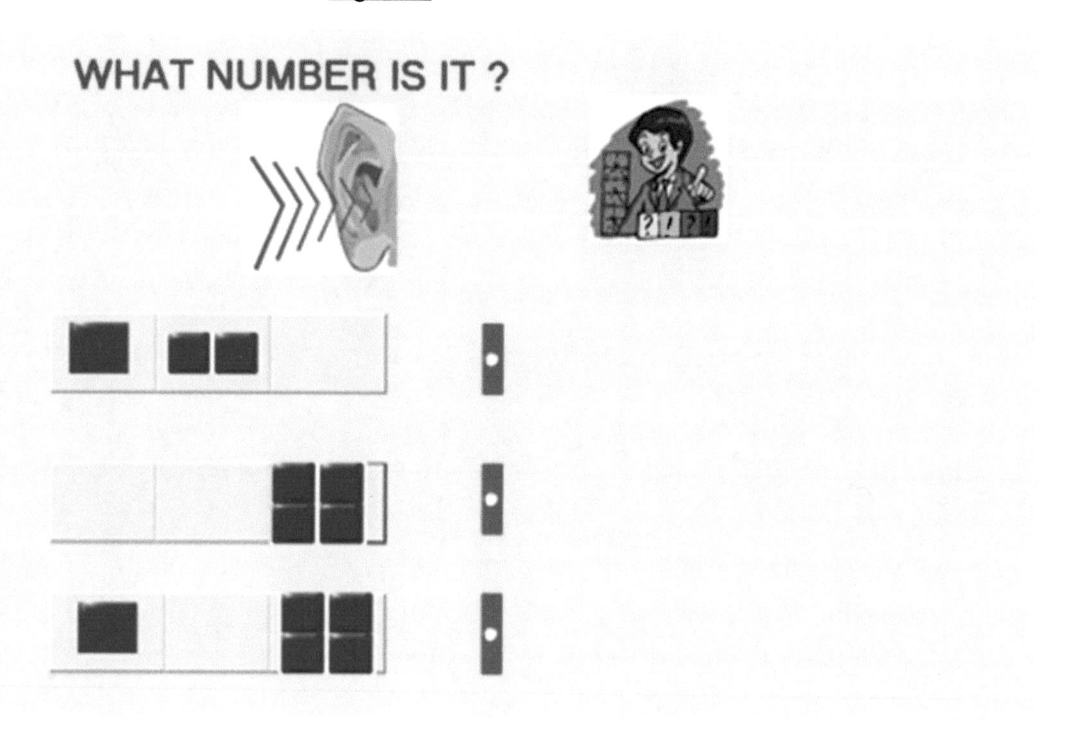

A knock represents a filled up position and a double knock represents an empty position. The clenching of the right hand represents a filled up position and the clenching of the left represents an empty position. The ***kinesthetic*** representation for the answer which corresponds to the third picture is the clenching of the right hand, left hand, and right hand. The kinesthetic representation for the first picture is the clenching of the right hand twice and left hand. The kinesthetic representation for the second picture is the clenching of the left hand twice and right hand.

BLUE LESSONS, EXERCISE, FIGURE 6

In Figure 6, we have three ***visual*** combinations, filled up with symbols or empty positions. The ***audio*** representation for one of the pictures below is **double knock, knock, and knock**.

Figure 6

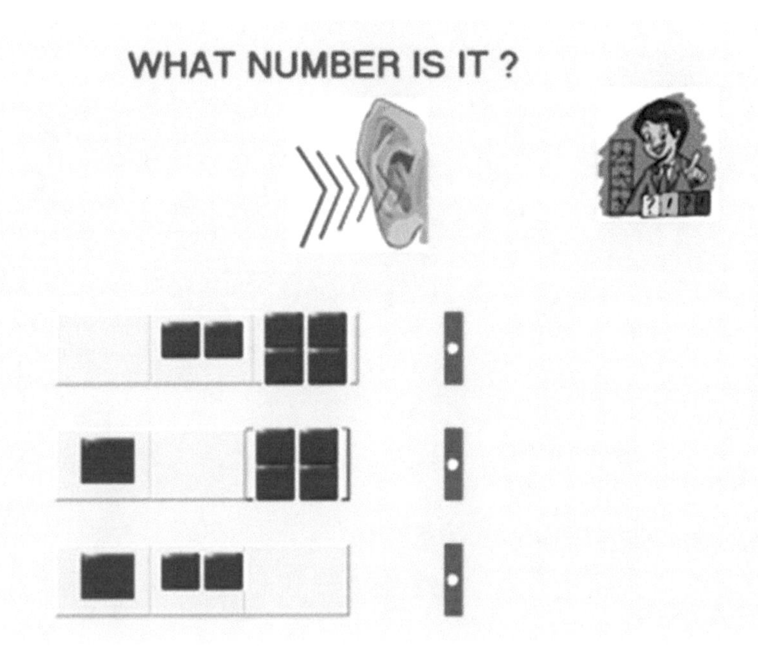

A knock represents a filled up position and a double knock represents an empty position. The clenching of the right hand represents a filled up position and the clenching of the left hand represents an empty position. The ***kinesthetic*** representation for the answer which corresponds to the first picture is the clenching of the left hand and right hand twice. The kinesthetic representation for the second picture is the clenching of the right hand, left hand, and right hand. The kinesthetic representation for the third picture is the clenching of the right hand twice and left hand.

MATHEMATICS

BLUE LESSONS, EXERCISE, FIGURE 7

In Figure 7, we have three **visual** combinations that are filled up with symbols or empty positions. The ***audio*** representation of one of the pictures below is **knock, knock, and knock**.

Figure 7

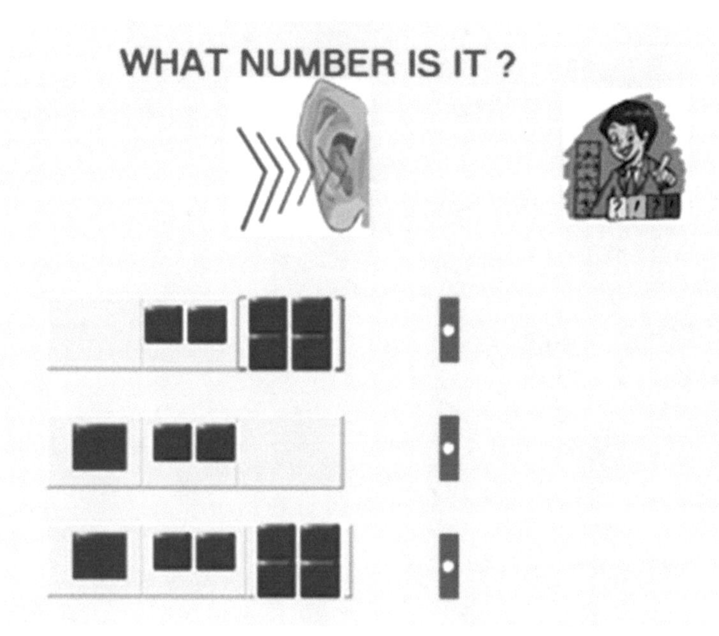

A knock represents a filled up position and a double knock represents an empty position. The clenching of the right hand represents a filled up position and the clenching of the left hand represents an empty position. The ***kinesthetic*** representation for the answer which corresponds to the third picture is the clenching of the right hand thrice. The kinesthetic representation for the first picture is the clenching of the left hand and right hand twice. The kinesthetic representation for the second picture is the clenching of the right hand twice and left hand.

BLUE LESSONS, EXERCISE, FIGURE 8

In Figure 8, we have three ***visual*** combinations that are filled up with symbols or empty positions. The ***audio*** representation of one of the pictures below is **double knock, double knock, double knock, and knock**.

Figure 8

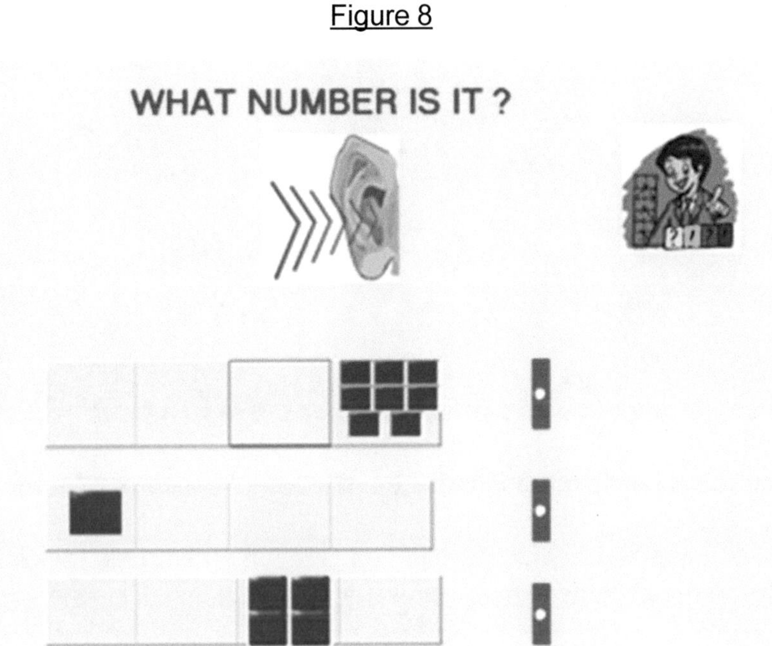

A knock represents a filled up position and a double knock represents an empty position. The clenching of the right hand represents a filled up position and the clenching of the left hand represents an empty position. The ***kinesthetic*** representation for the answer which corresponds to the first picture is the clenching of the left hand thrice and right hand. The kinesthetic representation for the second picture is the clenching of the right hand and left hand thrice. The kinesthetic representation for the third picture is the clenching of the left hand twice, right hand, and left hand.

In Figure 10 we have a **visual** display of two pictures with four positions. Positions one and four are filled up with symbols but positions two and three are empty. The **audio** representation of this figure is **knock, double knock, double knock, and knock.**

Figure 9

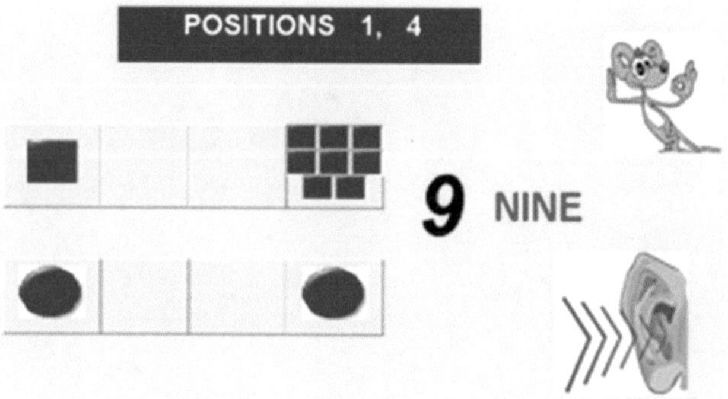

POSITIONS 1, 4

9 NINE

The pictures of boxes correspond to the mathematical value of number nine.

The **audio** signal of knock represents the filled up position and double knock represents an empty position. In our example knock, double knock, double knock, and knock correspond to both pictures.

The **visual** display of the first picture of boxes has a square symbol in the first place and in the fourth place, but the second place and third place both have empty spaces. There is a circle in the first and second positions of the second picture. The different symbols in the first and fourth positions of the pictures do not change the mathematical representation of both configurations. The shape of the symbols also does not have any effect on the mathematical meaning. The symbols exist just to show that positions are already filled up. The mathematical value of the first configuration as well as the second is equal to nine.

The **kinesthetic** representation for the knock is the clenching of the right hand and for double knock is the tightening of the left hand into a fist. The filled up position is represented by the clenching of the right hand and an empty position represented by the clenching of the left hand. The **kinesthetic** representation for the first and second configurations of boxes is the clenching of the right hand once, left hand twice, and then the right hand once.

BLUE LESSONS, LESSON 2, FIGURE 10

In Figure 10 we have the **visual** display of two pictures with four positions each. Positions two and four are filled up with symbols. Positions one and three are empty. The **audio** representation of those figures is **double knock, knock, double knock and knock**.

Figure 10

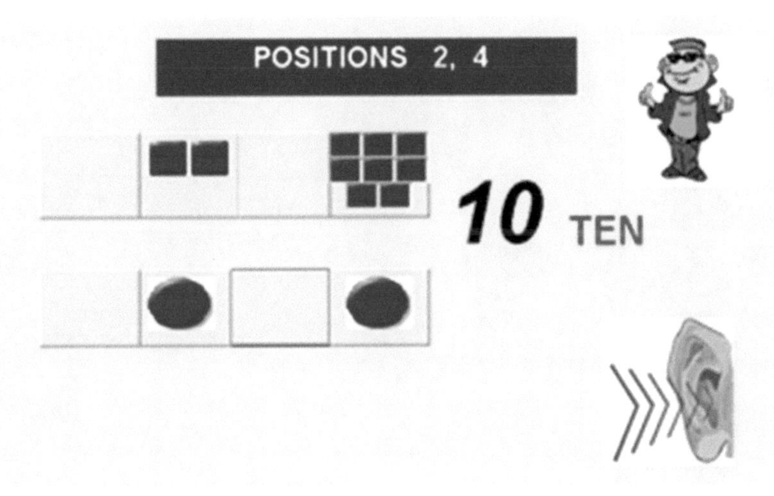

The pictures of boxes correspond to the mathematical value of number ten.

The **audio** signal of knock represents the filled up position and double knock represents an empty position. In our example double knock, knock, double knock, and knock correspond to both pictures. The **video** display of the first picture of boxes has square symbols in the second place and fourth place. First place and third place have empty spaces. In the second and third position of the second picture we see a circle. In the first and second positions we have empty spaces. The shape of the symbols does not affect the mathematical meaning. The symbols exist just to show that positions are already filled up. The mathematical value of the first configuration as well as the second is equal to ten.

The **kinesthetic** representation of the knock is the clenching of the right hand and for double knock is the tightening of the left hand into a fist. The filled up position is represented by the clenching of the right hand and the empty position represented by the clenching of the left hand. The **kinesthetic** representation for the first and second configurations of boxes is the clenching of the left, right, left and right hands.

MATHEMATICS

BLUE LESSONS, LESSON 2, FIGURE 11

In Figure 11 we have a ***visual*** display of the two pictures with four positions. The positions one, two and four are filled up with symbols but position three is empty. The ***audio*** representation of those figures is **knock, knock, double knock and knock**.

Figure 11

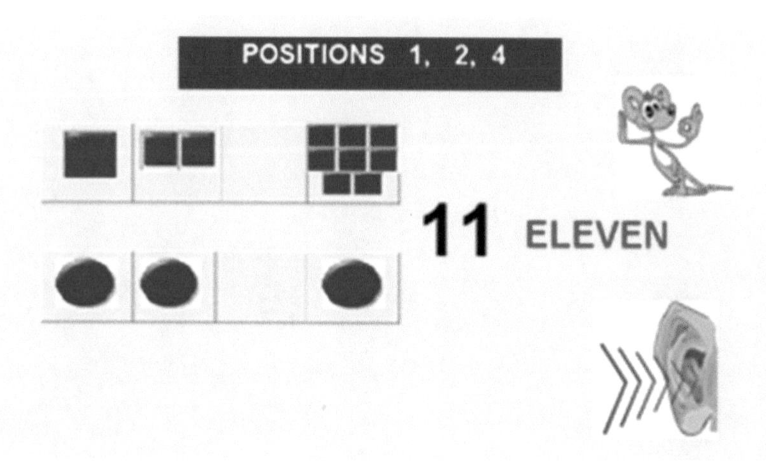

The pictures of boxes correspond to the mathematical value of eleven.

In our example the **audio** signals of knock and double knock correspond to both pictures.

The **visual** display of the first picture of boxes has a square symbol in the first, second and fourth places, and empty space in the second place. In the first, second and fourth positions of the second picture we see a circle. The different symbols in the first, second and fourth positions of the pictures do not change the mathematical representation of both configurations. The shape of the symbols also does not have any effect on the mathematical meaning. The symbols exist just to show that positions are already filled up. The mathematical value of the first and second configurations is equal to eleven.

The **kinesthetic** representation of the knock is the clenching of the right hand and for double knock is the tightening of the left hand into a fist. The filled up position is represented by the clenching of the right hand and the empty position by the clenching of the left hand. The **kinesthetic** representation for the first and second configurations of boxes is the clenching of the right hand twice, left once and the then the right hand one last time.

BLUE LESSONS, LESSON 2, FIGURE 12

In Figure 12, we have a ***visual*** display of two pictures containing four positions each. Positions three and four are filled up with symbols, but the first and second positions are empty.

The ***audio*** representation of this figure is **double knock, double knock, knock, and knock.**

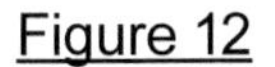

Figure 12

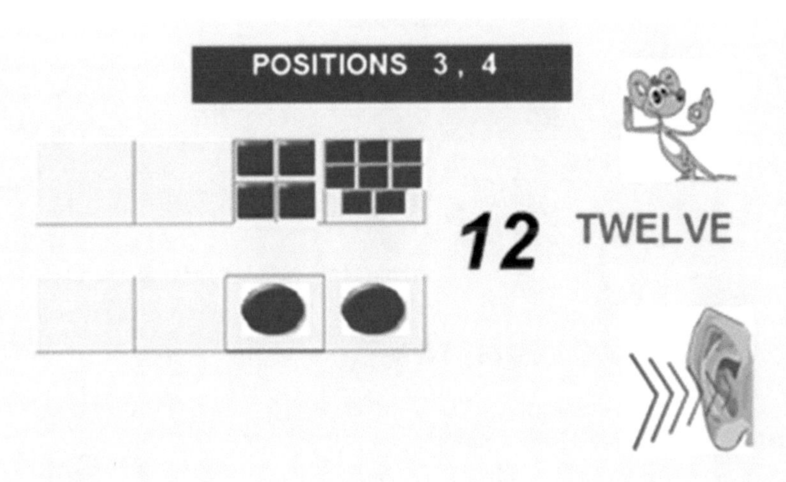

The pictures of boxes correspond to the mathematical value of number twelve.

The **audio** signal of knock represents the filled up position and double knock represents an empty position. In our example double knock, double knock, knock and knock correspond to both pictures. The **visual** display of the first picture of boxes has a square symbol in the third and fourth places but empty spaces in the first and second place. In the third and fourth positions of the second picture, we see a circle. The different symbols in the individual positions do not change the mathematical representation of both configurations. The symbols exist just to show that the positions are already filled up. The mathematical value of the first configuration as well as the second is equal to twelve.

The **kinesthetic** representation of the knock is the clenching of the right hand and for double knock is the tightening of the left hand into a fist. The filled up position is represented by the clenching of the right hand and the empty position represented by the clenching of the left hand. The **kinesthetic** representation for the first and second configurations of boxes is the clenching of the left hand twice, then the right hand twice.

MATHEMATICS BLUE LESSONS, LESSON 2, FIGURE 13

In Figure 13, we have a **visual** display of two pictures with four positions in each. Positions one, three, and four are filled with symbols and the second position is empty.

The **audio** representation of those figures is **knock, double knock, knock and knock**.

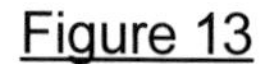

Figure 13

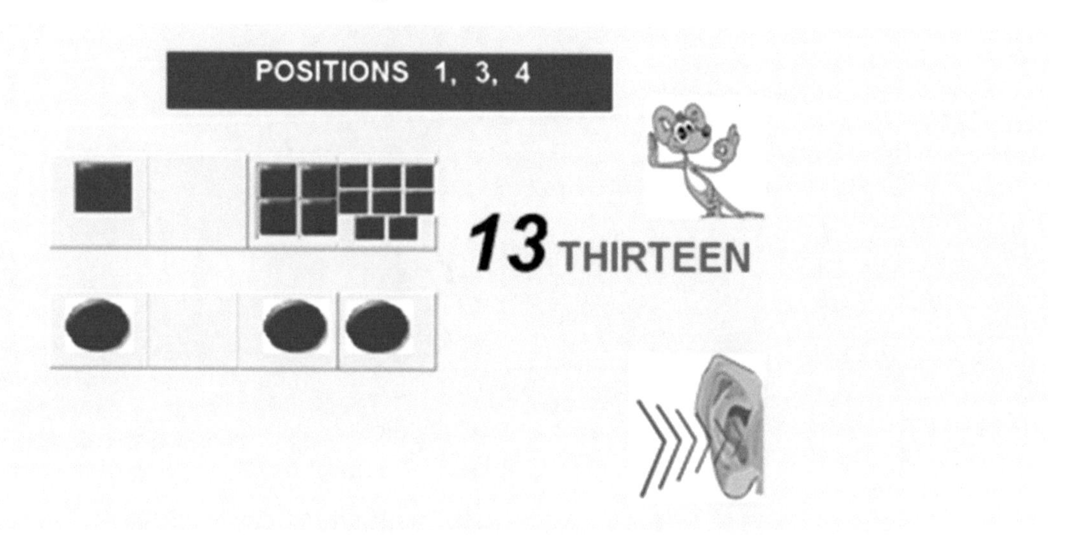

The pictures of boxes correspond to the mathematical value of number thirteen.

The **audio** signal of knock represents the filled position and the double knock represents the empty position. In our example knock, double knock, knock, and knock correspond to both pictures.

The **visual** display of the first picture of boxes has a square symbol in the first, third and fourth places, but there is empty space in the second place. In the first, third and fourth positions of the second picture we see circles. The mathematical value of the both the first and second configurations is equal to thirteen.

The **kinesthetic** representation of a knock is the clenching of the right hand and for a double knock is the tightening of the left hand into a fist. The **kinesthetic** representation for the first and second configurations of boxes is the clenching of the right, left, and the right twice.

BLUE LESSONS, LESSON 2, FIGURE 14

In Figure 14, we have a **visual** display of two pictures with four positions each. Position one is empty and the other three positions are filled up with symbols.

The **audio** representation of this figure is **double knock, knock, knock, and knock**.

Figure 14

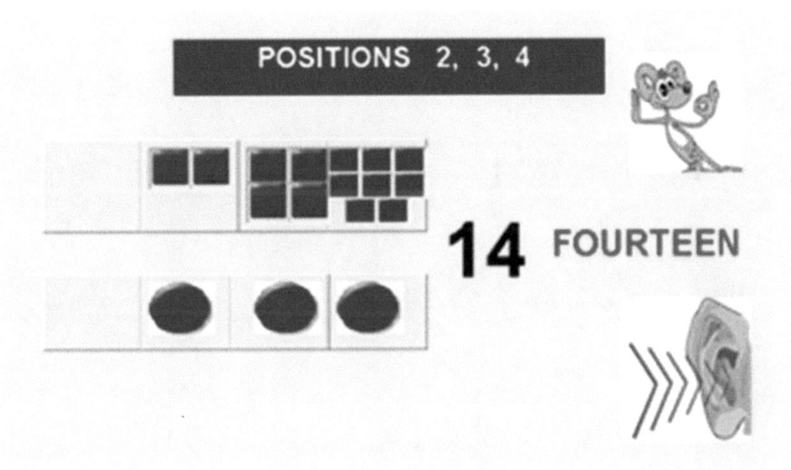

In the pictures above the **audio** signal of knock represents the filled up position and double knock represents an empty position. In our example, double knock, knock, knock, and knock correspond to both pictures.

The **video** display of the first picture of boxes has a square symbol in the second, third, forth place and empty space in the first. In the second, third and forth position of the second picture we have a circle. The symbols and shapes do not change the mathematical representation of both configurations. The symbols are there just to illustrate that positions are already filled up. The mathematical value of the first and second configurations is equal to fourteen.

A filled up position is represented by the clenching of the right hand and an empty position represented by the clenching of the left hand. In conclusion, the **kinesthetic** representation for the first and second configurations of boxes is the clenching of the left hand once and then the right hand three times.

MATHEMATICS

BLUE LESSONS, FIGURE 15

In Figure 15, in a **visual** display of both pictures we have four positions on each picture. All four positions are filled up with symbols.

The **audio** representation of this figure is **knock, knock, knock and knock**.

Figure 15

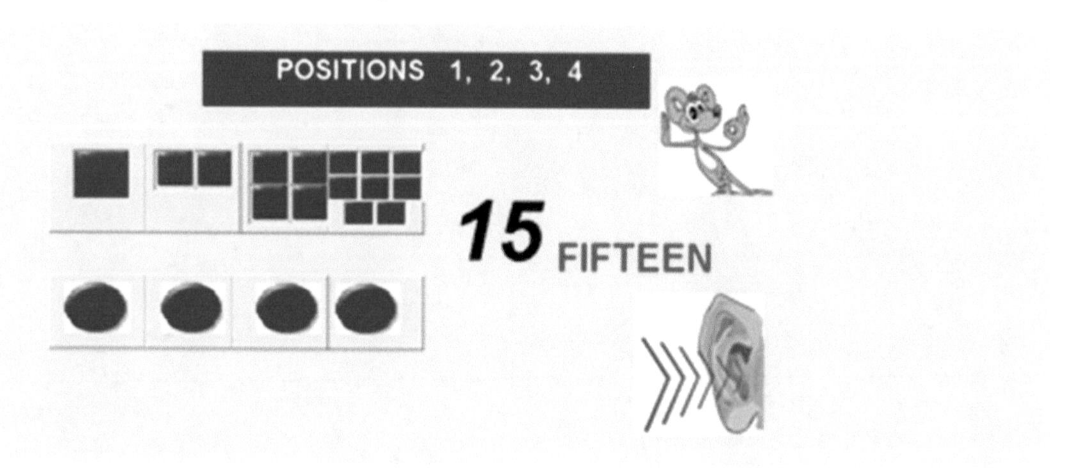

The **audio** signal of knock represents a filled up position. In our example we have four knocks that represent both pictures. In the **video** display of the first picture of boxes, we have a square symbol in all four places. Similarly, we have all places filled up in the second picture, but this time with a circle instead of a square. The symbols are there to indicate that the positions are already filled up. The mathematical value of the first and second configurations is equal to fifteen.

The filled up position is represented by the clenching of the right hand and the empty position is represented by the clenching of the left hand.

The **kinesthetic** representation for the first and second configurations of boxes is the clenching of the right hand four times.

BLUE LESSONS, FIGURE 16

In Figure 16, in a ***visual*** display of both pictures we have five positions in each picture. Position five is filled up with symbols and first four positions are empty.

The ***audio*** representation of this figure is **double knock, double knock, double knock, double knock and knock.**

Figure 16

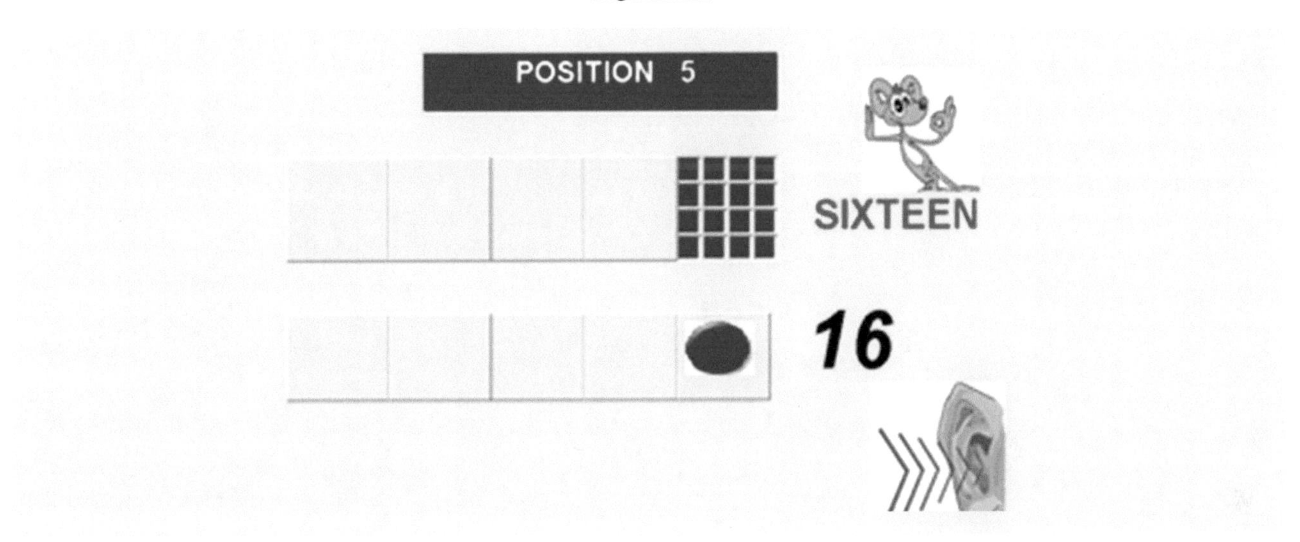

In the first picture of boxes we see square symbols in the fifth place and empty spaces before it. In the fifth position of the second picture we have a circle and empty spaces in the first four positions. The mathematical value of both configurations is equal to sixteen. A filled up position is represented by the clenching of the right hand. An empty position is represented by the clenching of the left hand.

The **kinesthetic** representation for the first and second configurations of boxes is clenching of the left hand four times and then the right hand once.

MATHEMATICS

BLUE LESSONS, EXERCISE, FIGURE 9

In Figure 9, we have three ***visual*** combinations that are filled up with symbols or empty positions. The ***audio*** representation of one of the pictures below is **knock, double knock, double knock and knock.**

Figure 9

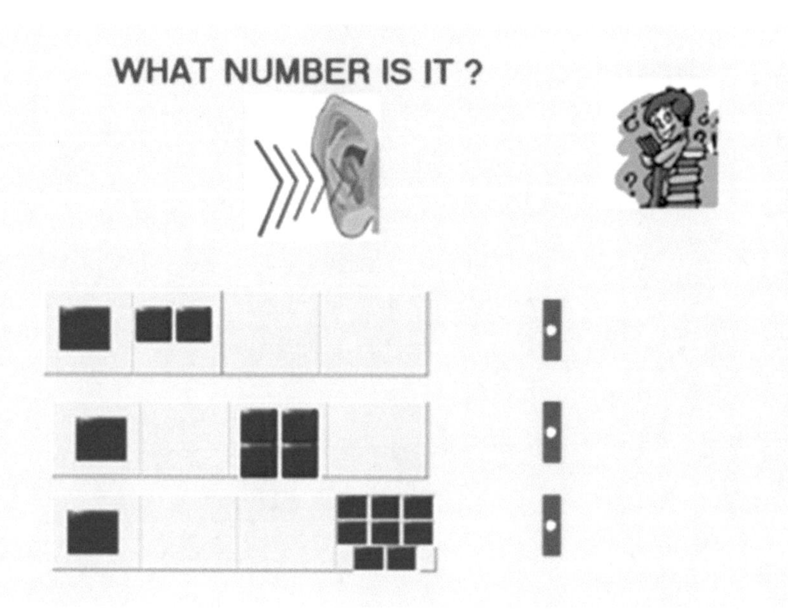

A knock represents a filled up position and a double knock represents an empty position. The clenching of the right hand represents a filled up position and the clenching of the left hand represents an empty position. The ***kinesthetic*** representation for the answer which corresponds to the third picture is the clenching of the right hand, left hand twice, and right hand. The kinesthetic representation for the first picture is the clenching of the right hand twice and left hand twice. The kinesthetic representation for the second picture is the clenching of the right hand, left hand, right hand, and left hand.

BLUE LESSONS, EXERCISE, FIGURE 10

In Figure 10, we have three ***visual*** combinations that are filled up with symbols or empty positions. The ***audio*** representation of one of the pictures below is **double knock, knock, double knock, and knock**.

Figure 10

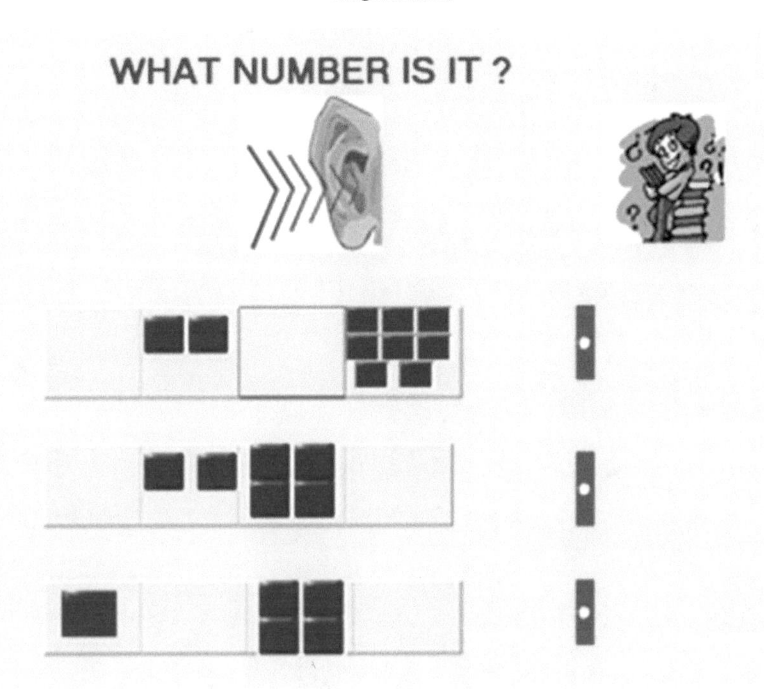

A knock represents a filled up position and a double knock represents an empty position. The clenching of the right hand represents a filled up position and the clenching of the left hand represents an empty position. The ***kinesthetic*** representation for the answer which corresponds to the first picture is the clenching of the left hand, right hand, left hand, and right hand. The kinesthetic representation for the second picture is the clenching of the left hand, right hand twice, and left hand. The kinesthetic representation for the third picture is the clenching of the right hand, left hand, right hand, and left hand.

MATHEMATICS BLUE LESSONS, EXERCISE, FIGURE 11

In Figure 11 we have three ***visual*** combinations that are filled up with symbols or empty positions. The ***audio*** representation of one of the pictures below is **knock, knock, double knock, and knock**

Figure 11

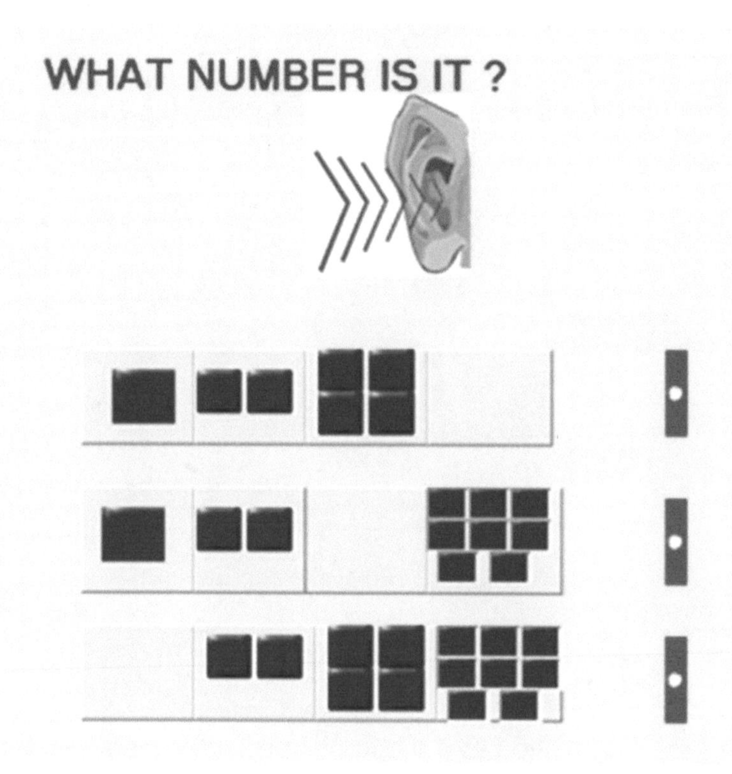

A knock represents a filled up position and a double knock represents an empty position. The clenching of the right hand represents a filled up position and the clenching of the left hand represents an empty position. The ***kinesthetic*** representation for the answer which corresponds to the second picture is the clenching of the right hand twice, left hand, and right hand. The kinesthetic representation for the first picture is the clenching right hand thrice and left hand. The kinesthetic representation for the third picture is the clenching of the left hand and right hand thrice.

BLUE LESSONS, EXERCISE, FIGURE 12

In Figure 12 we have three ***visual*** combinations that are filled up with symbols or empty positions. The ***audio*** representation of one of the pictures below is **double knock, double knock, knock, and knock**.

Figure 12

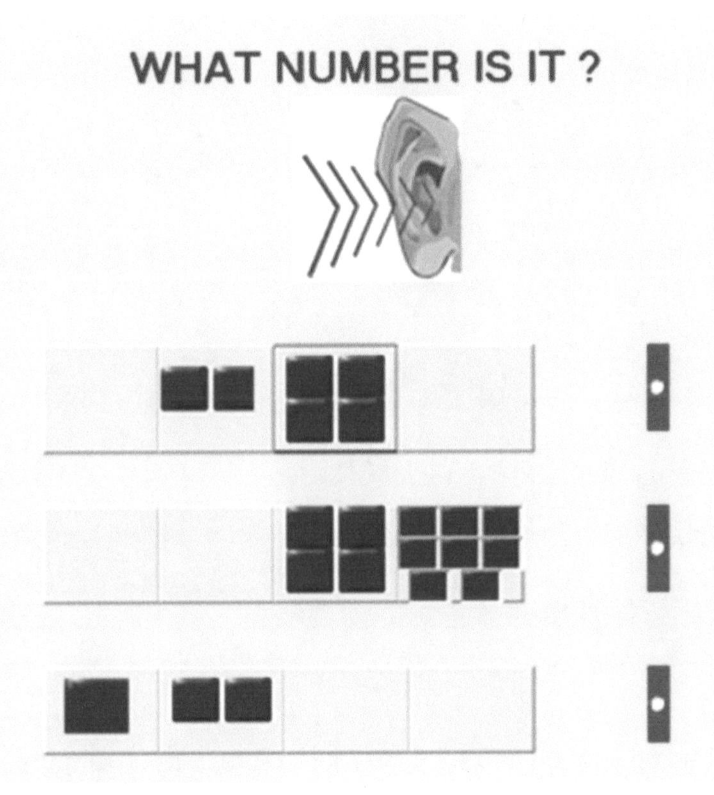

A knock represents a filled up position and a double knock represents an empty position. The clenching of the right hand represents a filled up position and the clenching of the left an empty position. The ***kinesthetic*** representation for the answer corresponding to the second picture is the clenching of the left hand twice and right hand twice. The kinesthetic representation for the first picture is the clenching of the left hand, right hand twice, and left hand. The kinesthetic representation for the third picture is the clenching of the right hand twice and left hand twice.

MATHEMATICS BLUE LESSONS, EXERCISE, FIGURE 13

In Figure 13, we have three ***visual*** combinations that are filled up with symbols or empty positions. The ***audio*** representation of one of the pictures below is **knock, double knock, knock, and knock**.

Figure 13

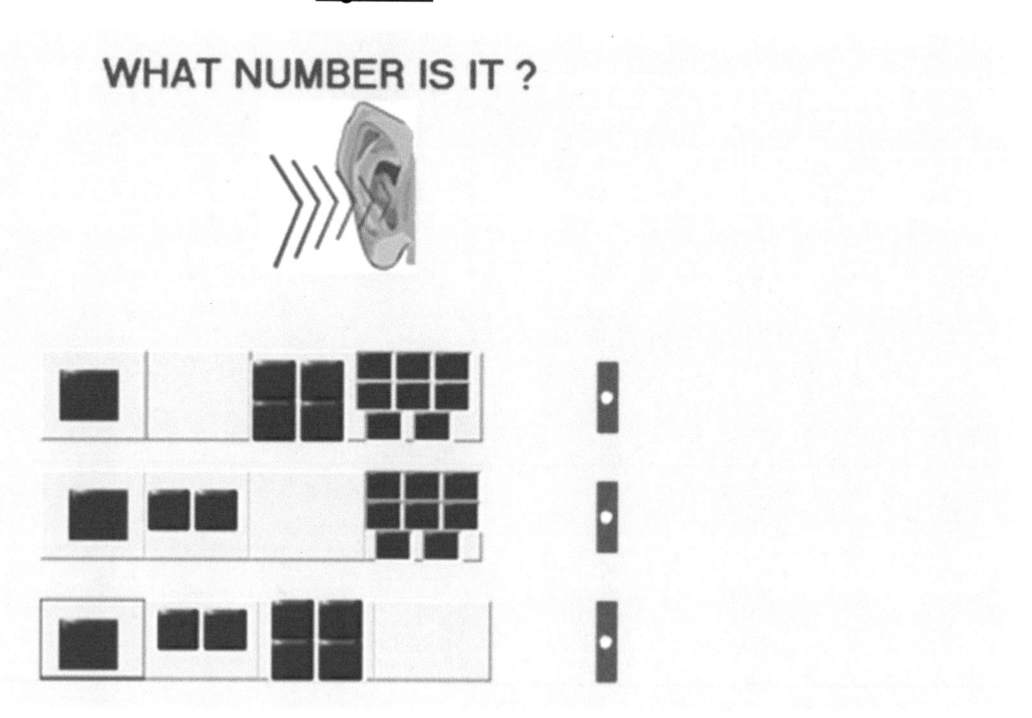

A knock represents a filled up position and a double knock represents an empty position. The clenching of the right hand represents a filled up position and the clenching of the left hand represents an empty position. The ***kinesthetic*** representation for the answer which corresponds to the first picture is the clenching of the right hand, left hand, and right hand twice. The kinesthetic representation for the second picture is the clenching of the right hand twice, left hand, and right hand. The kinesthetic representation for the third picture is the clenching of the right hand three times and left hand once.

BLUE LESSONS, EXERCISE 14, FIGURE 14

In Figure 14, we have three ***visual*** combinations that are filled up with symbols or empty positions. The ***audio*** representation of one of the pictures below is **double knock, knock, knock, and knock**.

Figure 14

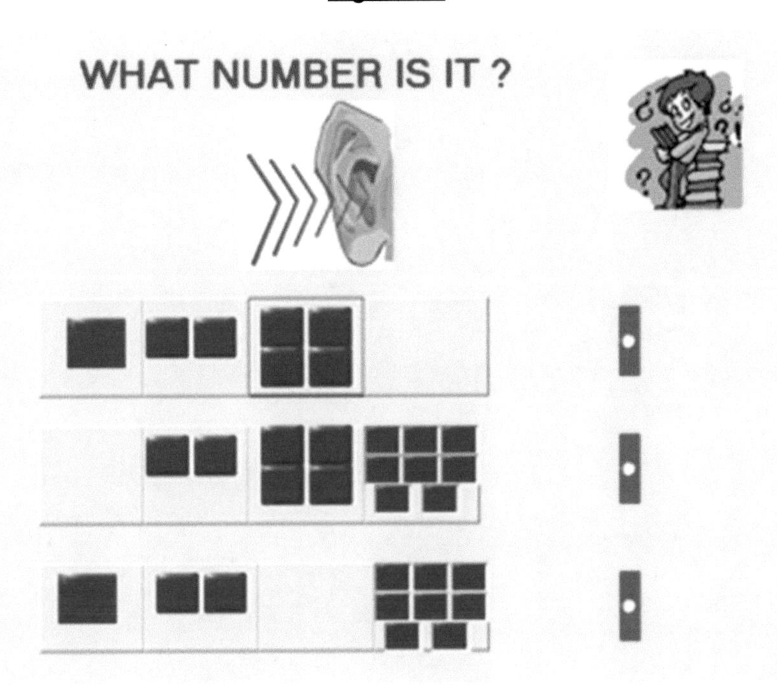

A knock represents a filled up position and a double knock represents an empty position. The clenching of the right hand represents a filled up position and the clenching of the left hand represents an empty position. The ***kinesthetic*** representation for the answer which corresponds to the second picture is the clenching of the left hand, followed by the right hand thrice. The kinesthetic representation for the first picture is the clenching of the right hand thrice and left hand. The kinesthetic representation for the third picture is the clenching of the right hand twice, left hand, and right hand.

MATHEMATICS BLUE LESSONS, EXERCISE 15, FIGURE 15

In Figure 15, we have three ***visual*** combinations that are filled with symbols or empty positions. The ***audio*** representation of one of the pictures below is **knock, knock, knock, and knock**.

Figure 15

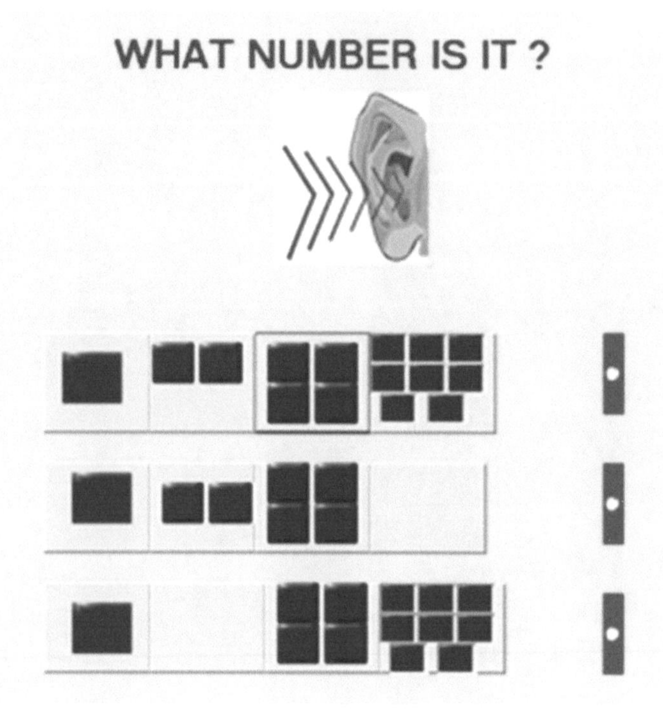

A knock represents a filled position and a double knock represents an empty position. The clenching of the right hand represents a filled position and the clenching of the left hand represents an empty position. The ***kinesthetic*** representation for the answer which corresponds to the first picture is the clenching of right hand four times. The kinesthetic representation for the second picture is the clenching of the right hand thrice and left hand. The kinesthetic representation for the third picture is the clenching of the right hand, left hand, and right hand twice.

BLUE LESSONS, EXERCISE 16, FIGURE 16

In Figure 16 we have three ***visual*** combinations that are filled up with symbols or empty positions. The ***audio*** representation of one of the pictures below is **double knock, double knock, double knock, double knock and knock**.

Figure 16

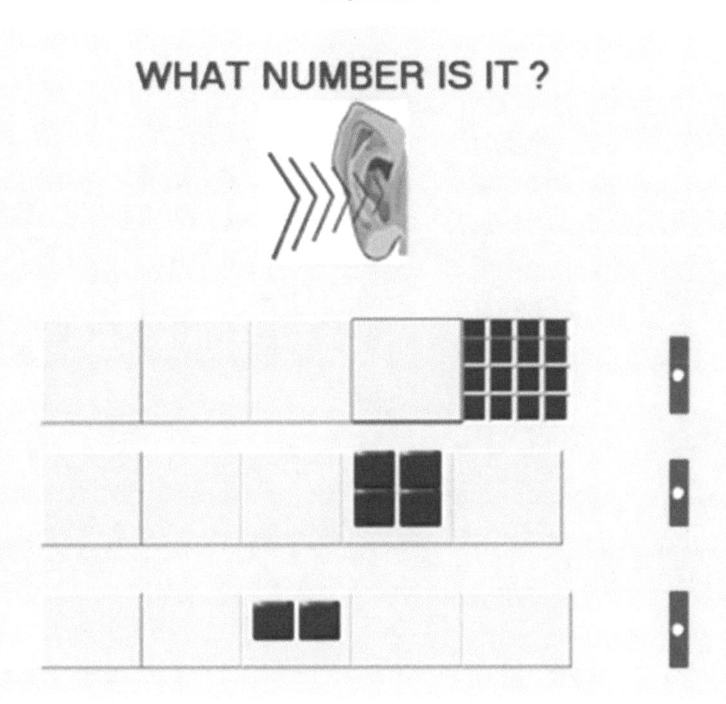

A knock represents a filled up position and a double knock represents an empty position. The clenching of the right hand represents a filled up position and the clenching of the left hand represents an empty position. The ***kinesthetic*** representation for the answer which corresponds to the first picture is the clenching of the left hand four times and right hand. The kinesthetic representation for the second picture is the clenching of the left hand thrice, right hand, and left hand. The kinesthetic representation for the third picture is the clenching of the left hand twice, right hand, and left hand twice.

MATHEMATICS BLUE LESSONS, LESSON, FIGURE 17

In Figure 17, in a ***visual*** display both pictures have five positions, where position one and five are filled up with symbols, and positions two, three and four are empty.

The ***audio*** representation of this figure is a **knock, double knock, double knock, double knock and knock.**

Figure 17

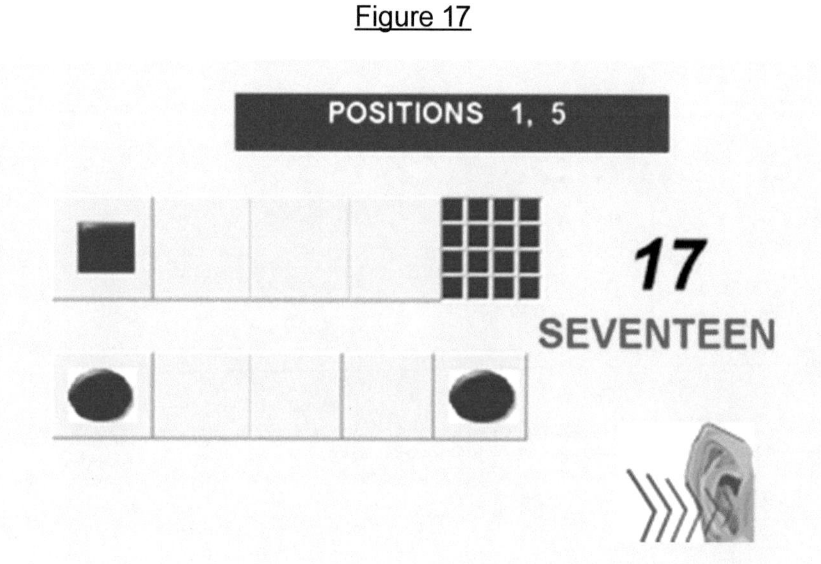

From the diagram we have the **visual** representation of a filled up position represented by a knock and the empty position represented by a double knock. The corresponding **audio** signals are knock, three double knocks and one single knock. In the first picture of boxes there are square symbols in the first and fifth places. In the same positions of the second picture, there are circles. The different symbols and shapes in both pictures do not change the mathematical representation of both configurations. The mathematical value of both configurations is equal to seventeen.

A filled up position is represented by the clenching of the right hand. An empty position is represented by the clenching of the left hand. The ***kinesthetic*** representation for the first and second configurations of boxes is the clenching of the right once, left hand three times and then the right hand once, signifying that our mathematical value is equal to seventeen.

BLUE LESSONS, LESSON, FIGURE 18

In Figure 18, in a ***visual*** display both pictures have five positions, where positions two and five are filled with symbols, and positions one, three, and four are empty.

The ***audio*** representation of this figure is a **double knock, knock, double knock, double knock and knock**.

Figure 18

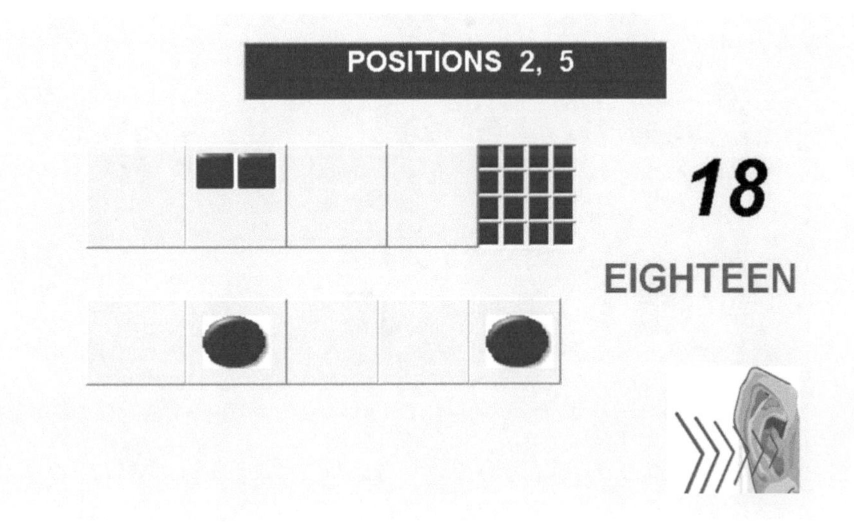

In the first picture of boxes we see square symbols in the second and fifth places. In the next picture we have circles filling up the second and fifth positions. The different symbols and shapes in both pictures do not change the mathematical representation of both configurations. The mathematical value of both configurations is equal to eighteen.

The ***kinesthetic*** representation for the first and second configurations of boxes is the clenching of the left, right, left twice and then the right hand once.

MATHEMATICS BLUE LESSONS, LESSON, FIGURE 19

In Figure 19, in both ***visual*** pictures we have five positions, where positions one, two and five are filled with symbols, and positions three to four are empty.

The ***audio*** representation of this figure is **knock, knock, double knock, double knock and knock**

Figure 19

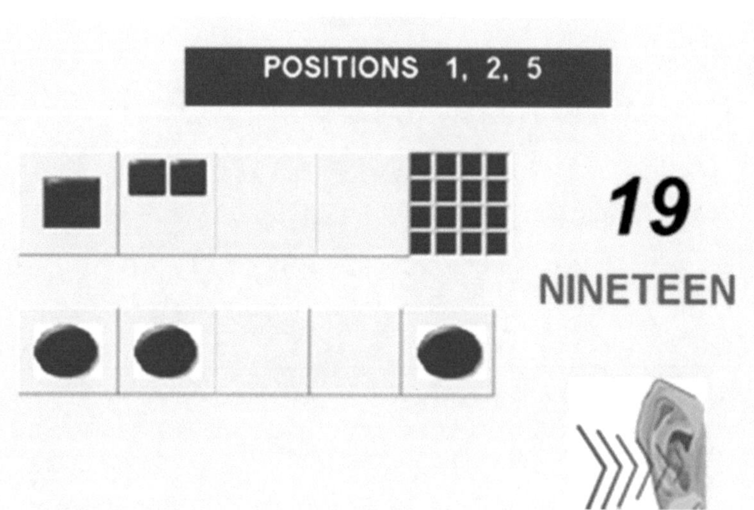

In the first picture of boxes we see square symbols in the first, second and fifth places. In the second picture the first, second and fifth positions are filled with circles. The different symbols and shapes in the pictures do not change the mathematical representation of both configurations. The mathematical value of both configurations is equal to nineteen.

The ***kinesthetic*** representation for the first and second configurations of boxes is the clenching of the right hand twice, left twice and then the right hand once.

BLUE LESSONS, LESSON, FIGURE 20

In Figure 20, in both ***visual*** pictures we have five positions, where positions three and five are filled with symbols, and positions one, two, and four are empty.

The ***audio*** representation of this figure is **double knock, double knock, knock, double knock and one single knock**.

Figure 20

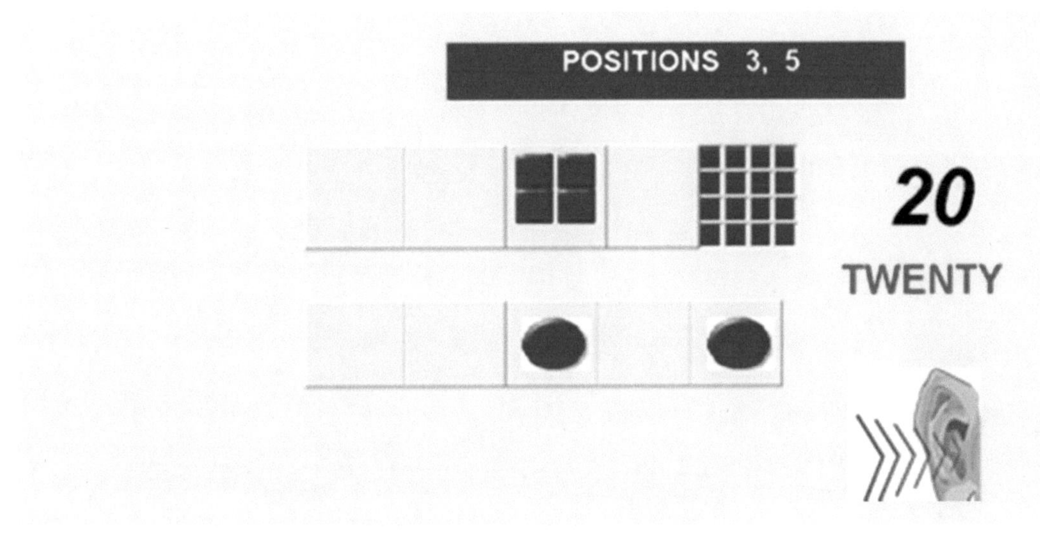

In the first picture of boxes we see square symbols in the third and fifth place. In the second picture the third and fifth positions are filled with circles. The different symbols and shapes in the pictures do not change the mathematical representation of both configurations. As a result, the mathematical value of both configurations is equal to twenty.

The ***kinesthetic*** representation for the first and second configurations of boxes is the clenching of the left twice, right once, left once and then the right hand once.

MATHEMATICS BLUE LESSONS, LESSON, FIGURE 21

In Figure 21, in both ***visual*** pictures we have five positions, where positions one, three and five are filled with symbols. In positions two and four we have empty spaces.

In the ***audio*** representation of both figures we have a **knock, double knock, knock, double knock and knock**.

Figure 21

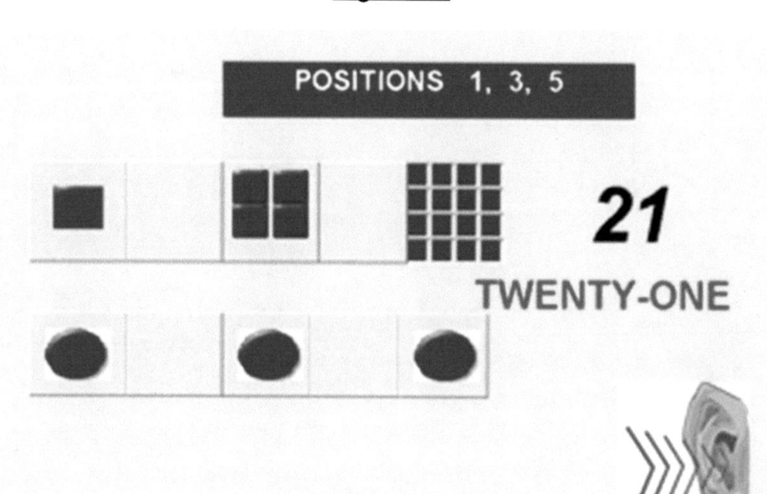

In the first picture of boxes we see square symbols in the first, third and fifth places. In the second picture positions three and five are all filled with circles. The different symbols and shapes in the pictures do not change the mathematical representation of both configurations. As a result the mathematical value of both configurations is equal to twenty one.

The ***kinesthetic*** representation for the first and second configurations of boxes is the clenching of the right, left, right, left and right hand once.

BLUE LESSONS, LESSON 3, FIGURE 22

In Figure 22, in both ***visual*** pictures we have five positions, where positions two, In the first picture of boxes we see square symbols in the second, third and ***audio*** representation of both figures is **double knock, knock, knock, double knock and single knock**.

Figure 22

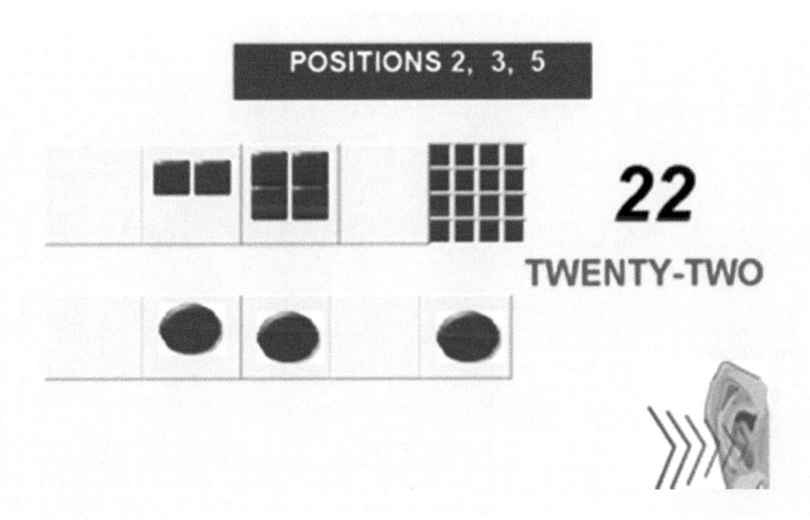

In the second pictures positions two, three and five are filled with circles. The different symbols and shapes in the pictures do not change the mathematical representation of both configurations. The mathematical value of both configurations is equal to twenty two.

The ***kinesthetic*** representation for the first and second configurations of the boxes is the clenching of the left, right twice, left and right hand once.

MATHEMATICS BLUE LESSONS, LESSON, FIGURE 23

In Figure 23, in both ***visual*** representations we have five positions, where positions one, two, three and five are filled with symbols. In position four we have an empty space.

In the ***audio*** representation of both figures we have a **knock, knock knock, double knock, and knock**.

Figure 23

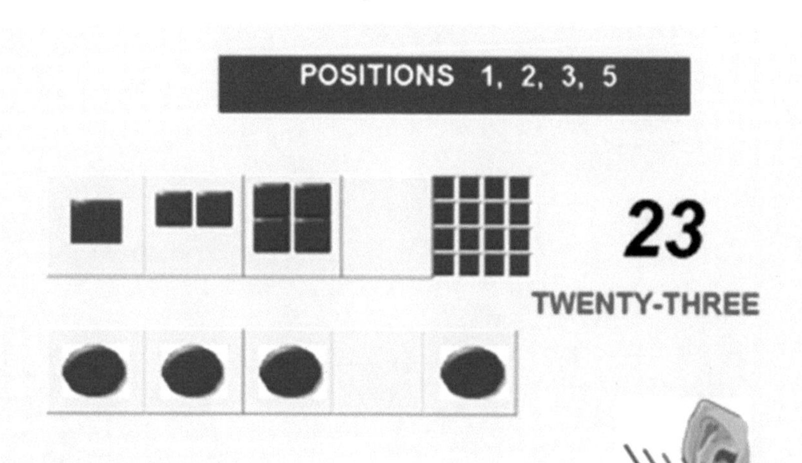

In the first picture of boxes we see square symbols in the first, second, third and fifth places. In the second picture, positions one, two, three and five are filled with circles. The different symbols and shapes in the pictures do not change the mathematical representation of both configurations. The mathematical value of both configurations is equal to twenty three.

The ***kinesthetic*** representation for the first and second configurations of the boxes is the clenching of the right hand three times, left hand once and right hand once.

BLUE LESSONS, LESSON, FIGURE 24

In Figure 24, in both ***visual*** representations we have five positions, where positions four and five are filled with symbols. Positions one, two, and three have empty spaces in the boxes.

In the ***audio*** representation of both figures we have **three double knocks, and two single knocks**

Figure 24

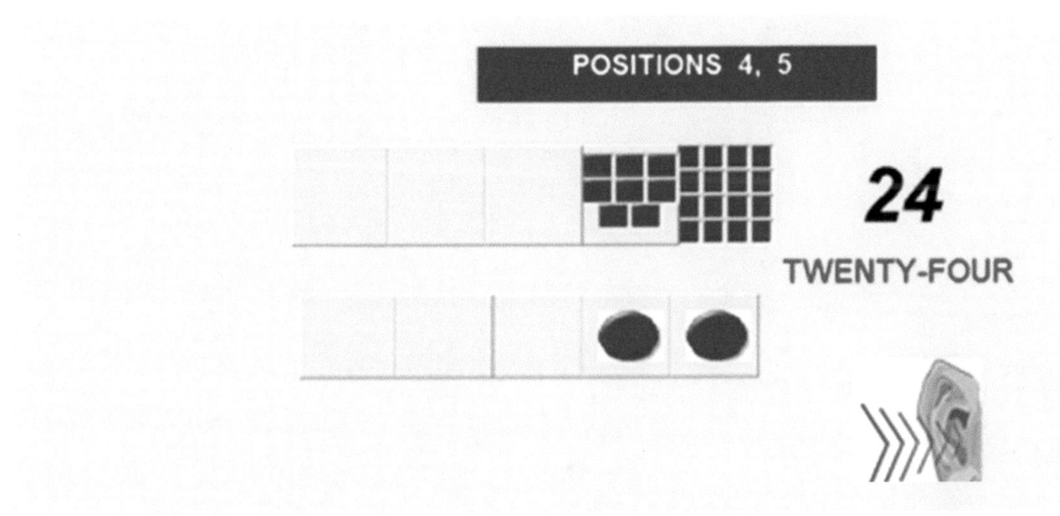

In the first picture of boxes we see square symbols in the fourth and fifth places. In the second picture, positions four and five are filled with circles. The different symbols and shapes in the pictures do not change the mathematical representation of both configurations. The mathematical value of both configurations is equal to twenty four.

The ***kinesthetic*** representation for the first and second configurations of the boxes is the clenching of the left hand three times and right hand twice.

MATHEMATICS BLUE LESSONS, EXERCISE, FIGURE 17

In Figure 17, we have three ***visual*** combinations that are filled up with symbols or empty positions. The ***audio*** representation of one of the pictures below is **knock, double knock, double knock, double knock and knock**.

Figure 17

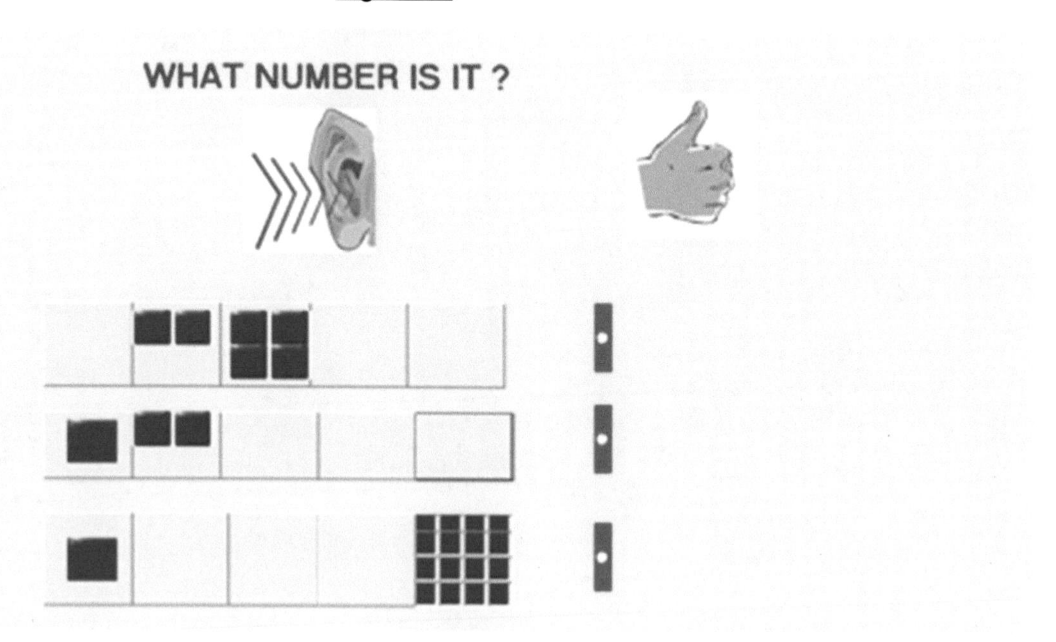

A knock represents a filled up position and a double knock represents an empty position. The clenching of the right hand represents a filled up position and the clenching of the left hand represents an empty position. The ***kinesthetic*** representation for the answer which corresponds to the third picture is the clenching of right hand, left hand thrice, and right hand. The kinesthetic representation for the second picture is the clenching of the right hand twice and left hand thrice. The kinesthetic representation for the first picture is the clenching of the left hand, right hand twice, and left hand twice.

BLUE LESSONS, EXERCISE, FIGURE 18

In Figure 18, we have three ***visual*** combinations that are filled up with symbols or empty positions. The ***audio*** representation of one of the pictures below is **double knock, knock, double knock, double knock and knock**.

Figure 18

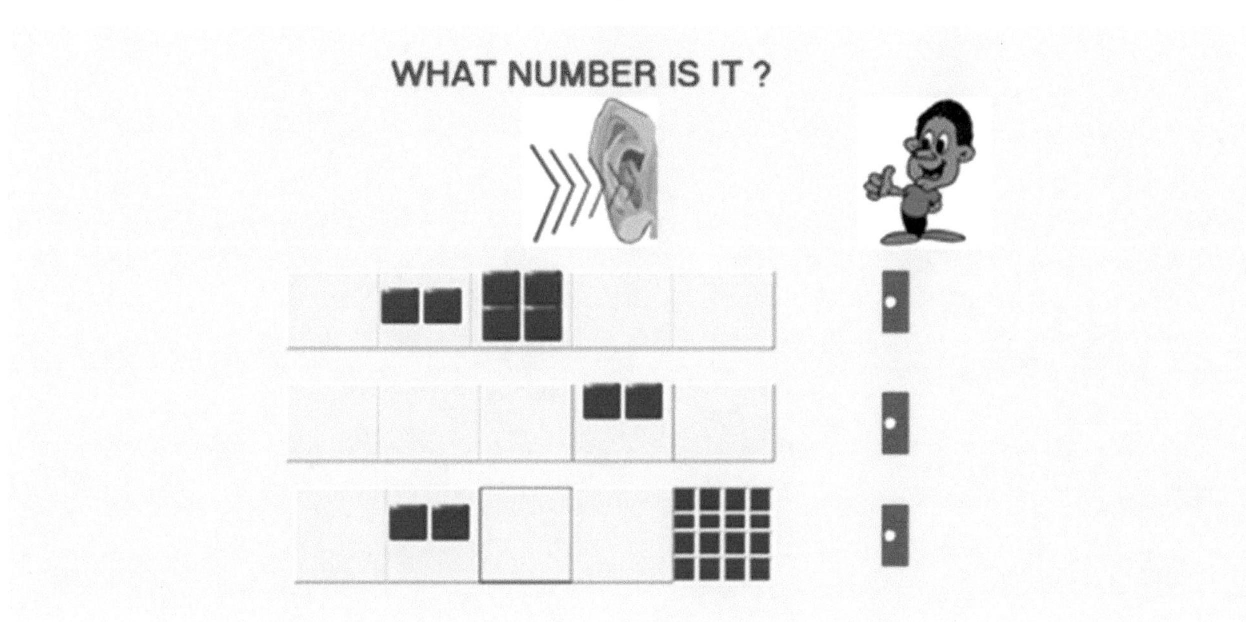

A knock represents a filled up position and a double knock represents an empty position. The clenching of the right hand represents a filled up position and the clenching of the left hand represents an empty position. The ***kinesthetic*** representation for the answer which corresponds to the third picture is the clenching of the left hand, right hand, left hand twice, and right hand. The kinesthetic representation for the second picture is the clenching of the left hand thrice, right hand, and left hand. The kinesthetic representation for the first picture is the clenching of the left hand, right hand twice, and left hand twice.

MATHEMATICS BLUE LESSONS, EXERCISE, FIGURE 19

In Figure 19, we have three ***visual*** combinations that are filled up with symbols or empty positions. The ***audio*** representation of one of the pictures below is **knock, knock, double knock, double knock and knock**.

Figure 19

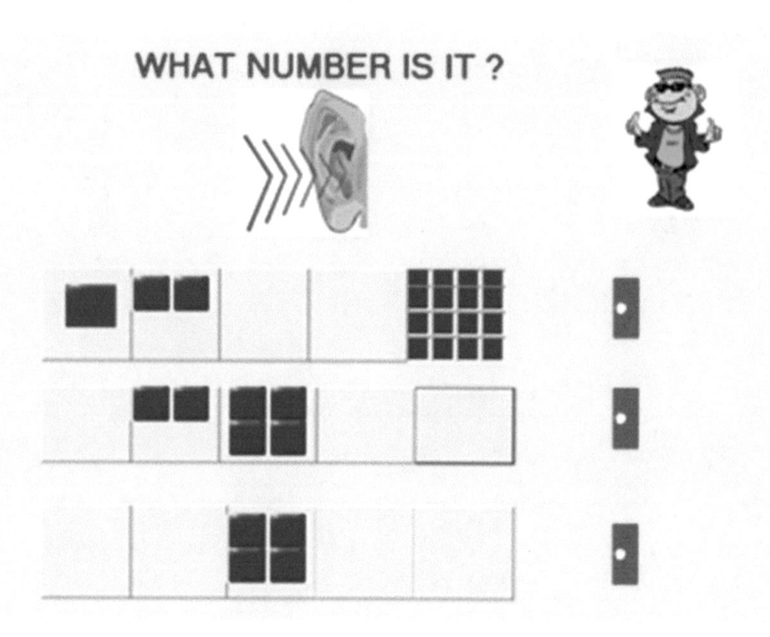

A knock represents a filled up position and a double knock represents an empty position. The clenching of the right hand represents a filled up position and the clenching of the left hand represents an empty position. The ***kinesthetic*** representation for the answer which corresponds to the first picture is the clenching of the right hand twice, left hand twice, and right hand. The kinesthetic representation for the second picture is the clenching left hand, right hand twice, and left hand twice. The kinesthetic representation for the third picture is the clenching of the left hand twice, right hand, and left hand twice.

BLUE LESSONS, EXERCISE, FIGURE 20

In Figure 20, we have three ***visual*** combinations that are filled up with symbols or empty positions. The ***audio*** representation of one of the pictures below is **double knock, double knock, knock, double knock, and knock**.

Figure 20

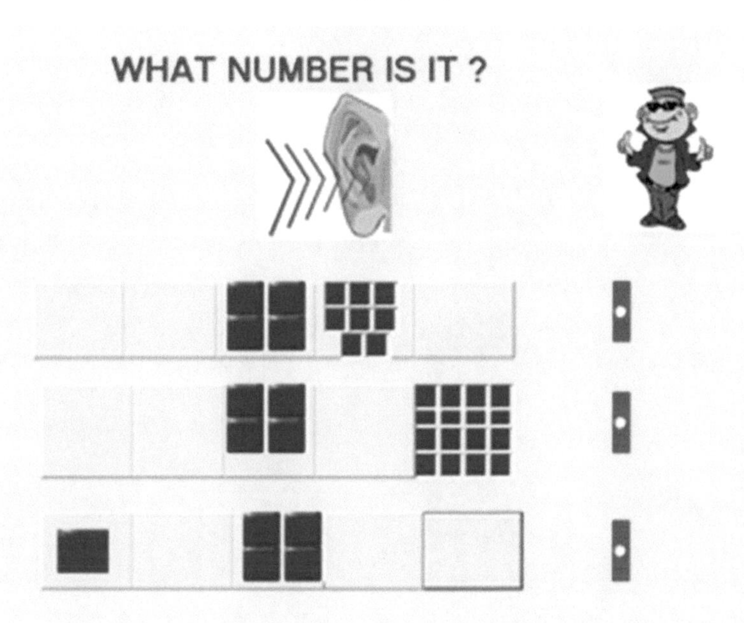

A knock represents a filled up position and a double knock represents an empty position. The clenching of the right hand represents a filled up position and the clenching of the left hand represents an empty position. The ***kinesthetic*** representation for the answer which corresponds to the second picture is the clenching of the left hand twice, right hand, left hand, and right hand. The kinesthetic representation for the first picture is the clenching of the left hand twice, right hand twice, and left hand. The kinesthetic representation for the third picture is the clenching of the right hand, left hand, right hand, and left hand twice.

MATHEMATICS

BLUE LESSONS, EXERCISE 21, FIGURE 21

In Figure 21, we have three ***visual*** combinations that are filled up with symbols or empty positions. The ***audio*** representation of one of the pictures below is **knock, double knock, knock, double knock, and knock**.

Figure 21

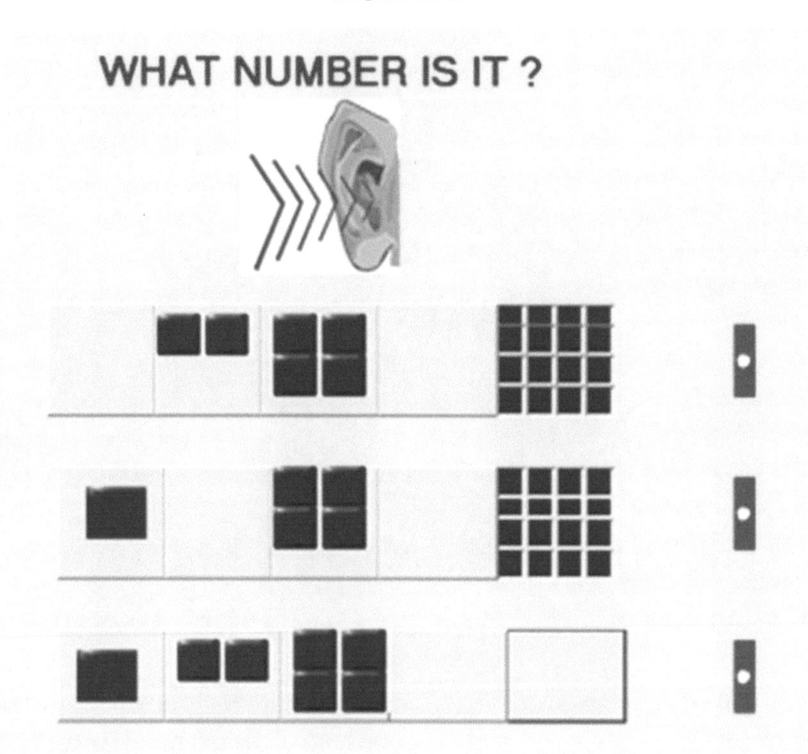

A knock represents a filled up position and a double knock represents an empty position. The clenching of the right hand represents a filled up position and the clenching of the left hand represents an empty position. The ***kinesthetic*** representation for the answer which corresponds to the second picture is the clenching of right hand, left hand, right hand, left hand, and right hand. The kinesthetic representation for the first picture is the clenching of the left hand, right hand twice, left hand, and right hand. The kinesthetic representation for the third picture is the clenching of the right hand thrice and left hand twice.

BLUE LESSONS, EXERCISE 22, FIGURE 22

In Figure 22, we have three ***visual*** combinations that are filled up with symbols or empty positions. The ***audio*** representation of one of the pictures below is **double knock, knock, knock, double knock, and knock**.

Figure 22

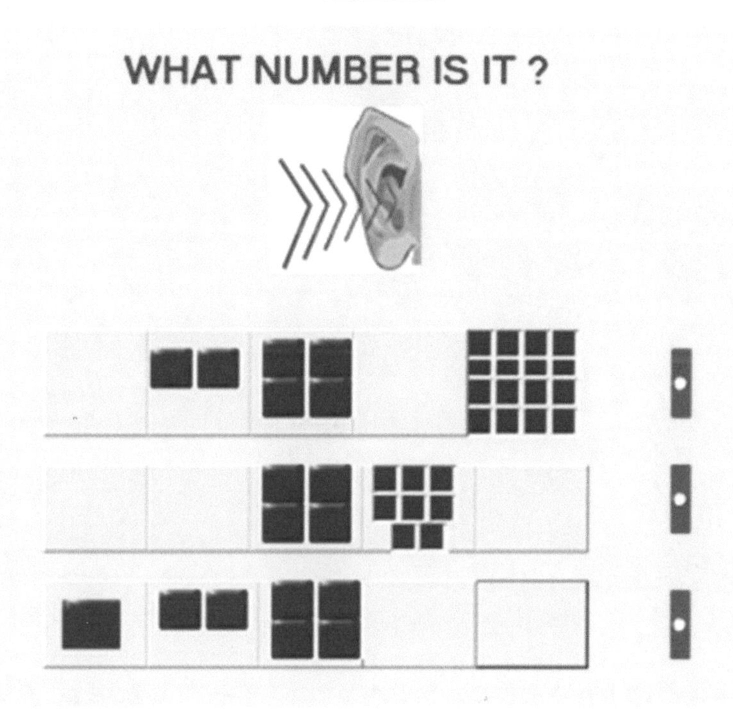

A knock represents a filled up position and a double knock represents an empty position. The clenching of the right hand represents a filled up position and the clenching of the left hand represents an empty position. The ***kinesthetic*** representation for the answer which corresponds to the first picture is the clenching of the left hand, right hand twice, left hand, and right hand. The kinesthetic representation for the second picture is the clenching of the left hand twice, right hand twice, and left hand. The kinesthetic representation for the third picture is the clenching of the right hand thrice and left hand twice.

MATHEMATICS BLUE LESSONS, EXERCISE 23, FIGURE 23

In Figure 23, we have three ***visual*** combinations that are filled up with symbols or empty positions. The ***audio*** representation of one of the pictures below is **knock, knock, knock, double knock, and knock**.

Figure 23

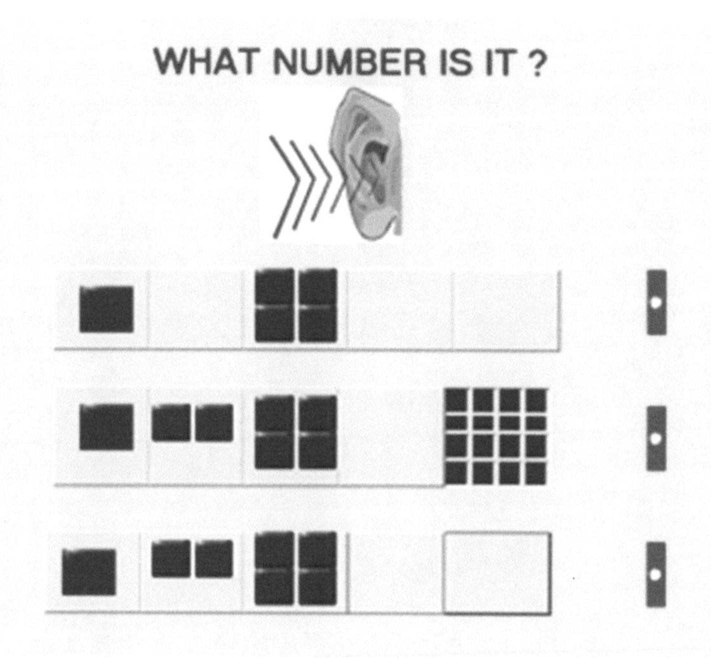

A knock represents a filled up position and a double knock represents an empty position. The clenching of the right hand represents a filled up position and the clenching of the left hand represents an empty position. The ***kinesthetic*** representation for the answer which corresponds to the second picture is the clenching of right hand thrice, left hand, and right hand. The kinesthetic representation for the first picture is the clenching of the right hand, left hand, right hand, and left hand twice. The kinesthetic representation for the third picture is the clenching of the right hand thrice and left hand twice.

BLUE LESSONS, FIGURE 24

In Figure 24, we have three ***visual*** combinations that are filled up with symbols or empty positions. The ***audio*** representation of one of the pictures below is **double knock, double knock, double knock, knock, and knock**.

Figure 24

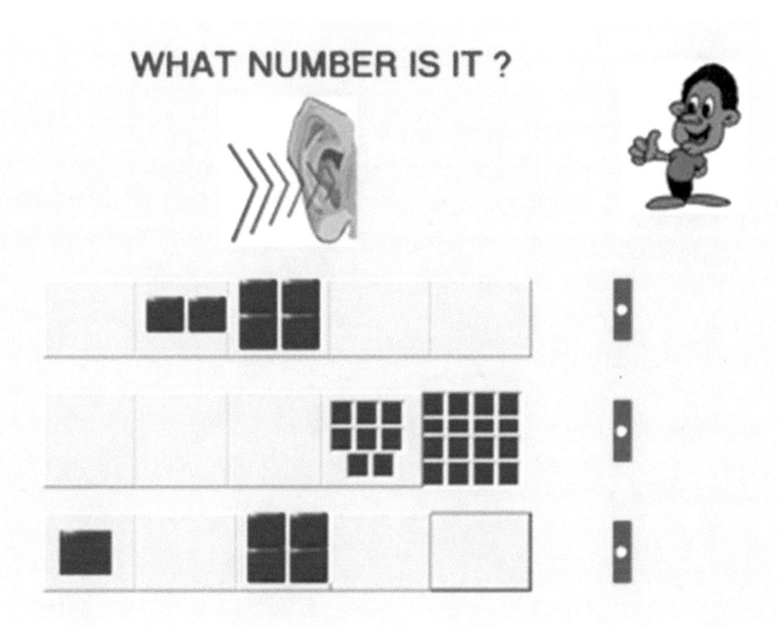

A knock represents a filled up position and a double knock represents an empty position. The clenching of the right hand represents a filled up position and the clenching of the left hand represents an empty position. The ***kinesthetic*** representation for the answer which corresponds to the second picture is the clenching of the left hand thrice and right hand twice. The kinesthetic representation for the first picture is the clenching of the left hand, right hand twice, and left hand twice. The kinesthetic representation for the third picture is the clenching of the right hand, left hand, right hand, and left hand twice.

MATHEMATICS BLUE LESSONS, LESSON 4, FIGURE 25

In Figure 25, in a ***visual*** display in both pictures we have five positions each, where positions one, four and five are filled with symbols. In positions two and three, we have empty spaces in the boxes.

In the **audio** representation of both figures we have a **knock, double knock, double knock, knock and knock**.

Figure 25

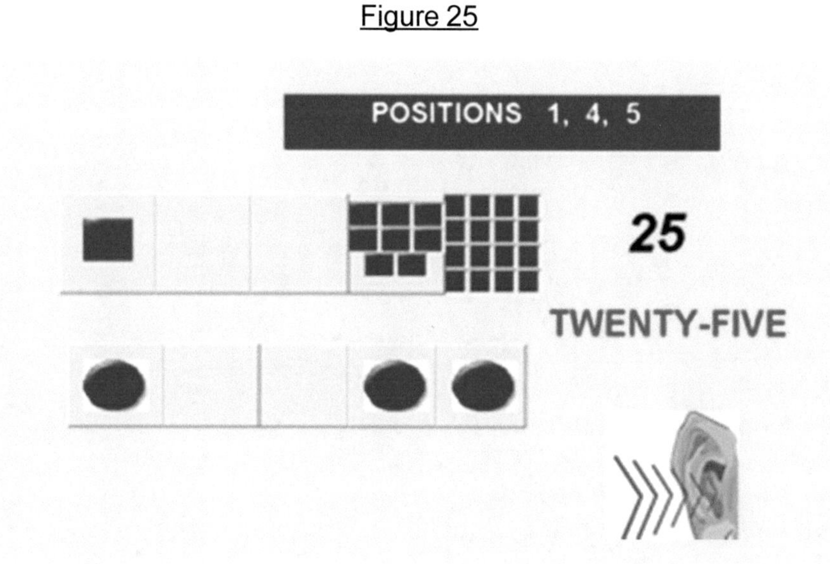

In the first picture of boxes we see square symbols in the first, fourth and fifth places. In the second picture, positions one, four and five are filled with circles. The different symbols and shapes in the pictures do not change the mathematical representation of both configurations. The mathematical value of both configurations is equal to twenty five.

The **kinesthetic** representation for the first and second configurations of boxes is the clenching of the right hand once, left twice and right hand twice.

BLUE LESSONS, LESSON 4, FIGURE 26

In Figure 26, we have a **visual** display of two pictures. In both pictures we have five positions, where positions two, four and five are filled up with symbols, and positions one and three are empty.

In the **audio** representation of both figures we have a **double knock, knock, double knock, knock and knock**.

Figure 26

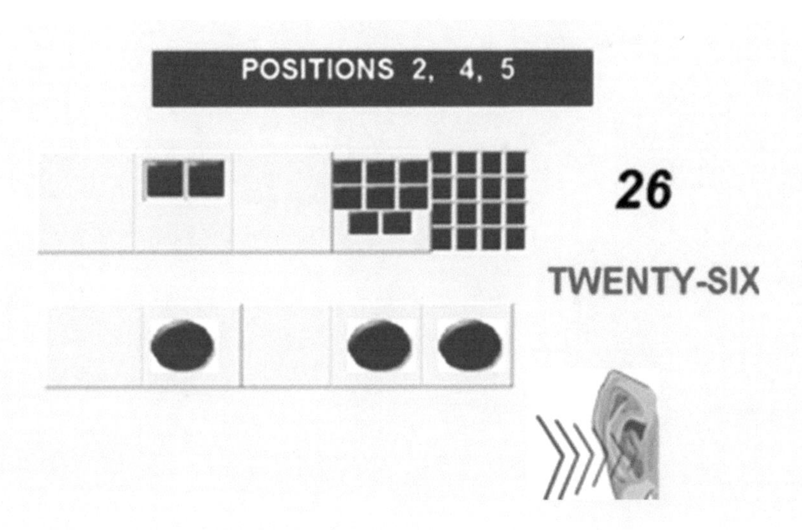

While the first picture has its positions filled with twenty six squares, the same positions in the second picture are filled with symbols. The different symbols and shapes in the pictures do not change the mathematical representation of both configurations, which is equal to twenty six.

The **kinesthetic** representation for the first and second configurations of boxes is *the clenching of the left, right, left and right hand twice.*

MATHEMATICS

BLUE LESSONS, LESSON 4, FIGURE 27

In Figure 27, we have a ***visual*** display of two pictures. In both pictures we have five positions, where positions one two, four and five are filled with symbols, and position three is empty.

In the **audio** representation of both figures we have a **knock, knock, double knock, knock and knock.**

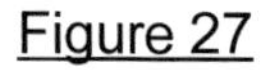

Figure 27

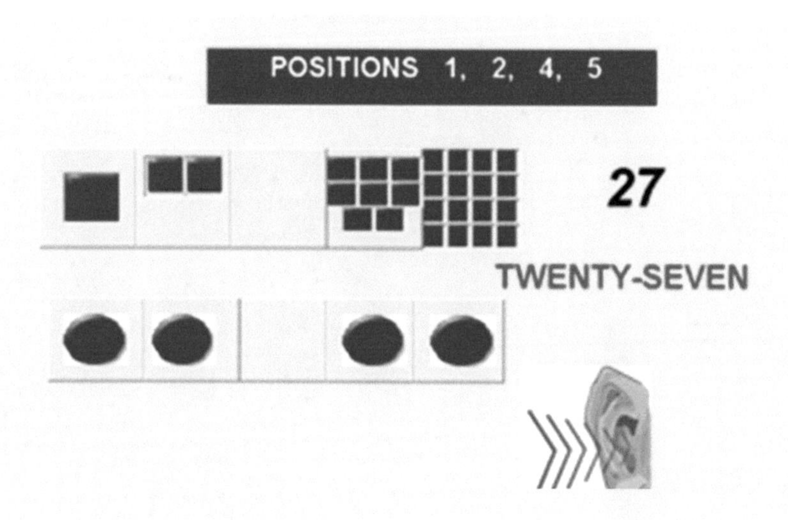

While the first picture has its positions filled with squares, the second picture has the same positions filled with circles. The different symbols and shapes in the pictures do not change the mathematical representation of both configurations. The mathematical value of both configurations is equal to twenty seven.

The ***kinesthetic*** representation for the first and second configurations of boxes is *the clenching of the right twice, left once and right hand twice.*

BLUE LESSONS, LESSON 4, FIGURE 28

In Figure 28, in a ***visual*** display of the pictures below, positions one and two are empty, while positions three four and five are filled with symbols. In the ***audio*** representation of both figures we have a **double knock, double knock, knock, knock and knock**.

Figure 28

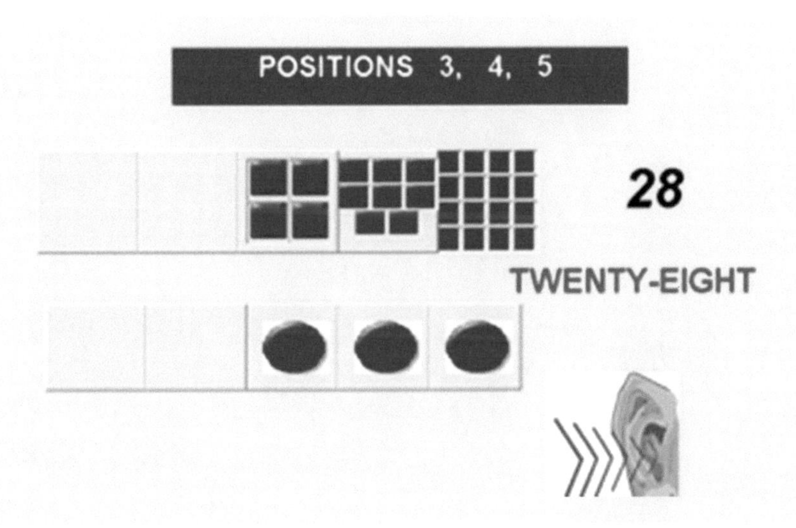

While the first picture has its last three positions filled with square symbols, the same positions in the second picture are filled with circles. The different symbols and shapes in the pictures do not change the mathematical representation of both configurations. The mathematical value of both configurations is equal to twenty eight.

The ***kinesthetic*** representation for the first and second configurations of boxes is *the clenching of the left hand twice and right hand three times.*

MATHEMATICS

BLUE LESSONS, LESSON 4, FIGURE 29

In Figure 29, in a ***visual*** display of the pictures below, positions one, three, four, and five are filled with symbols, while position two is empty.

In the ***audio*** representation of both figures we have a **knock, double knock, knock, knock and knock**.

Figure 29

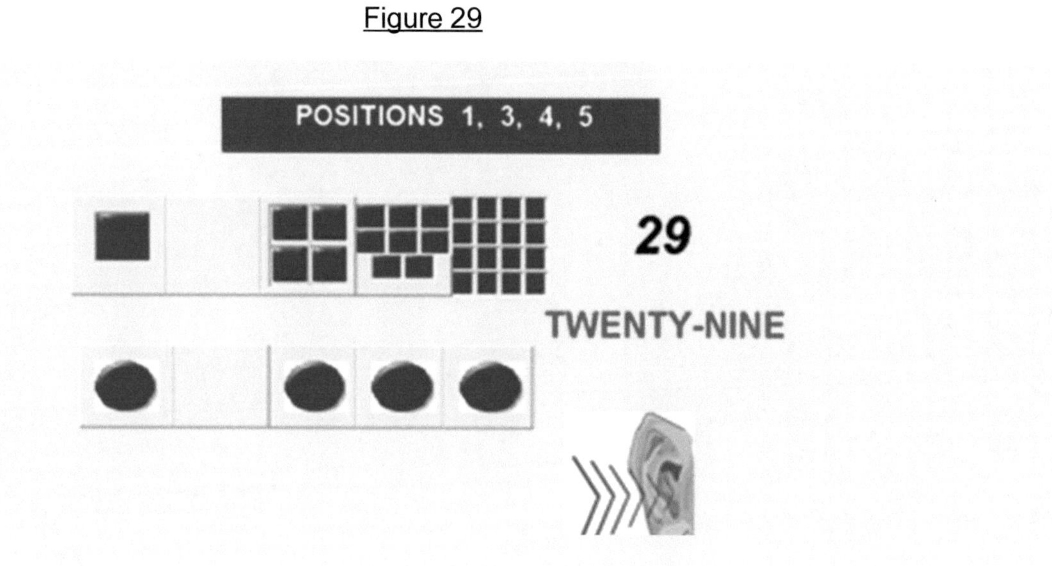

While the first picture has its positions filled with squares, the same positions in the second picture are filled with circles. The different symbols and shapes in the pictures do not change the mathematical representation of both configurations, which is equal to twenty nine.

The ***kinesthetic*** representation for the first and second configurations of boxes is *the clenching of the right hand once, left once and right hand three times.*

BLUE LESSONS, EXERCISE, FIGURE 25

In Figure 25, we have three ***visual*** combinations that are filled up with symbols or empty positions.

The ***audio*** representation of one of the pictures below is **knock, double knock, double knock, knock, and knock**.

Figure 25

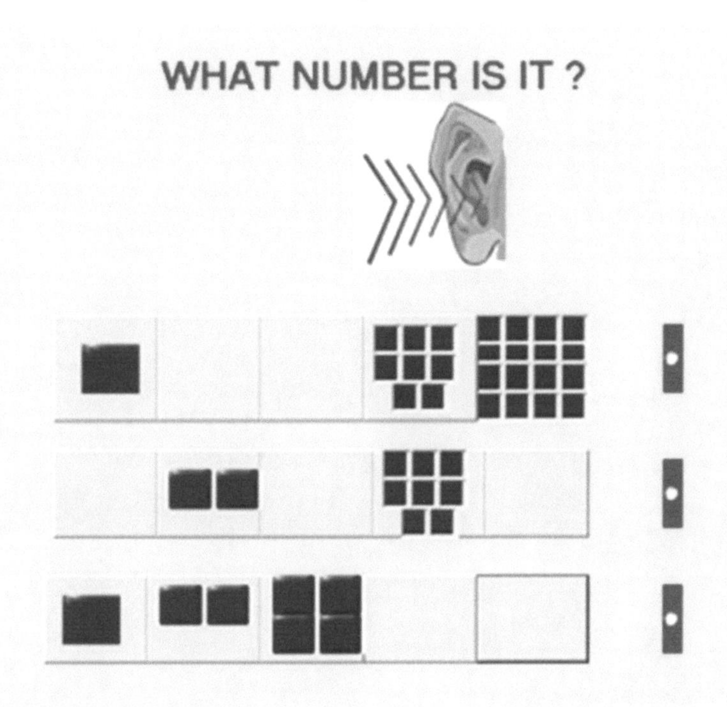

A knock represents a filled up position and a double knock represents an empty position. The clenching of the right hand represents a filled up position and the clenching of the left hand represents an empty position. The ***kinesthetic*** representation for the answer which corresponds to the first picture is the clenching of the right hand, left hand twice, and right hand twice. The kinesthetic representation for the second picture is the clenching of the left hand, right hand, left hand, right hand, and left hand. The kinesthetic representation for the third picture is the clenching of the right hand thrice and left hand twice.

MATHEMATICS

BLUE LESSONS, EXERCISE, FIGURE 26

In Figure 26, we have three ***visual*** combinations that are filled with symbols or empty positions.

The ***audio*** representation of one of the pictures below is **double knock, knock, double knock, knock, and knock**

Figure 26

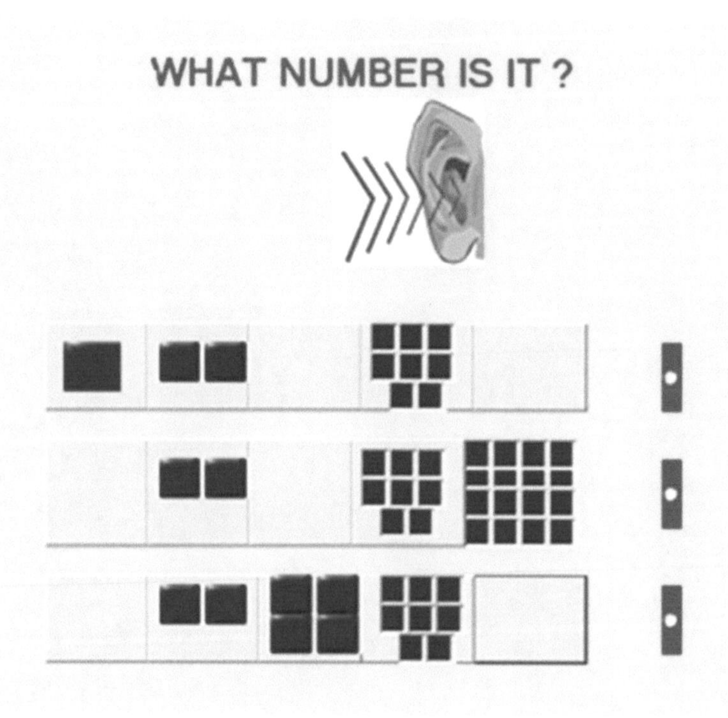

A knock represents a filled up position and a double knock represents an empty position. The clenching of the right hand represents a filled up position and the clenching of the left hand represents an empty position. The ***kinesthetic*** representation for the answer which corresponds to the second picture is the clenching of the left hand, right hand, left hand, and right hand twice. The kinesthetic representation for the first picture is the clenching of the left hand twice, right hand, left hand, and right hand. The kinesthetic representation for the third picture is the clenching of the left hand, right hand thrice, and left hand.

BLUE LESSONS, EXERCISE, FIGURE 27

In Figure 27, we have three ***visual*** combinations that are filled with symbols or empty positions. The ***audio*** representation of one of the pictures below is **knock, knock, double knock, knock, and knock**.

Figure 27

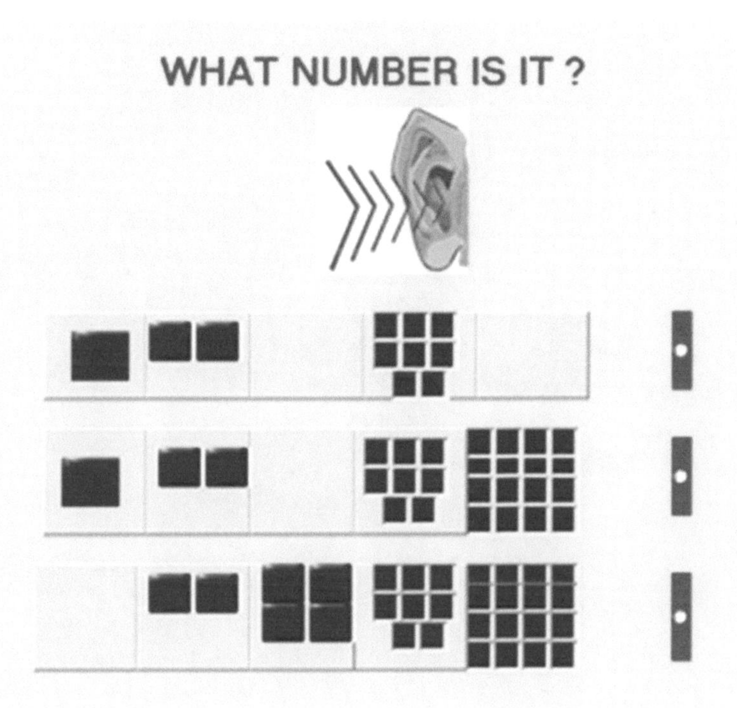

A knock represents a filled up position and a double knock represents an empty position. The clenching of the right hand represents a filled up position and the clenching of the left hand represents an empty position. The ***kinesthetic*** representation for the answer which corresponds to the second picture is the clenching of right hand twice, left hand, and right hand twice. The kinesthetic representation for the first picture is the clenching of the right hand twice, left hand, right hand, and left hand. The kinesthetic representation for the third picture is the clenching of the left hand and right hand four times.

MATHEMATICS

BLUE LESSONS, EXERCISE, FIGURE 28

In Figure 28, we have three ***visual*** combinations that are filled with symbols or empty positions. The ***audio*** representation of one of the pictures below is **double knock, double knock, knock, knock, and knock**.

Figure 28

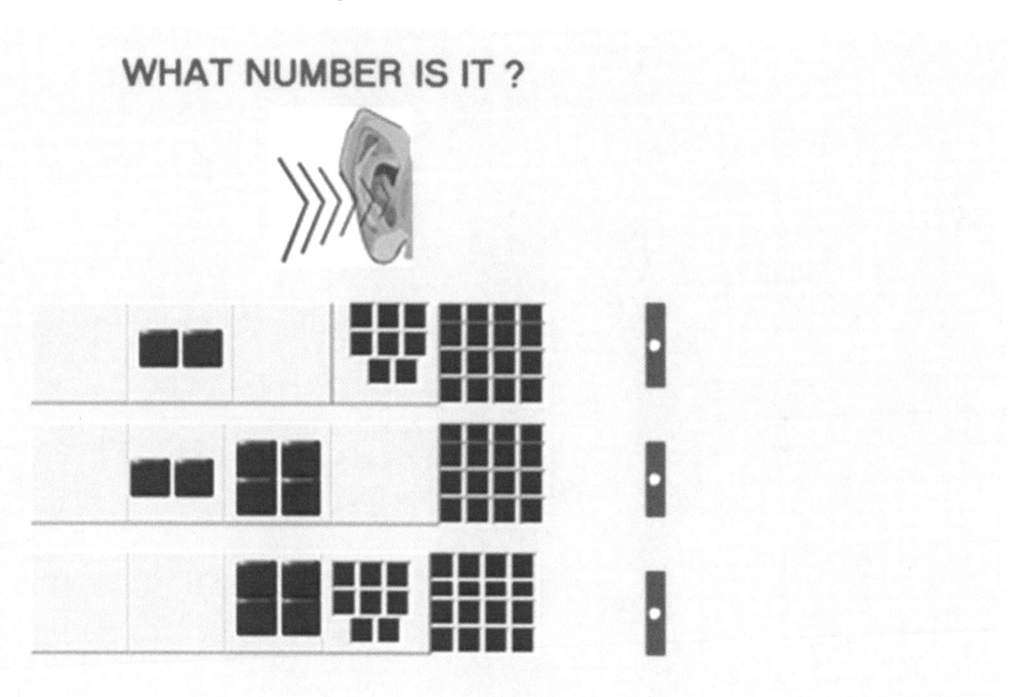

A knock represents a filled up position and a double knock represents an empty position. The clenching of the right hand represents a filled up position and the clenching of the left hand represents an empty position. The ***kinesthetic*** representation for the answer which corresponds to the third picture is the clenching of the left hand twice and right hand thrice. The kinesthetic representation for the first picture is the clenching of the left hand, right hand, left hand, and right hand. The kinesthetic representation for the second picture is the clenching of the left hand, right hand twice, left hand, and right hand.

BLUE LESSONS, EXERCISE, FIGURE 29

In Figure 29, we have three ***visual*** combinations that are filled with symbols or empty positions. The ***audio*** representation of one of the pictures below is **knock, double knock, knock, knock, and knock**.

Figure 29

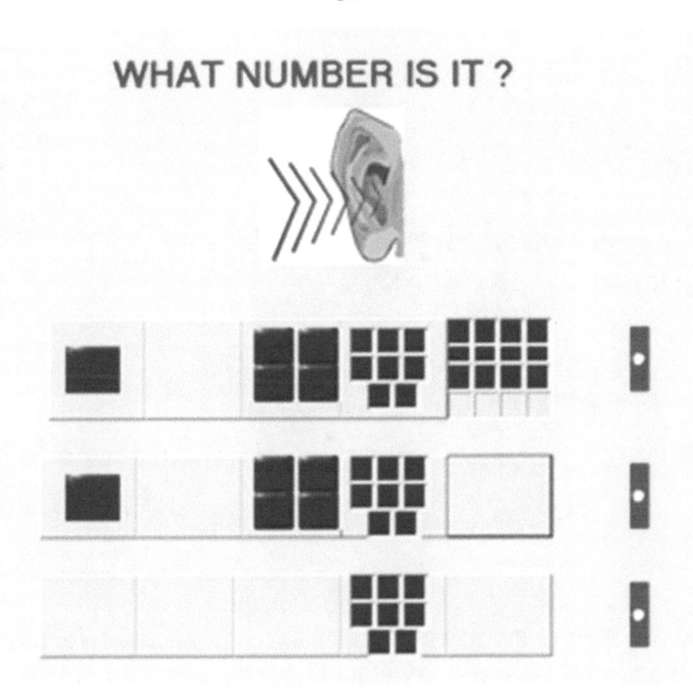

A knock represents a filled up position and a double knock represents an empty position. The clenching of the right hand represents a filled up position and the clenching of the left hand represents an empty position. The ***kinesthetic*** representation for the answer which corresponds to the first picture is the clenching of the right hand, left hand, and right hand thrice. The kinesthetic representation for the second picture is the clenching of the right hand, left hand, right hand twice, and left hand. The kinesthetic representation for the third picture is the clenching of the left hand thrice, right hand, and left hand.

MATHEMATICS BLUE LESSONS, LESSON, FIGURE 30

In Figure 30, in both ***visual*** pictures we have five positions, where positions two through to five are filled with symbols. Position one has an empty space. In the ***audio*** representation of both figures we have a **double knock, knock, knock, knock and knock**.

Figure 30

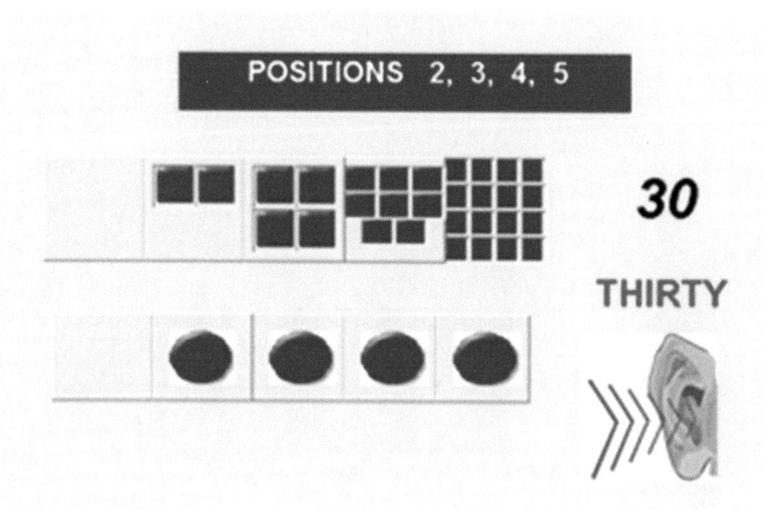

In the first picture of boxes we see square symbols filled in positions two, three, four and five. In the second picture positions two through five are also filled with circles. The different symbols and shapes in the pictures do not change the mathematical representation of both configurations. The mathematical value of both configurations is equal to thirty.

The ***kinesthetic*** representation for the first and second configurations of the boxes is the clenching of the left hand once and the right hand four times.

BLUE LESSONS, LESSON, FIGURE 31

In Figure 31, in both ***visual*** pictures we have five positions, where all positions are filled with symbols.

In the ***audio*** representation of both figures we have **knock, knock, knock, knock and knock**.

Figure 31

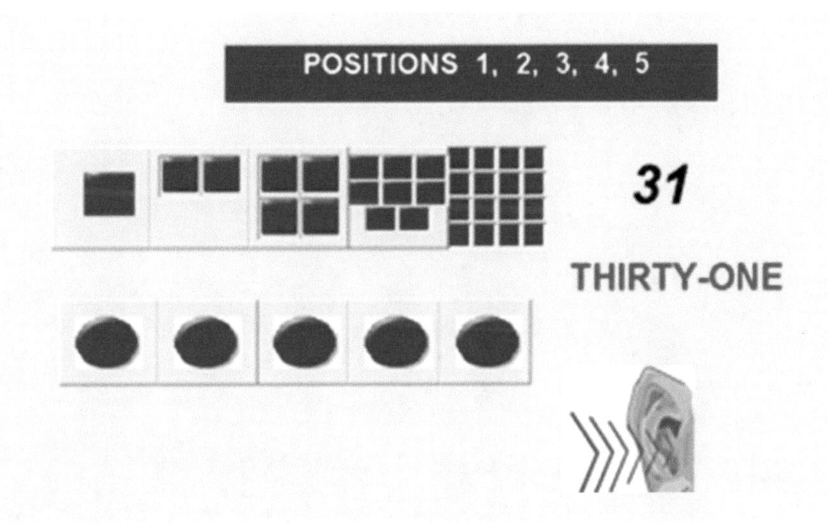

In both pictures of boxes we have all positions filled with both square and circle symbols. The different symbols and shapes in the pictures do not change the mathematical representation of both configurations. The mathematical value of both configurations is equal to thirty one.

The ***kinesthetic*** representation for the first and second configurations of the boxes is the clenching of the right hand five times.

MATHEMATICS

BLUE LESSONS, EXERCISE 30, FIGURE 30

In Figure 30, we have three ***visual*** combinations that are filled with symbols or empty positions.

The ***audio*** representation of one of the pictures below is **double knock, knock, knock, knock, and knock**.

Figure 30

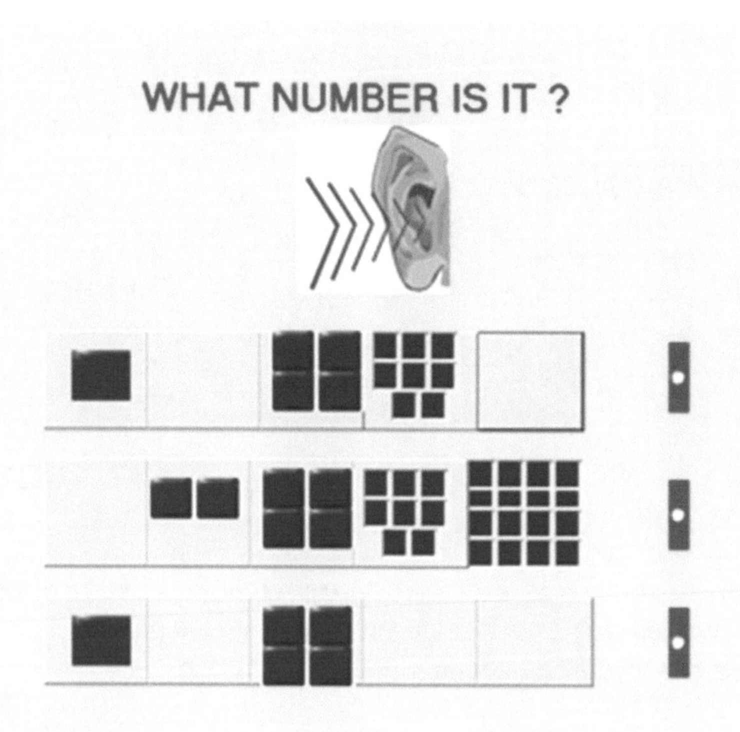

A knock represents a filled position and a double knock represents an empty position. The clenching of the right hand represents a filled position and the clenching of the left hand represents an empty position. The ***kinesthetic*** representation for the answer which corresponds to the second picture is *the clenching of the left hand and right hand four times.* The kinesthetic representation for the first picture is *the clenching of the right hand, left hand, right hand twice, and left hand.* The kinesthetic representation for the third picture is *the clenching of the right hand, left hand, right hand, and left hand twice.*

BLUE LESSONS, EXERCISE 31, FIGURE 31

In Figure 31, we have three ***visual*** combinations that are filled with symbols or empty positions.

The ***audio*** representation of one of the pictures below is **knock, knock, knock, knock, and knock**.

Figure 31

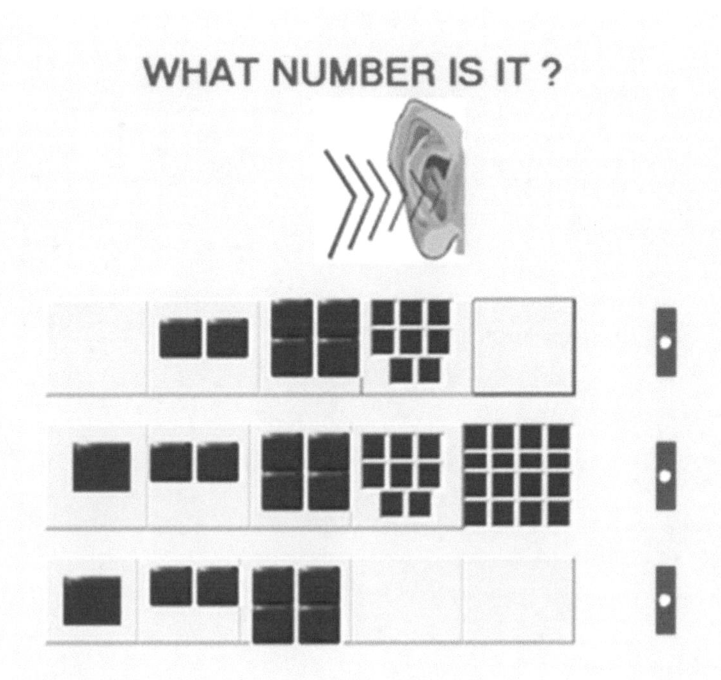

A knock represents a filled position and a double knock represents an empty position. The clenching of the right hand represents a filled position and the clenching of the left hand represents an empty position. The ***kinesthetic*** representation for the answer which corresponds to the second picture is *the clenching of the right hand five times.* The kinesthetic representation for the first picture is *the clenching of the left hand, right hand thrice, and left hand.* The kinesthetic representation for the third picture is *the clenching of the right hand thrice and left hand twice.*

INTRODUCTION TO POSITIONS

LESSON 1 INTRODUCTION TO POSITIONS

In Figure 1 there are two pictures of boxes. There is a combination of positions, filled with symbols or empty spaces. A knock represents a filled position and a double knock represents an empty position. We are balancing our visual and audio reaction on the symbol (knock). The audio representation of one of the pictures below is **knock and double knock**.

Figure 1

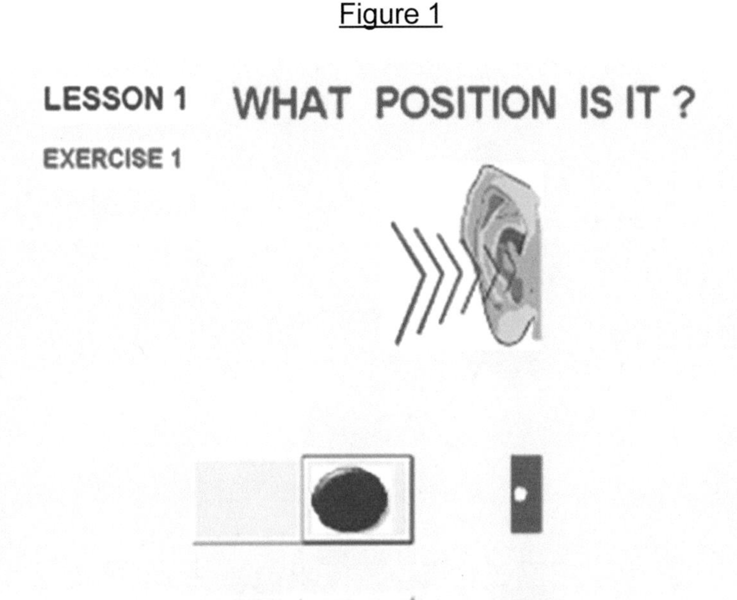

In our example knock and double knock correspond to the second picture. The first drawing of the boxes corresponds to double knock and knock.

The kinesthetic representation involves movement. The method of representation of a knock is clenching of the right fist, and double knock is clenching the left fist. Clenching of the right hand then represents the filled position and left hand clenching represents the empty space. In our example, kinesthetic representation for the answer is *to clench the right hand then the left*. For the first picture the kinesthetic representation is *clenching of the left hand then right hand.*

INTRODUCTION TO POSITIONS

LESSON 2

In Figure 2 we have three combinations of positions, filled with symbols or empty spaces. The audio representation of one of the pictures below is **double knock and knock.**

Figure 2

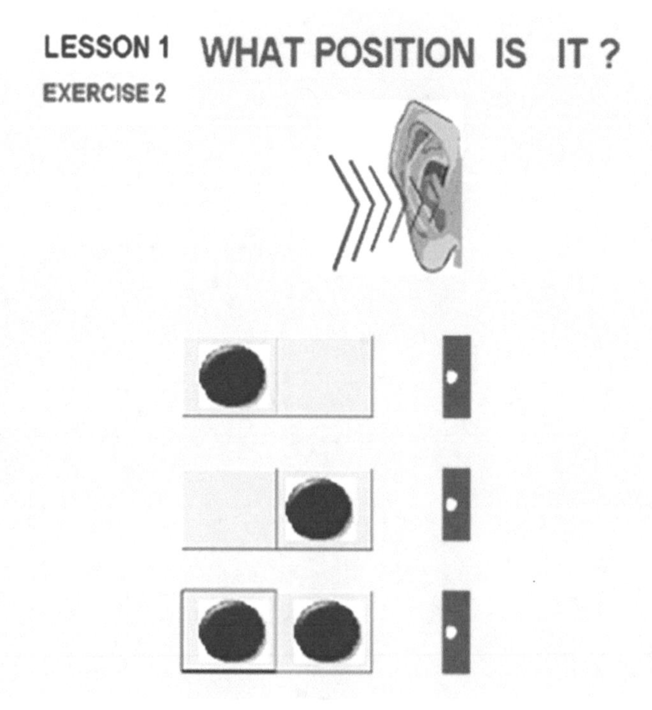

A double knock represents an empty space and a knock represents a filled space. In our example, double knock and knock correspond to the second picture. In accordance with the audio representations, the first picture of boxes corresponds to knock and double knock, and the third represents knock, knock.

The kinesthetic representation of the double knock is clenching of the left hand, and for knock is tightening the right hand into a fist. Clenching of the right hand represents a filled position, and left hand clenching represents an empty position. For the first picture of boxes, kinesthetic representation is clenching of the right then the left hand, and for the third picture kinesthetic representation will be clenching of the right hand twice.

LESSON 3 INTRODUCTION TO POSITIONS

In Figure 3 we have three different combinations of positions, filled with symbols or empty spaces. The audio representation of one of the pictures below is represented by **knock, knock.**

Figure 3

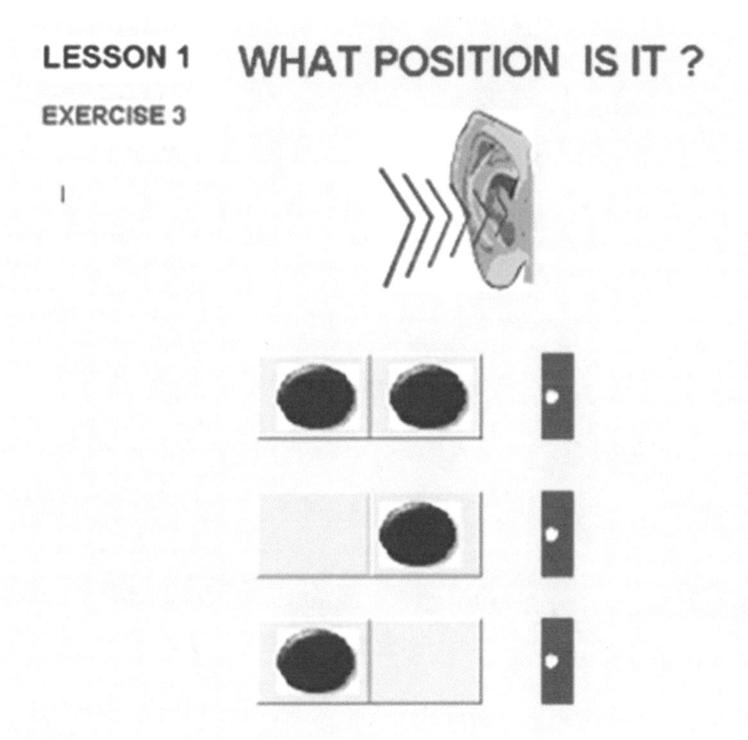

The knock, knock visually represents two filled spaces. In our example knock and knock corresponds to the first picture of boxes. The second picture represents double knock, and knock, and the third picture of boxes corresponds to knock and double knock.

The kinesthetic representation of the answer knock and knock is the *clenching of the right hand twice.* For the second picture of the boxes *clenching of the left hand* represents double knock *and clenching of the right hand* represents knock. A knock and then a double knock represent the third picture. The kinesthetic representation is *clenching of the right and left hand.*

INTRODUCTION TO POSITIONS

LESSON 4

In Figure 4 we have three combinations of filled or empty spaces. The audio representation of one of the pictures below is **knock, double knock, and double knock.**

Figure 4

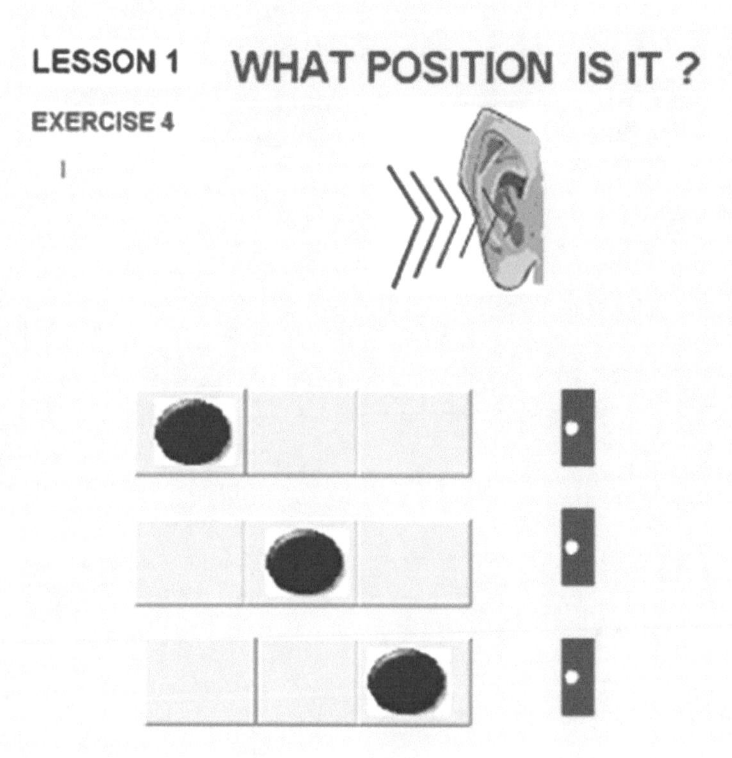

In this figure the first picture of the boxes with the filled space and two empty spaces, is the right answer.

The kinesthetic representation of the answer for knock is *clenching of the right hand, and making a fist with the left hand two times* represents the two empty positions.

For the second picture of boxes audio representation is double knock, knock, and double knock. Audio representation for the third picture of boxes is double knock, double knock, and knock.

LESSON 5 INTRODUCTION TO POSITIONS

In Figure 5 we have three combinations of filled or empty spaces. The audio representation of one of the pictures below is **double knock, knock, knock.**

Figure 5

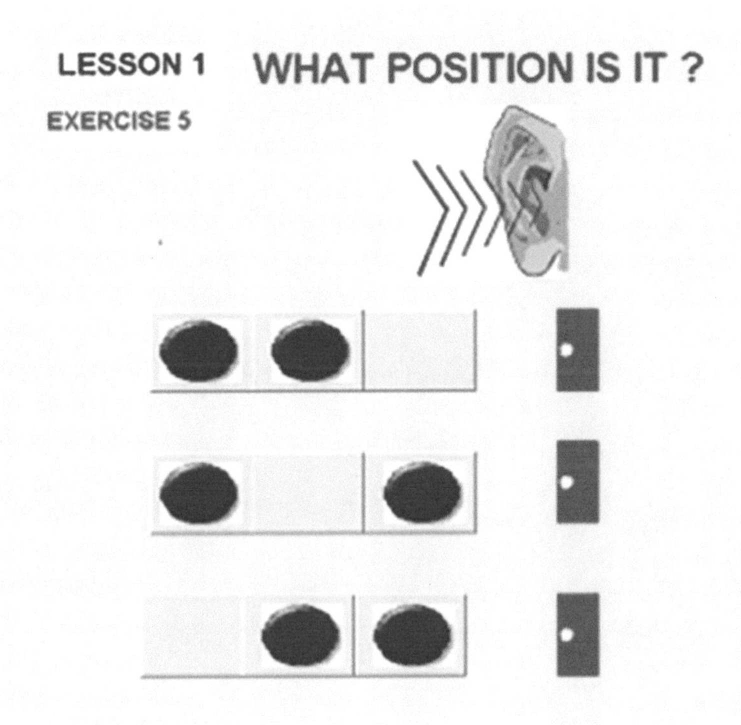

The visual representation of the answer is the third picture, where the first space is empty, and the second and third spaces are filled with a symbol.

The kinesthetic representation of the answer would be *clenching of the left hand once* for the empty position *and the right hand twice* for the second and third filled positions.

INTRODUCTION TO POSITIONS

LESSON 6

In Figure 6 we have three combinations of filled and empty spaces. The audio representation of one of the pictures below is **knock, double knock, knock.**

Figure 6

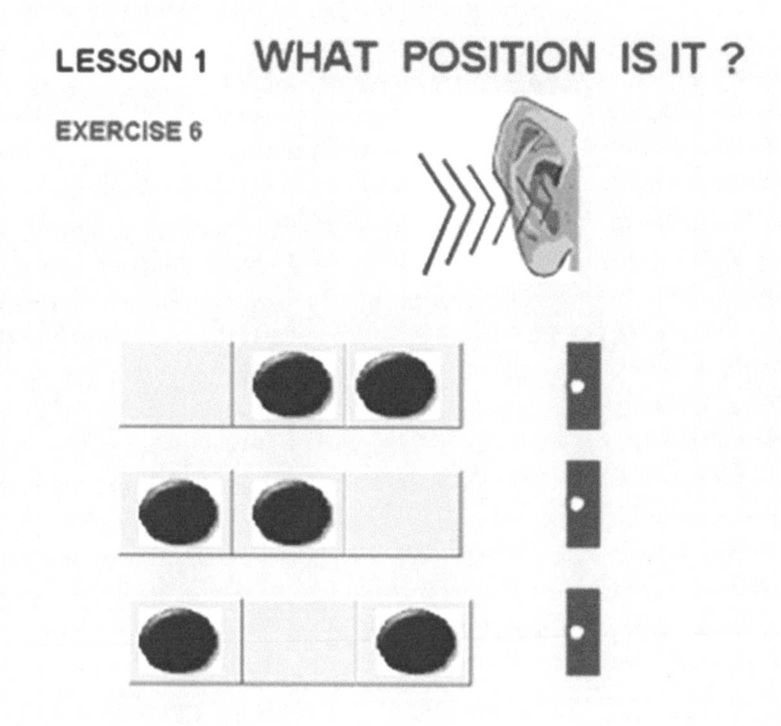

The answer is the third picture of boxes. The visual representation of the answer would be the first position filled with a symbol, the second position empty, and third position filled with a symbol.

The kinesthetic representation of the answer is *clenching of the right hand* for the first position, *clenching of the left* for the second position and *clenching of the right hand* to represent the third position.

LESSON 7 INTRODUCTION TO POSITIONS

In Figure 7 we have three combinations that are filled with symbols or empty spaces. The audio representation of one of the pictures below is **knock, knock, and knock.**

Figure 7

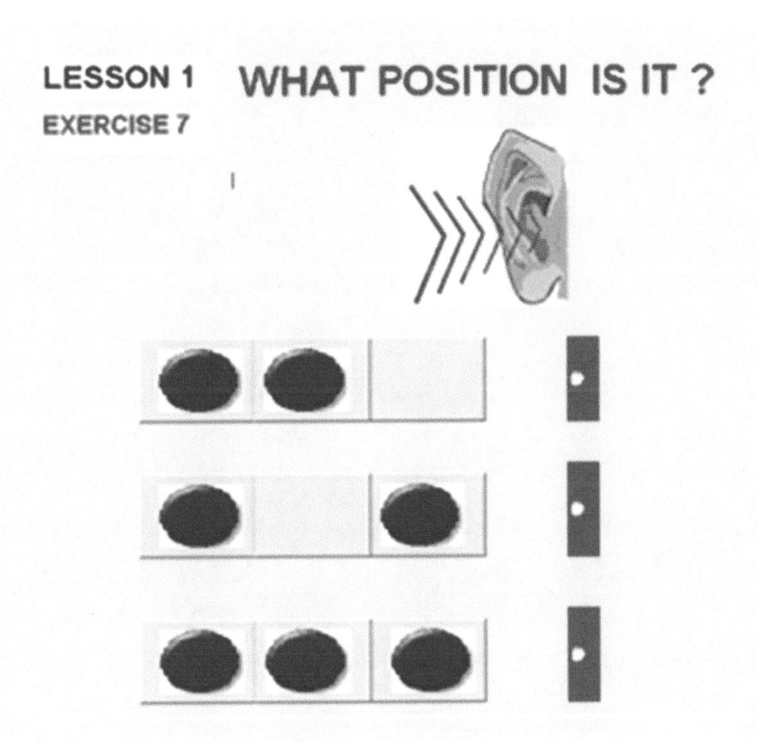

The visual representation of the answer is the third picture of the boxes, where all three spaces are filled with a symbol.

The kinesthetic representation of the answer is *clenching of the right hand three times* in a row to represent the filled positions with symbols.

INTRODUCTION TO POSITIONS LESSON 8

In Figure 8 we have three combinations of filled or empty spaces. The audio representation of one of the pictures below is **double knock, double knock, double knock, and knock.**

Figure 8

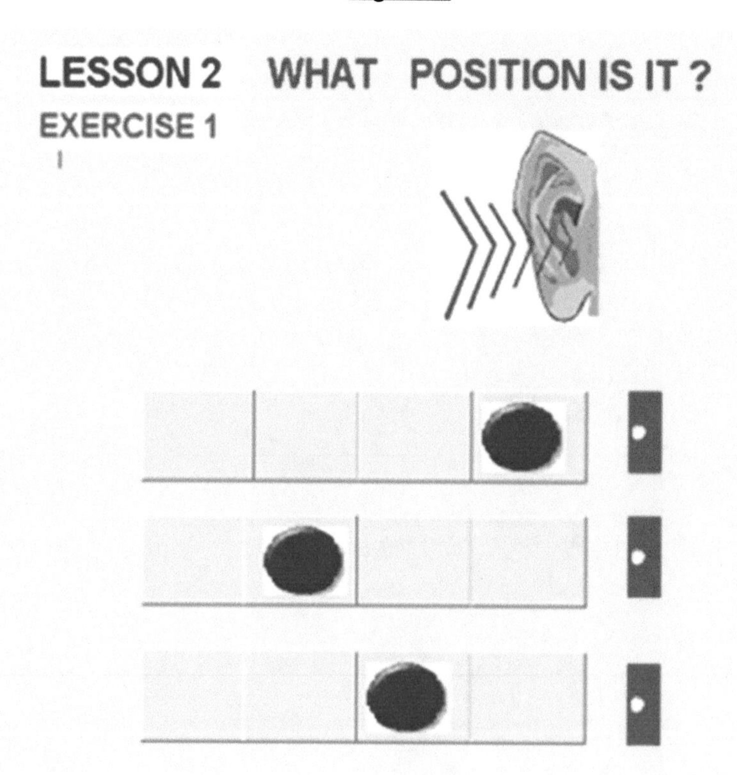

The answer is the first picture of the boxes. The visual representation of the answer has the first three spaces left empty. The fourth position is filled with a symbol.

The kinesthetic representation of the answer is *clenching of the left hand three times*, followed by the *clenching of the right hand once.*

LESSON 9 INTRODUCTION TO POSITIONS

In Figure 9 we have three combinations that are filled with symbols or empty spaces. The audio representation of one of the pictures below is **knock, double knock, double knock, knock.**

Figure 9

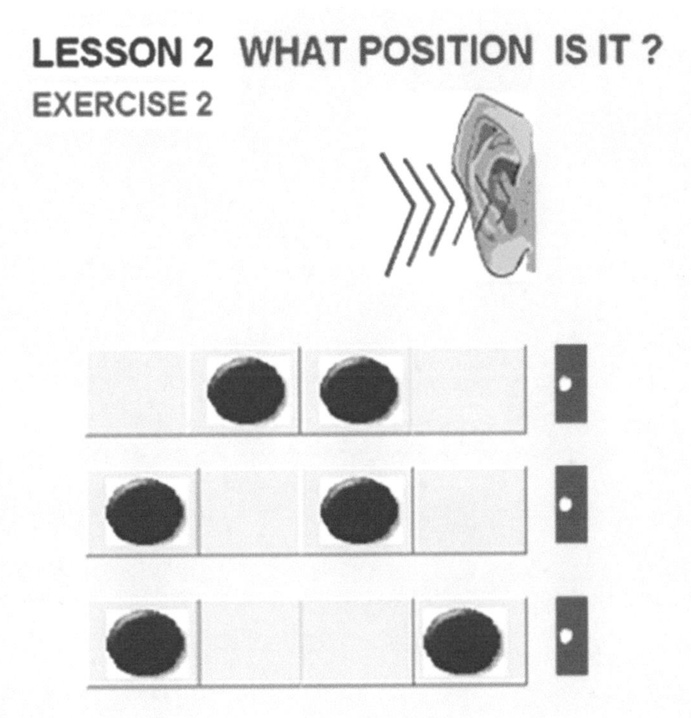

The answer for the audio representations is the third picture of boxes. The visual representation of the answer is a symbol in the first space followed by the second and third spaces being left empty and the fourth space filled with a symbol.

The kinesthetic representation of the answer is *clenching of the right hand* for the first position, followed by the second and third positions being represented by *clenching of the left hand twice and the clenching of the right hand* to represent the symbol in the fourth position.

INTRODUCTION TO POSITIONS

LESSON 10

In Figure 10 we have three combinations that are filled with symbols or empty spaces. The audio representation of one of the pictures below is **double knock, knock, knock, and double knock.**

Figure 10

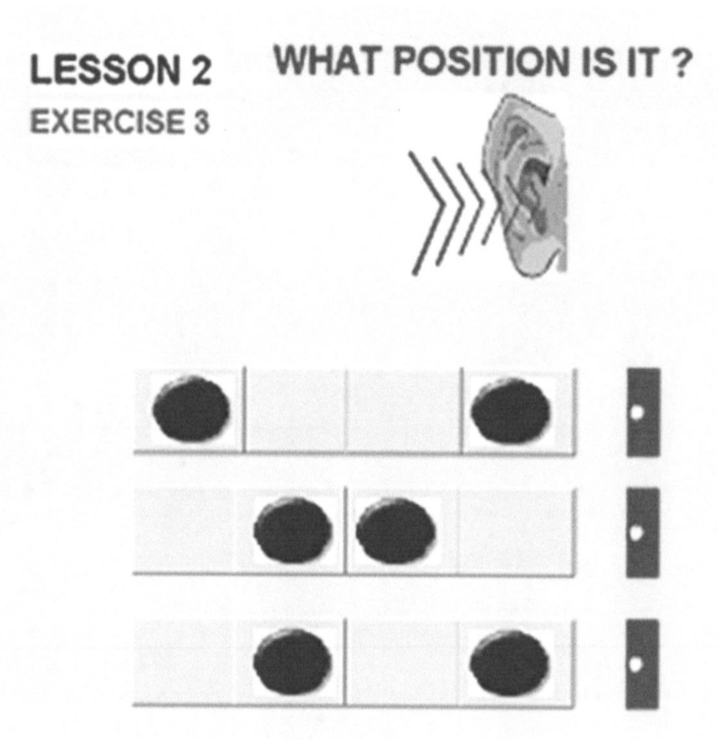

The answer is the second picture of boxes. In the visual representation of the answer, the first space is empty, followed by the second and third spaces filled with symbols and finally the fourth space being empty.

The kinesthetic representation of the answer is the *clenching of the left hand* for the first position, followed by *clenching of the right hand twice* to indicate the presence of symbols in the second and third positions. Finally, *clenching of the left hand* indicates the absence of a symbol in the fourth position.

LESSON 11 INTRODUCTION TO POSITIONS

In Figure 11 we have three combinations that are filled with symbols or empty spaces. The audio representation of one of the pictures below is **knock, knock, double knock, and knock.**

Figure 11

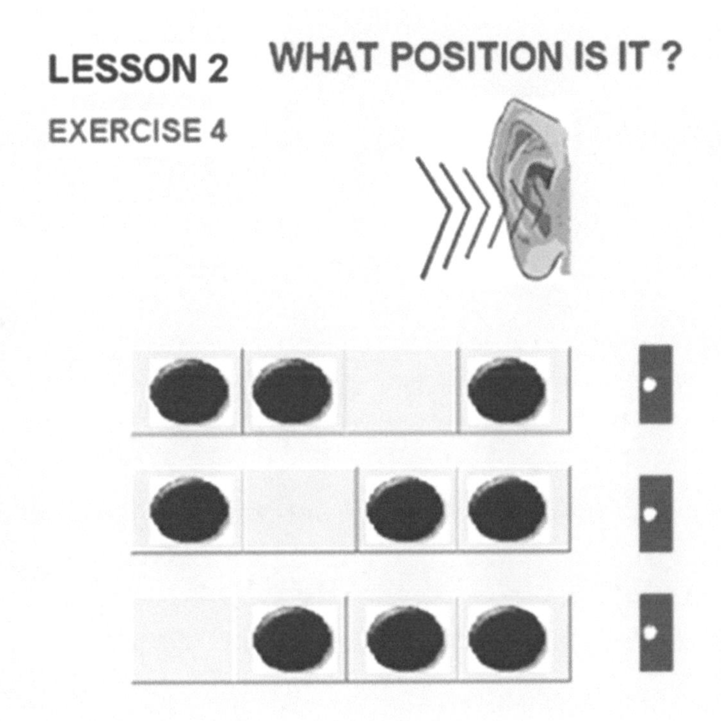

The correct answer in audio representation is the first picture of boxes. In the visual representation of the answer the first and second spaces are filled with a symbol, followed by the third empty space and the fourth space that is filled with a symbol.

The kinesthetic representation of the answer is the *clenching of the right hand twice* for the first and second positions, followed by *clenching of the left hand once* to represent an empty space in the third position. Finally, the *clenching of the right hand* is to represent a symbol in the fourth position.

INTRODUCTION TO POSITIONS

LESSON 12

In Figure 12 we have three combinations that are filled with symbols or empty spaces. The audio representation of one of the pictures below is **double knock, double knock, knock, and knock.**

Figure 12

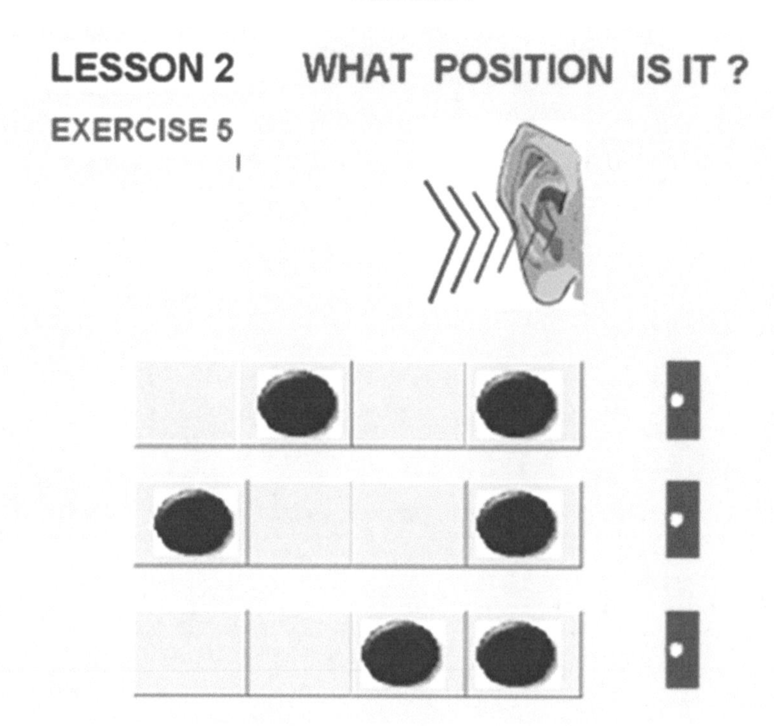

The answer in audio representations is the third picture of boxes. In the visual representation of the answer the first and second squares are empty, followed by third and fourth squares being filled with a symbol.

The kinesthetic representation of the answer is *clenching of the left hand twice* for the first and second positions, followed by *clenching of the right hand twice* to represent the presence of a symbol in the third and fourth positions.

LESSON 13 INTRODUCTION TO POSITIONS

In Figure 13 we have three combinations filled with symbols or empty squares. The audio representation of one of the pictures below is **knock, double knock, knock, and knock.**

Figure 13

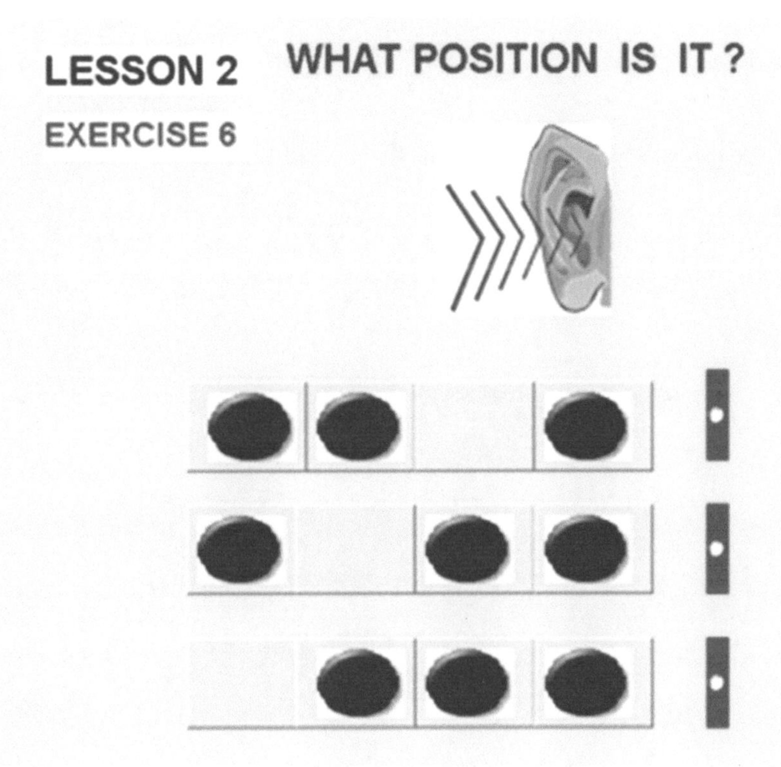

The answer for one of the pictures above in audio representations is the second picture of boxes. In the visual representation of the answer the first position contains a symbol, while the second square is empty, followed by a symbol in the third and fourth positions.

The kinesthetic representation of the answer is *clenching of the right hand* for the first position, *followed by clenching the left hand once* to represent a blank space, and the *clenching of the right hand twice* to represent a symbol in the third and fourth positions.

INTRODUCTION TO POSITIONS

LESSON 14

In Figure 14 we have three combinations filled with symbols or empty squares. The audio representation of one of the pictures below is **knock, knock, knock, and double knock.**

Figure 14

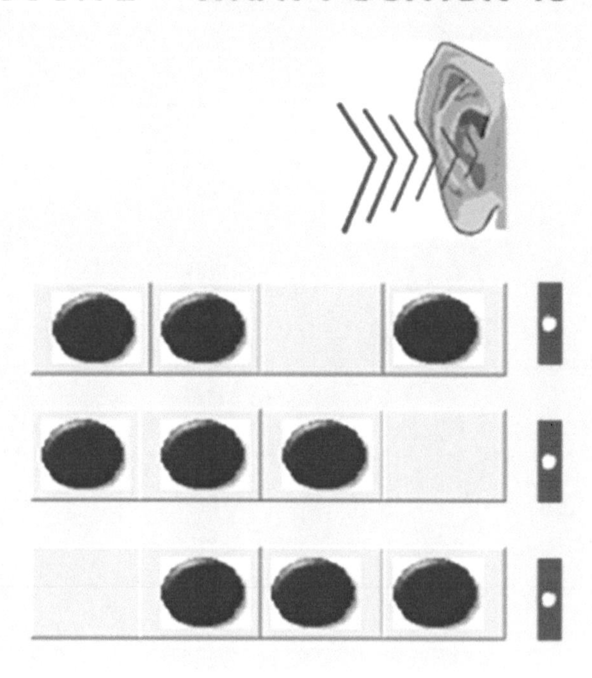

The answer for one of the pictures above in audio representations is the second picture of boxes. In the visual representation of the answer the first, second and third positions are filled with symbols, while the fourth square is empty.

The kinesthetic representation of the answer is the *clenching of the right hand three times* to represent the presence of a symbol in the first, second and third positions, followed by the *clenching of the left hand* to indicate an empty square.

LESSON 15 INTRODUCTION TO POSITIONS

In Figure 15 we have three combinations that are filled with symbols or empty squares. The audio representation of one of the pictures below is **knock, knock, knock, and knock.**

Figure 15

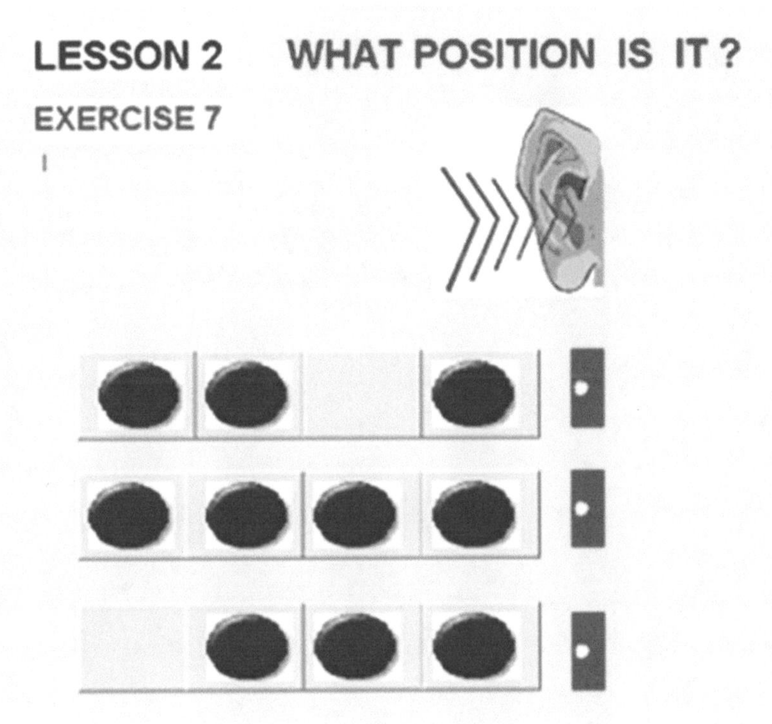

The answer for the picture above in audio representation is the second picture of the boxes. In the visual representation the answer requires all the positions to be filled with a symbol.

The kinesthetic representation of the answer is *clenching of the right hand four times* for the first, second, third and fourth positions to indicate the presence of a symbol.

INTRODUCTION TO POSITIONS

LESSON 16

In Figure 16 we have three combinations that are filled with symbols or empty positions. The audio representation of one of the pictures below is **double knock, knock, double knock, double knock and knock.**

Figure 16

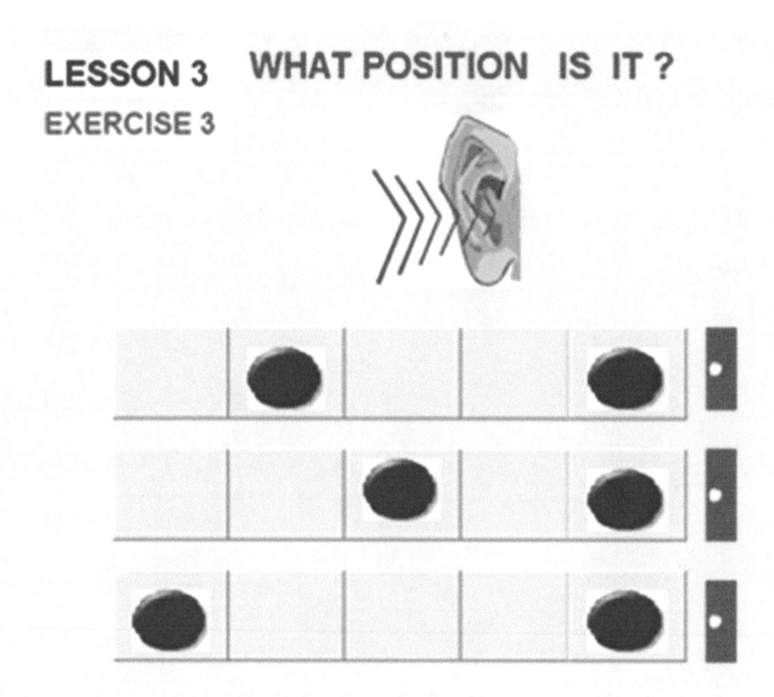

The answer for the assignment above in the audio representation is the first picture of boxes. In the visual representation of the answer the first box is empty, while the second contains a symbol, followed by the third and fourth positions, which are empty, while the fifth square is filled with a symbol.

The kinesthetic representation of the answer is *clenching of the left hand* for the first position, followed by *clenching the right hand* to represent presence of a symbol in the second square. The third and fourth positions are indicated by *clenching of the left hand twice, followed by clenching of the right hand* to indicate the presence of a symbol in the fifth square.

LESSON 17

INTRODUCTION TO POSITIONS

In Figure 17 we have three combinations that are filled up with symbols or empty positions. The audio representation of one of the pictures below is **double knock, knock, double knock, double knock and knock.**

Figure 17

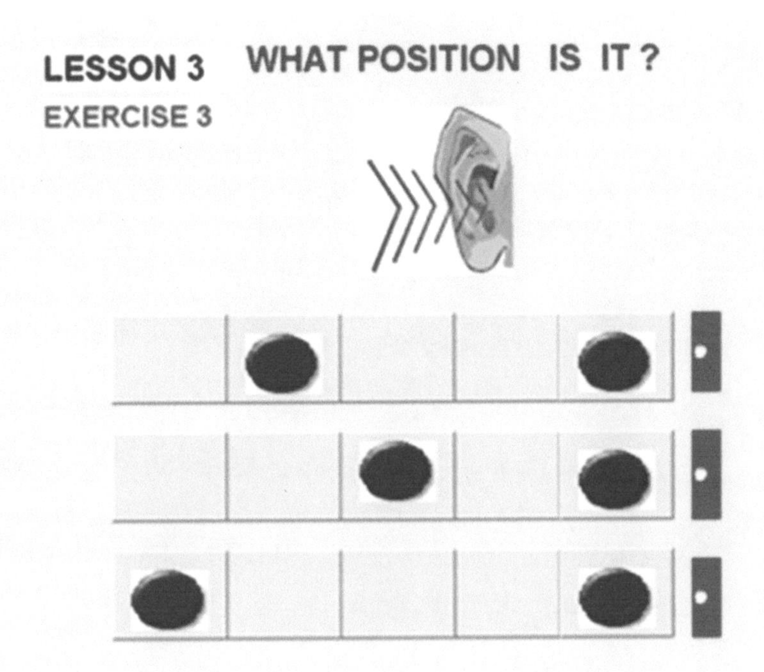

The answer for the assignment above in the audio representations is the first picture of boxes. In the visual representation of the answer the first position is empty and the second is filled with a symbol, followed by the third and fourth positions which are empty, while the fifth box is filled with a symbol.

The kinesthetic representation of the answer is the clenching *of the left hand once* to represent an empty box in the first position, followed by *clenching the right fist* to indicate the presence of a symbol. The third and fourth positions, which are empty, are represented by *clenching the left hand twice.* The *right hand is clenched* to indicate a symbol in the last box.

INTRODUCTION TO POSITIONS

LESSON 18

In Figure 18 we have three combinations that are filled up with symbols or empty positions. The audio representation of one of the pictures below is **knock, double knock, double knock, double knock and knock**.

Figure 18

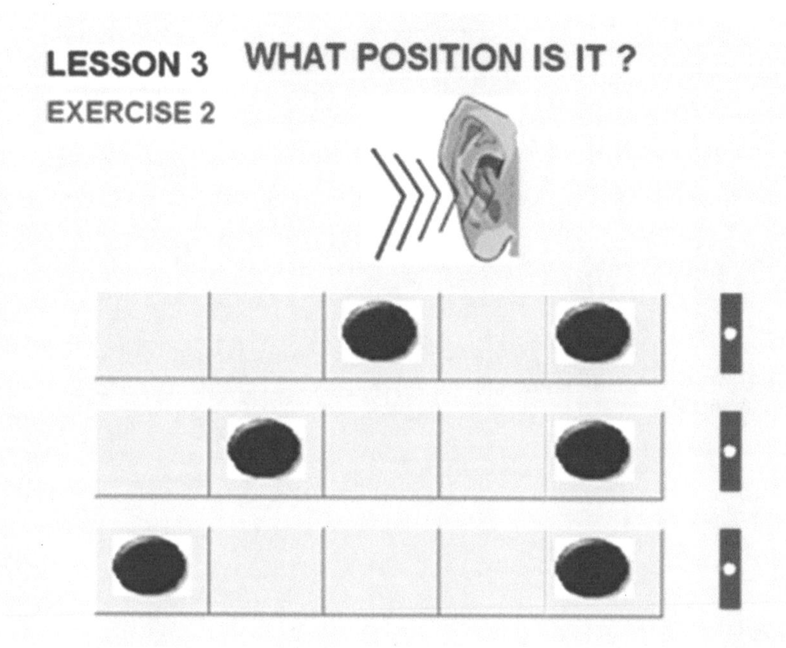

The answer for the audio representations is the third picture of boxes. The visual representation of the answer is for the first position that is filled with a symbol, followed by the second, third and fourth positions being empty, and finally the fifth box containing a symbol.

The kinesthetic representation of the answer is *clenching of the right hand* for the first position, followed by the second third and fourth positions being represented by *clenching of the left hand three times and finally clenching of the right hand* to represent a symbol in the fifth position.

LESSON 19 INTRODUCTION TO POSITIONS

In Figure 19 we have three combinations that contain symbols or empty boxes. The audio representation of one of the pictures below is **knock, knock, double knock, double knock and knock**.

Figure 19

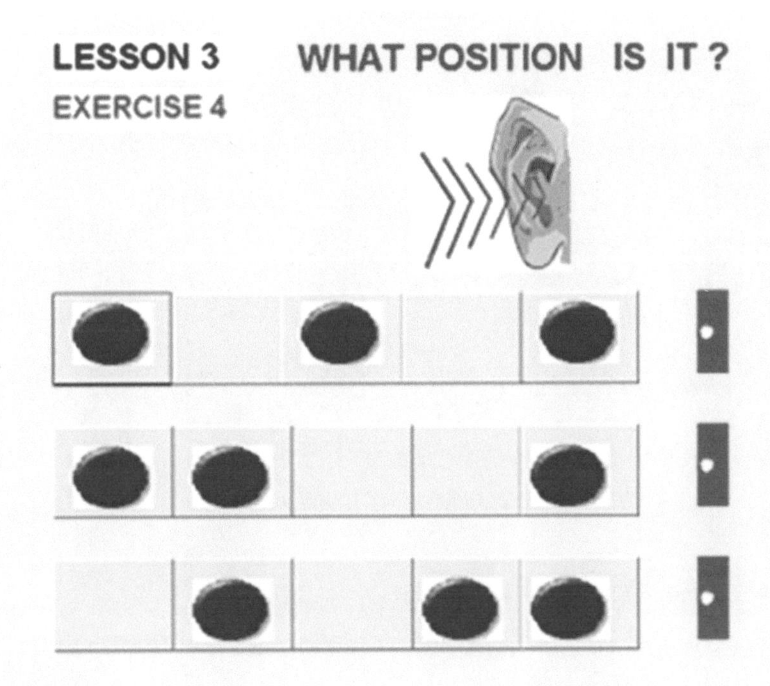

The answer for the audio representations is the second picture of boxes. The visual representation of the answer is indicated by a symbol in the first and second positions, followed by the third and fourth positions being left empty and finally the fifth box being filled with a symbol.

The kinesthetic representation of the answer is *clenching of the right hand* for the first and second position, followed by *clenching of the left hand* for the third and fourth positions and finally, *clenching the right hand* to represent a symbol in the fifth box.

INTRODUCTION TO POSITIONS

LESSON 20

In Figure 20 we have three combinations that contain symbols or empty squares. The audio representation of one of the pictures below is **double knock, double knock, knock, double knock, and knock.**

Figure 20

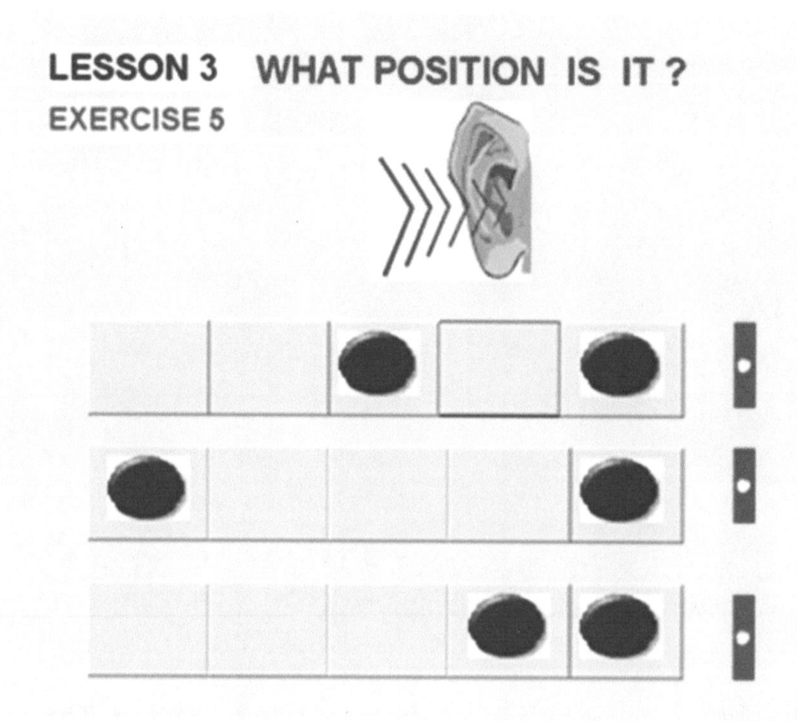

The answer for the audio representations is the first picture of boxes. The visual representation of the answer shows the first and second positions as empty, while the third contains a symbol. The fourth square is left empty and the fifth contains a symbol.

The kinesthetic representation of the answer is *clenching the left hand twice* for the first and second positions. The third square is represented by *clenching the right hand*, the fourth position with the *clenching of the left hand* and finally, *clenching of the right h*and to represent a symbol in the fifth square.

LESSON 21 INTRODUCTION TO POSITIONS

In Figure 21 we have three combinations that contain symbols or empty squares. The audio representation of one of the pictures below is **double knock, knock, double knock, knock, and knock.**

Figure 21

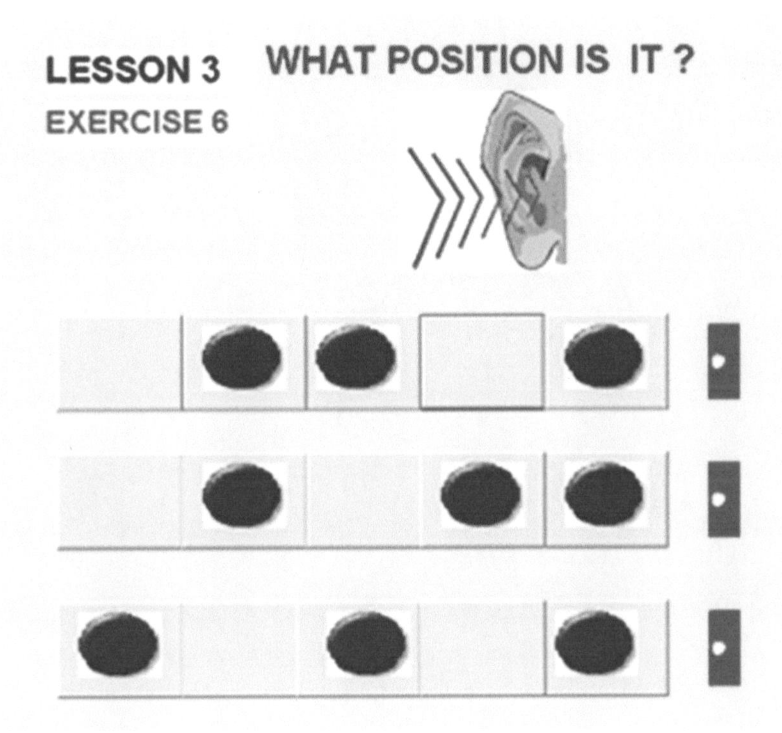

The answer for the audio representations is the second picture of boxes. The visual representation of the answer for the first position is an empty square followed by a symbol in the second square. The third box is empty while the fourth and fifth positions contain a symbol.

The kinesthetic representation of the answer is *clenching the left hand* for the first position, followed by the second position being represented by *clenching of the right hand.* The third square is described by *clenching of the left fist and clenching the right hand twice* represents a symbol in the fourth and fifth position.

INTRODUCTION TO POSITIONS

LESSON 22

In Figure 22 we have three combinations that contain symbols or empty squares. The audio representation of one of the pictures below is **knock, knock, double knock, double knock, and knock.**

Figure 22

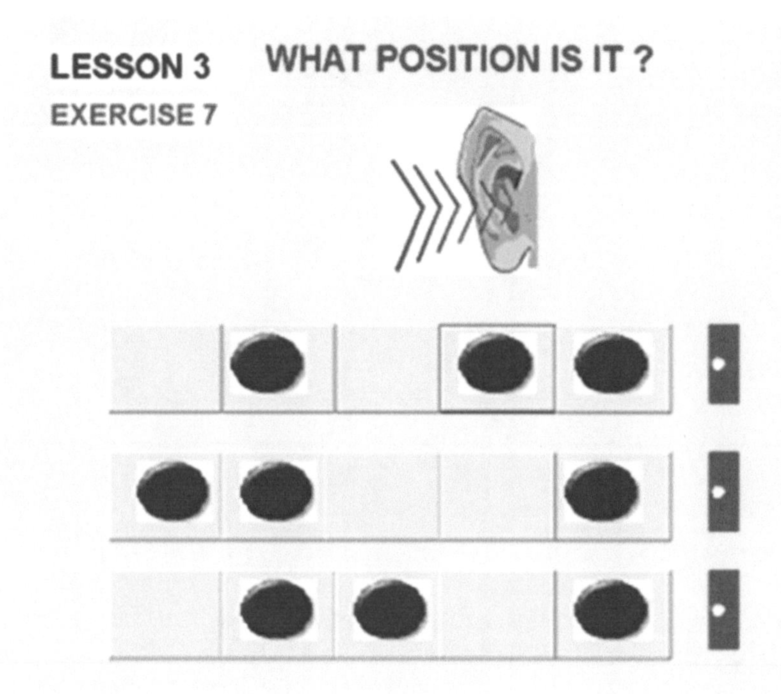

The answer for the audio representations is the second picture of the boxes. The visual description of the answer shows a symbol in the first and second position followed by empty squares in the third and fourth positions, and finally the fifth position containing a symbol.

The kinesthetic representation of the answer is *clenching of the right hand twice* for the first and second positions, followed by the third and fourth positions being represented by *clenching of the left hand twice.* Finally, *clenching of the right hand* represents a symbol in the fifth position.

LESSON 23 INTRODUCTION TO POSITIONS

In Figure 23 we have three combinations that contain symbols or empty squares. The audio representation of one of the pictures below is **knock, knock, knock, double knock, and knock.**

Figure 23

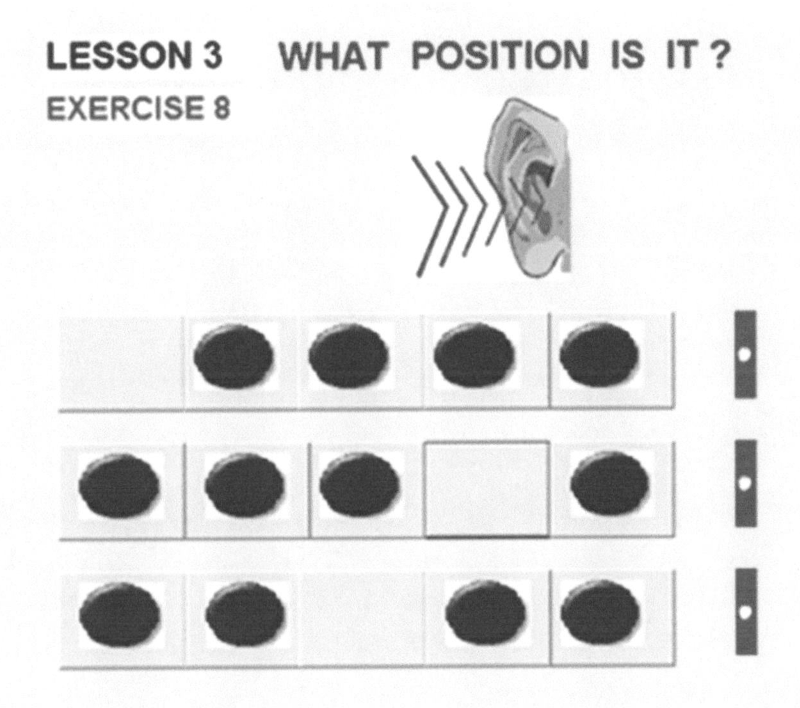

The answer for the audio descriptions is the second picture of boxes. The visual representation of the answer for the first, second and third positions contains a symbol, followed by an empty square for the fourth position while the fifth box contains a symbol.

The kinesthetic representation of the answer is *clenching of the right hand three times* for the first, second, and third position, followed by the fourth position being represented by *clenching of the left hand.* Finally *clenching of the right hand* represents a symbol in the fifth position.

INTRODUCTION TO POSITIONS

LESSON 24

In Figure 24 we have three combinations that contain symbols or empty squares. The audio representation of one of the pictures below is **double knock, double knock, double knock, knock, and knock.**

Figure 24

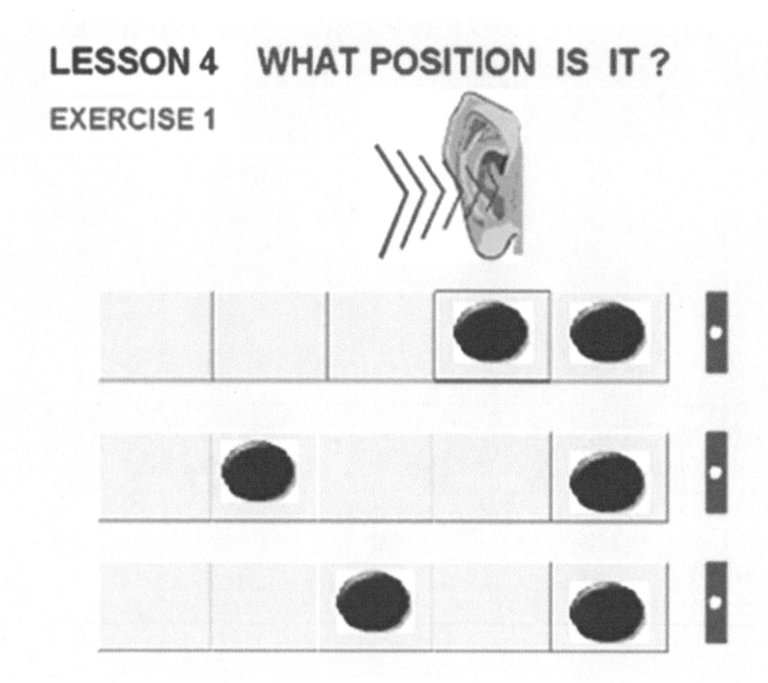

The answer for the audio representations is the first picture of boxes. The visual description of the answer is the first, second and third positions being empty while the fourth & fifth positions each contain a symbol.

The kinesthetic description of the answer is *clenching of the left hand three times* for the first, second, and third positions, followed by the fourth and fifth positions being described by *clenching of the right hand twice.*

LESSON 25 INTRODUCTION TO POSITIONS

In Figure 25 we have three combinations that contain symbols or empty squares. The audio representation of one of the pictures below is **double knock, double knock, knock, knock, and knock.**

Figure 25

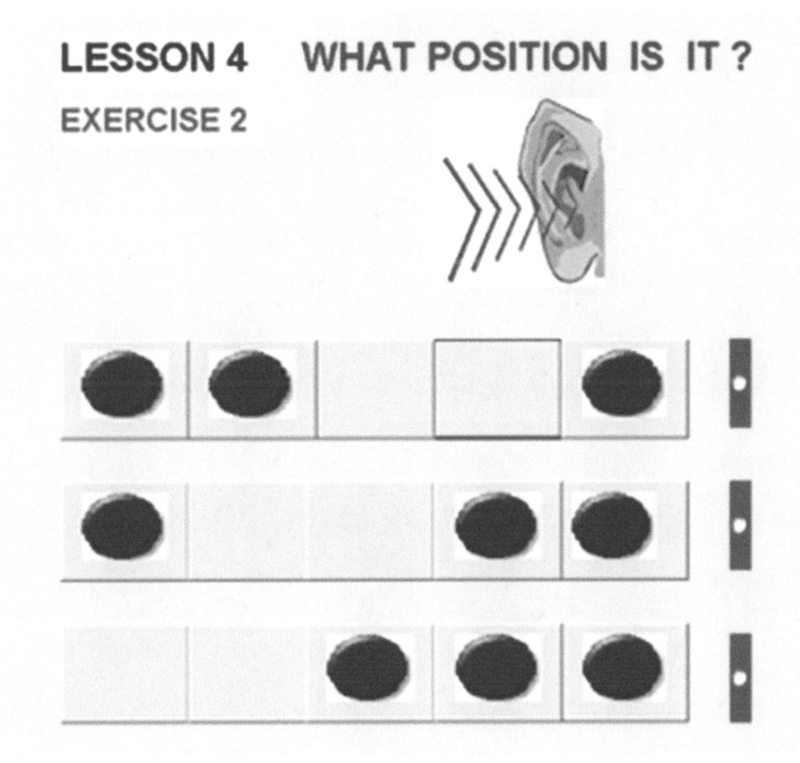

The answer for the audio representation is the third picture of boxes. The visual description of the answer are empty first and second positions, while the third, fourth & fifth boxes contain a symbol.

The kinesthetic representation of the answer requires *clenching of the left hand twice* for the first and second positions, while the third, fourth and fifth positions are represented by *clenching of the right hand three times.*

INTRODUCTION TO POSITIONS

LESSON 26

In Figure 26 we have three combinations that contain symbols or empty squares. The audio representation of one of the pictures below is **double knock, knock, double knock, knock, and knock.**

Figure 26

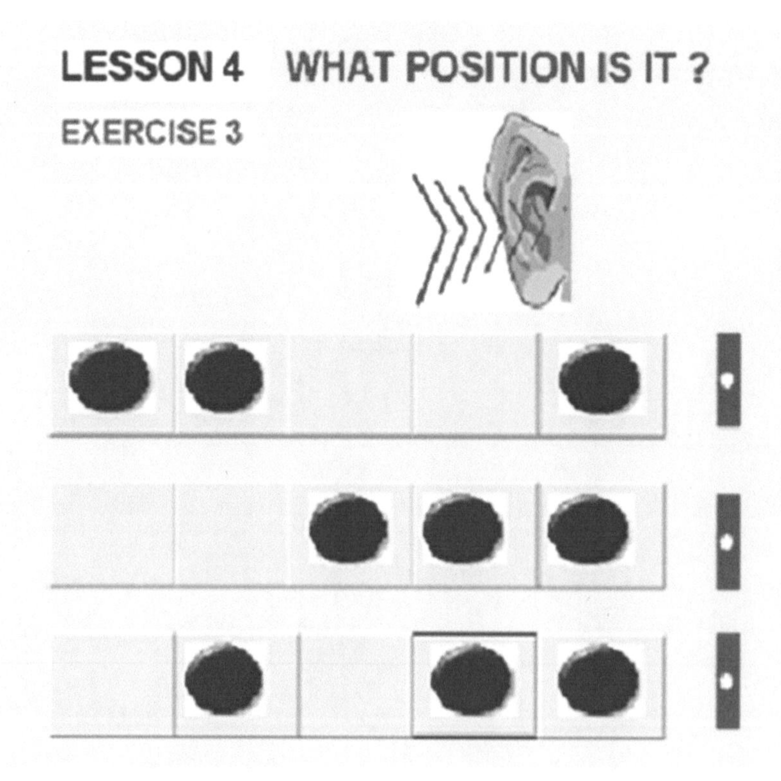

The answer for the audio representations is the third picture of boxes. The visual description of the answer for the first is an empty space followed by a symbol in the second position and a blank space for the third position, while the fourth & fifth positions contain symbols.

The kinesthetic representation of the answer requires *clenching of the left hand* for the first position, *clenching of the right hand* for the symbol in the second position. The third position is represented by *clenching the left hand* followed by the fourth and fifth positions being represented by *clenching the right hand twice.*

LESSON 27 INTRODUCTION TO POSITIONS

In Figure 27 we have three combinations that contain symbols or empty squares. The audio representation of one of the pictures below is **knock, double knock, knock, knock, and knock.**

Figure 27

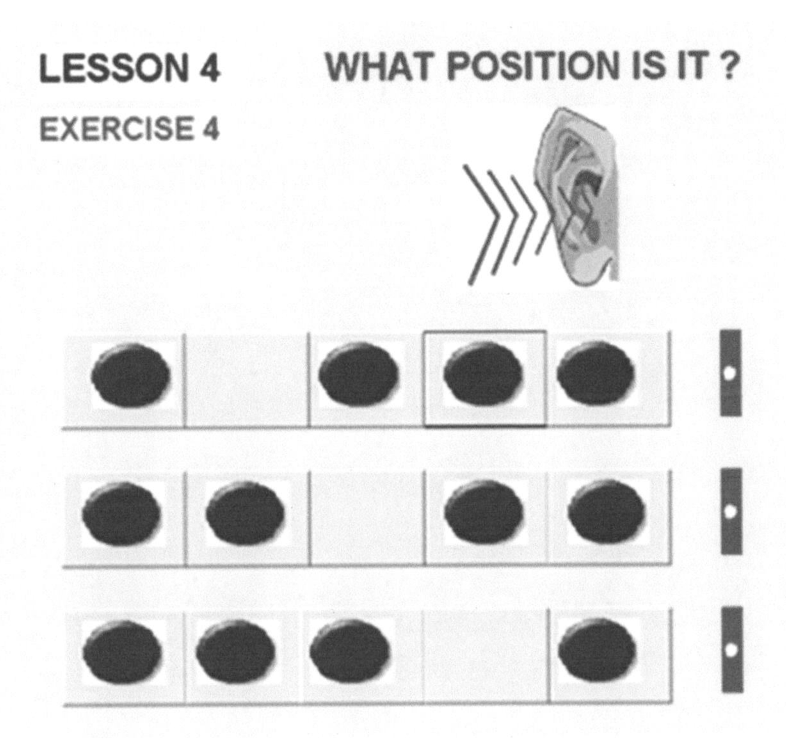

The answer for the audio representations is the first picture of boxes. The visual representation of the answer has a symbol in the first square followed by an empty box. The third, fourth and fifth squares each contain a symbol.

The kinesthetic representation of the answer is *clenching of the right hand* for the first position. The second position requires *clenching of the left hand* while the third, fourth and fifth are described by *clenching the right hand three times.*

INTRODUCTION TO POSITIONS

LESSON 28

In Figure 28 we have three combinations that contain symbols or empty positions. The audio representation of one of the pictures below is **double knock, double knock, knock, knock, and knock.**

Figure 28

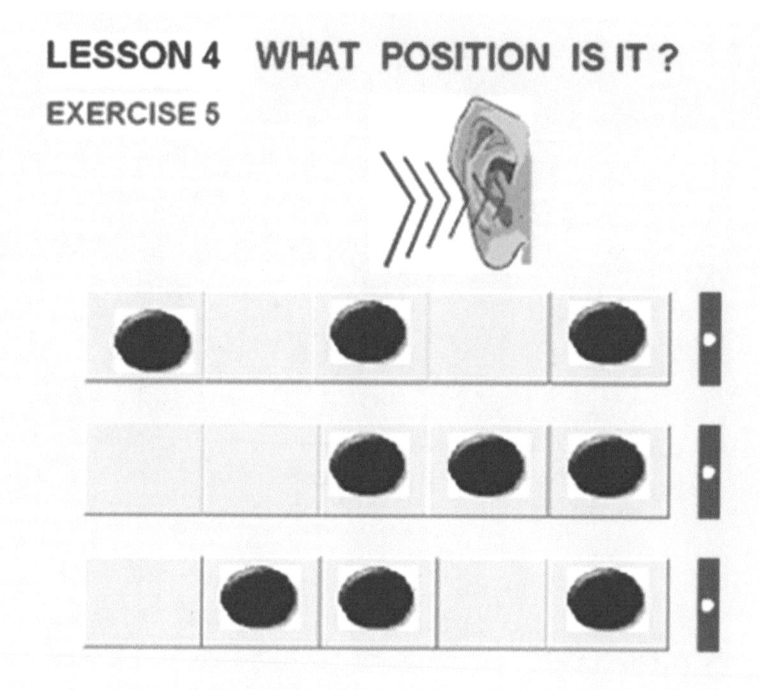

The answer for the audio representations is the second picture of boxes. In the visual representation of the answer the first and second positions are empty followed by a symbol in the third, fourth and fifth positions.

The kinesthetic representation of the answer is *clenching of the left hand twice* for the first, and second positions, while the third, fourth and fifth positions are represented by *clenching of the right hand three times.*

LESSON 29 INTRODUCTION TO POSITIONS

In Figure 29 we have three combinations that contain symbols or empty positions. The audio representation of one of the pictures below is **double knock, knock, knock, knock, and knock.**

Figure 29

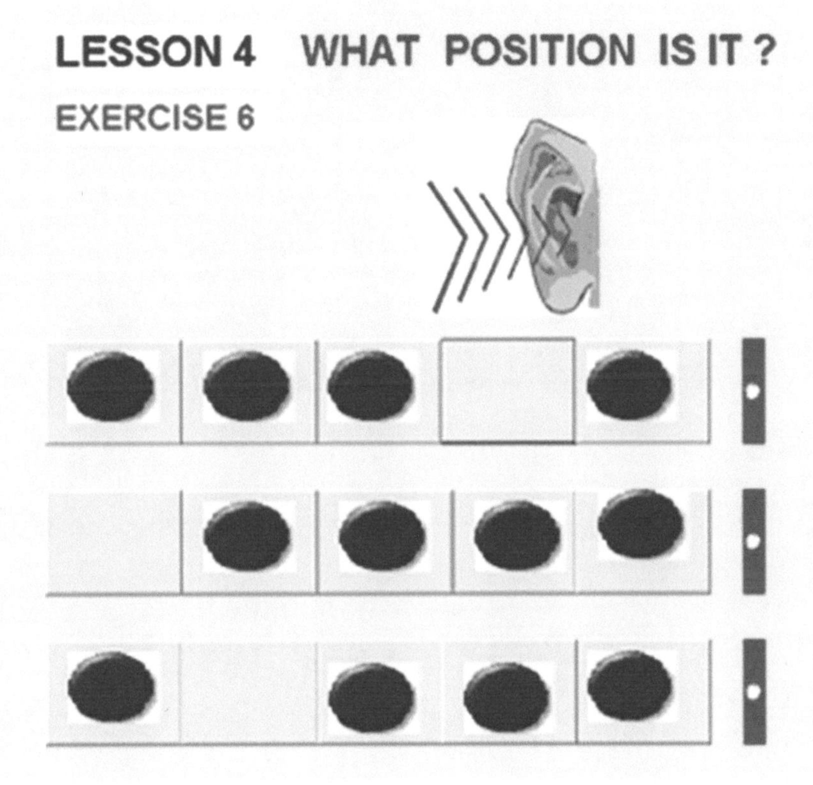

The answer for the audio representations is the second picture of boxes. The visual representation of the answer for the first position is an empty box, followed by symbols in the second, third, fourth, and fifth position.

The kinesthetic description of the answer is *clenching of the left hand* for the first position, while the second, third, fourth and fifth positions are represented by clenching of the right hand four times.

INTRODUCTION TO POSITIONS

LESSON 30

In Figure 30 we have three combinations that contain symbols or empty squares. The audio representation of one of the pictures below is **knock, knock, double knock, knock, and knock.**

Figure 30

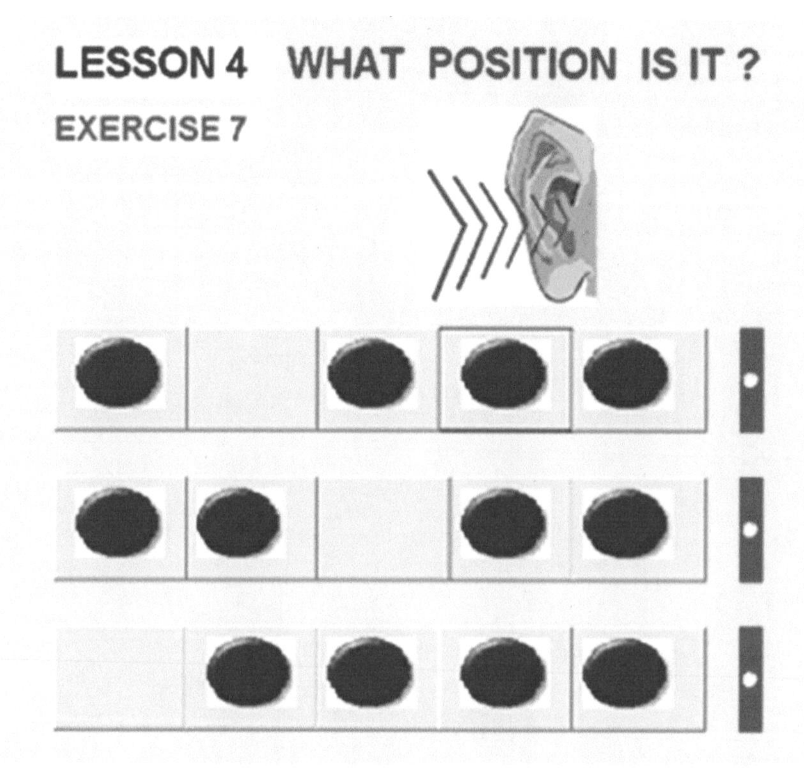

The answer for the audio representations is the second picture of boxes. The visual representation of the answer is a symbol in each of the first and second positions followed by an empty square in the third position. The fourth and fifth positions each contain a symbol.

The kinesthetic description of the answer is *clenching of the right hand twice* for the first and second positions, *clenching the left hand* for the third position, then *clenching of the right hand twice* for the fourth and fifth positions.

LESSON 31 INTRODUCTION TO POSITIONS

In Figure 31 we have three combinations that contain symbols or empty squares. The audio representation of one of the pictures below is **knock, knock, knock, knock, and knock.**

Figure 31

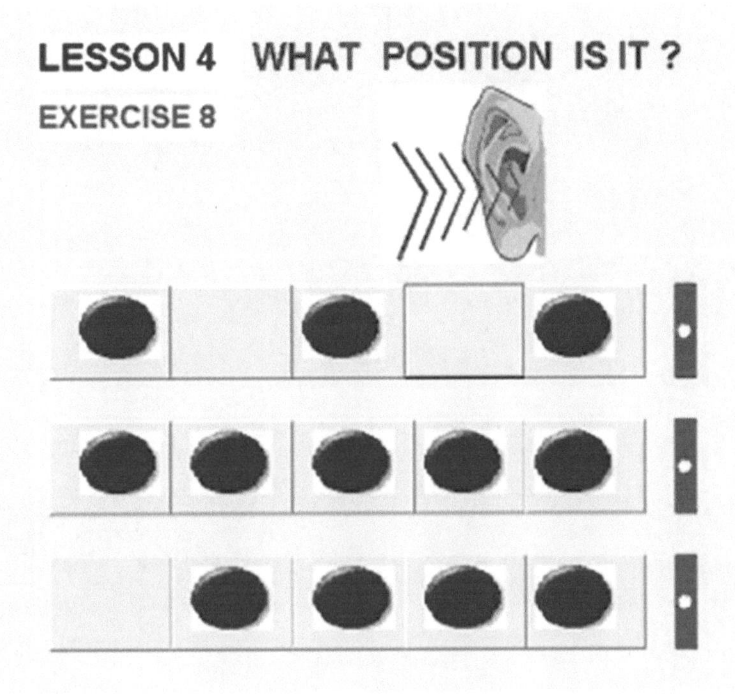

The answer for the audio representations is the second picture of boxes. The visual representation of the answer requires all of the squares to contain a symbol.

The kinesthetic representation of the answer is *clenching of the right hand five times* for all the positions.

INTRODUCTION TO SIGNS

LESSON ONE INTRODUCTION TO SIGNS

There are five different signs. There is an addition sign, division sign, subtraction sign, multiplication sign, and an equal sign. Addition is represented by the sound, **tram**. Division is represented by the sound, **double click**. Subtraction is represented by the sound, **double tram**. Multiplication is represented by the sound, **blick**. The equal sign is represented by the sound, **cling**.

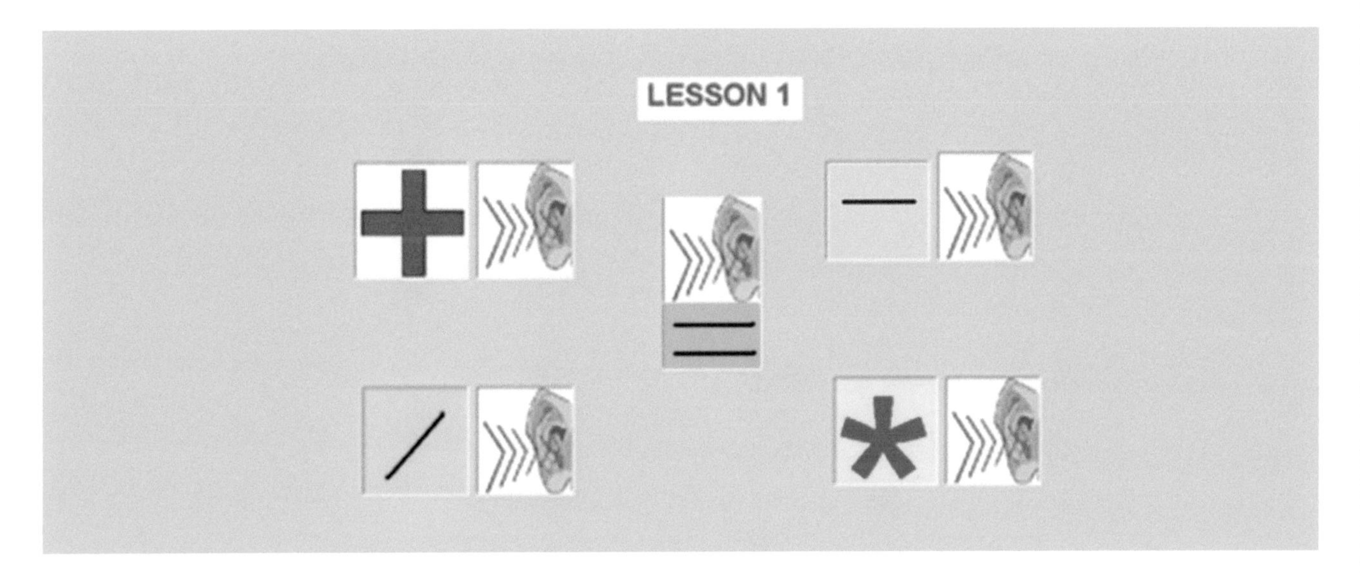

The addition sign is represented by one hand extended in front. The division sign is represented by both hands extended up. The subtraction sign is represented by both hands extended in front. The multiplication sign is represented by one hand raised up. The equal sign is represented by crossing both hands.

INTRODUCTION TO SIGNS

LESSON 1, EXERCISE 1

In Lesson 1 Ex. 1 there are two different signs below. The first picture is an addition sign. Addition is represented by the sound of a **tram**. The second picture is the multiplication sign. Multiplication is represented by the sound of a **blick**.

Exercise 1

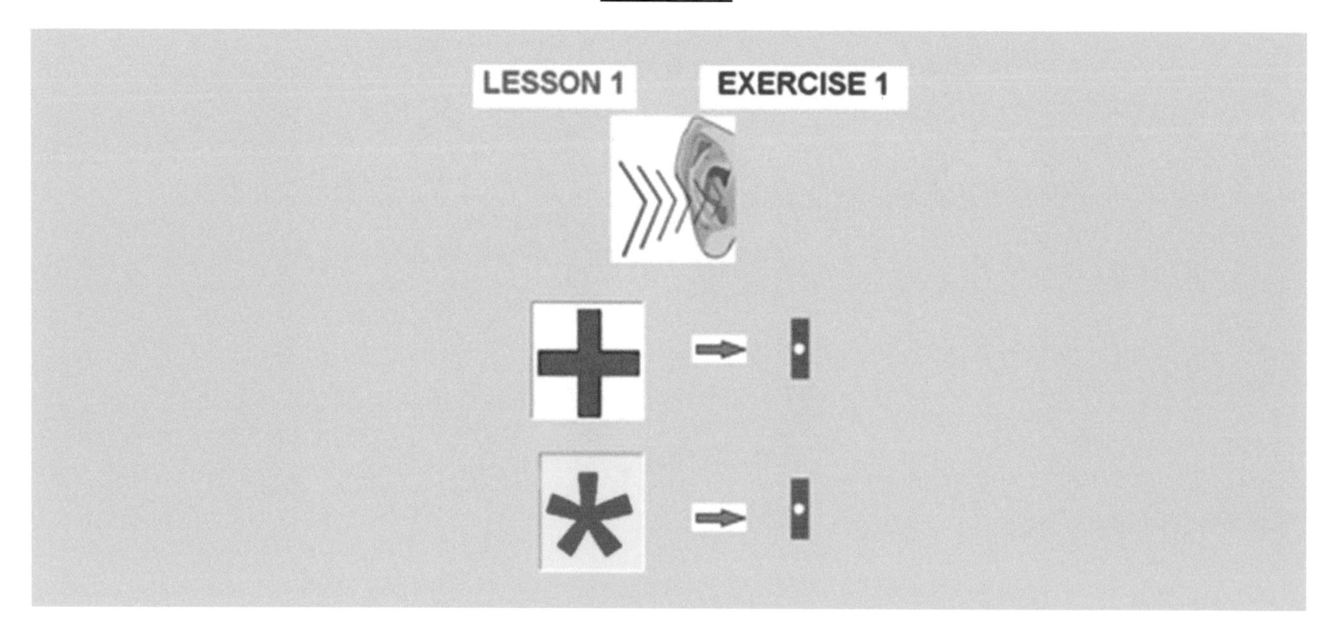

The addition sign is represented by one hand extended in front. The multiplication sign is represented by one hand raised up.

LESSON 1, EXERCISE 2 INTRODUCTION TO SIGNS

In Lesson 1 Ex. 2 there are two different signs below. The first picture is of a subtraction sign. The subtraction sign is represented by the sound of a **double tram**. The second picture is of a division sign. The division sign is represented by the sound of a **double click**.

Exercise 2

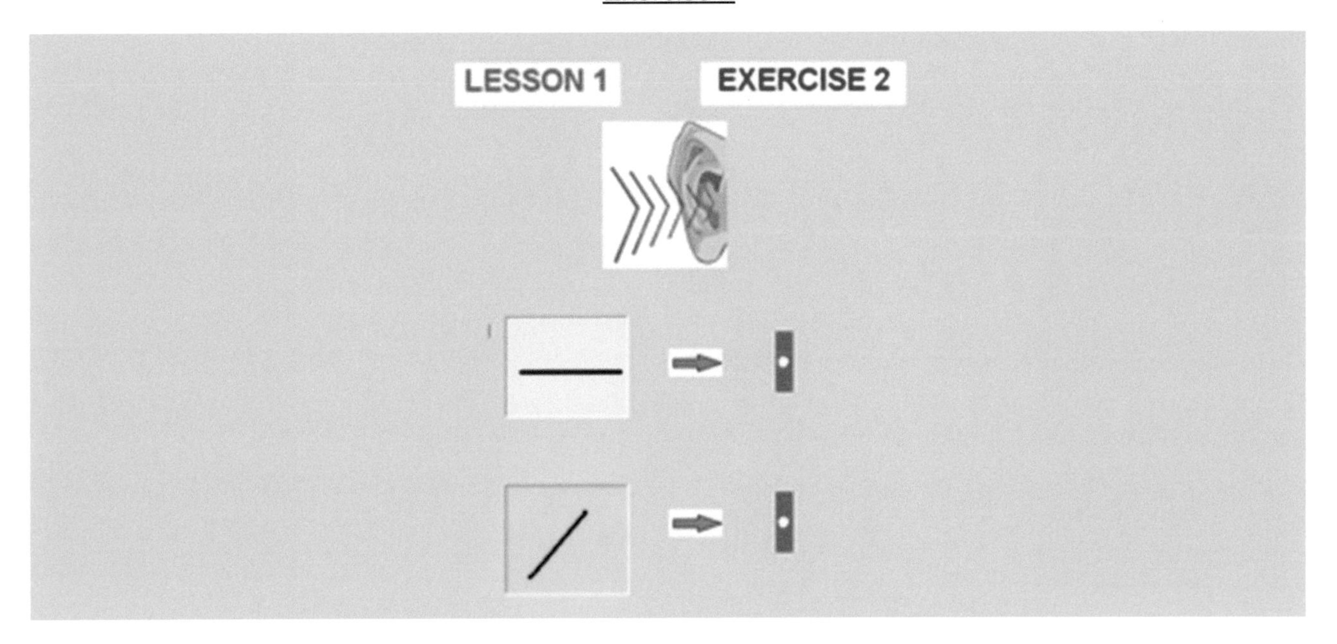

The subtraction sign is represented by both hands extended in front. The division sign is represented by both hands extended up.

INTRODUCTION TO SIGNS

LESSON 1, EXERCISE 3

In Lesson 1 Ex. 3 there are two different signs below. The first picture is an addition sign. Addition is represented by the sound of a **tram**. The second picture is of a multiplication sign. Multiplication is represented by the sound of a **blick**.

Exercise 3

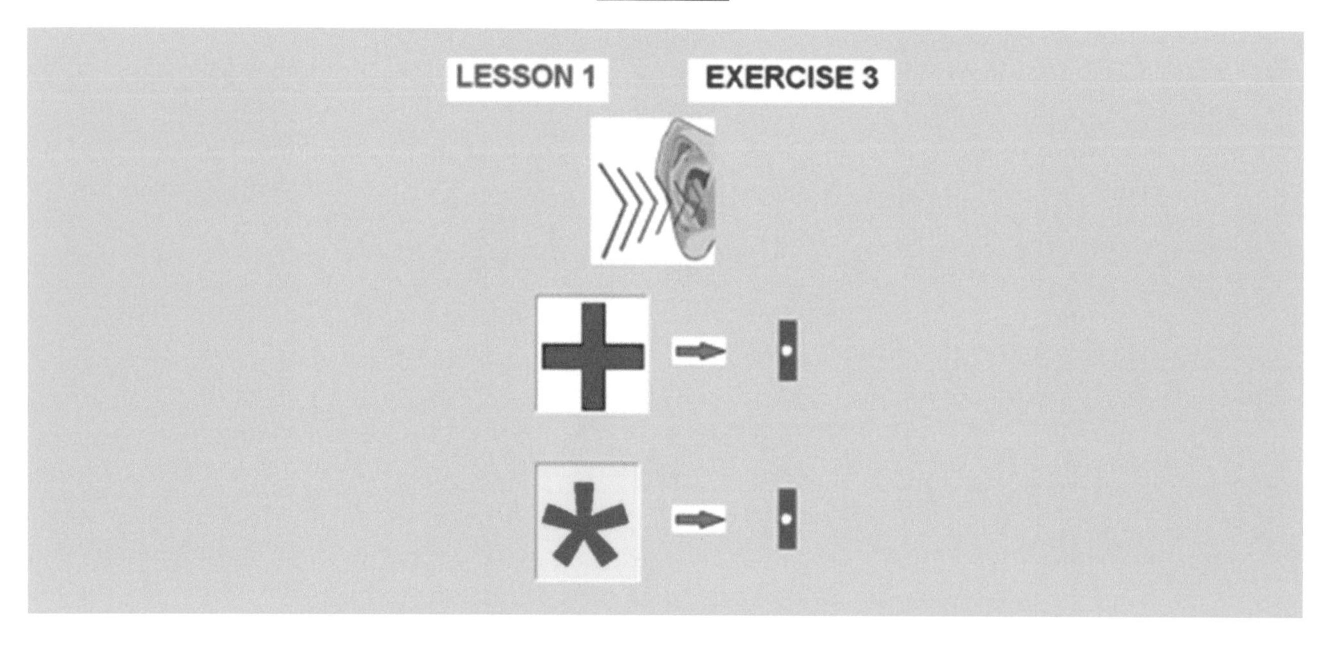

The addition sign is represented by one hand extended in front. The multiplication sign is represented by one hand raised up.

LESSON 1, EXERCISE 4 INTRODUCTION TO SIGNS

In Lesson 1 Ex. 4 there are two different signs below. The first picture is of a multiplication sign. Multiplication is represented by the sound of a **blick**. The second picture is of a division sign. Division is represented by the sound of a **click**.

Exercise 4

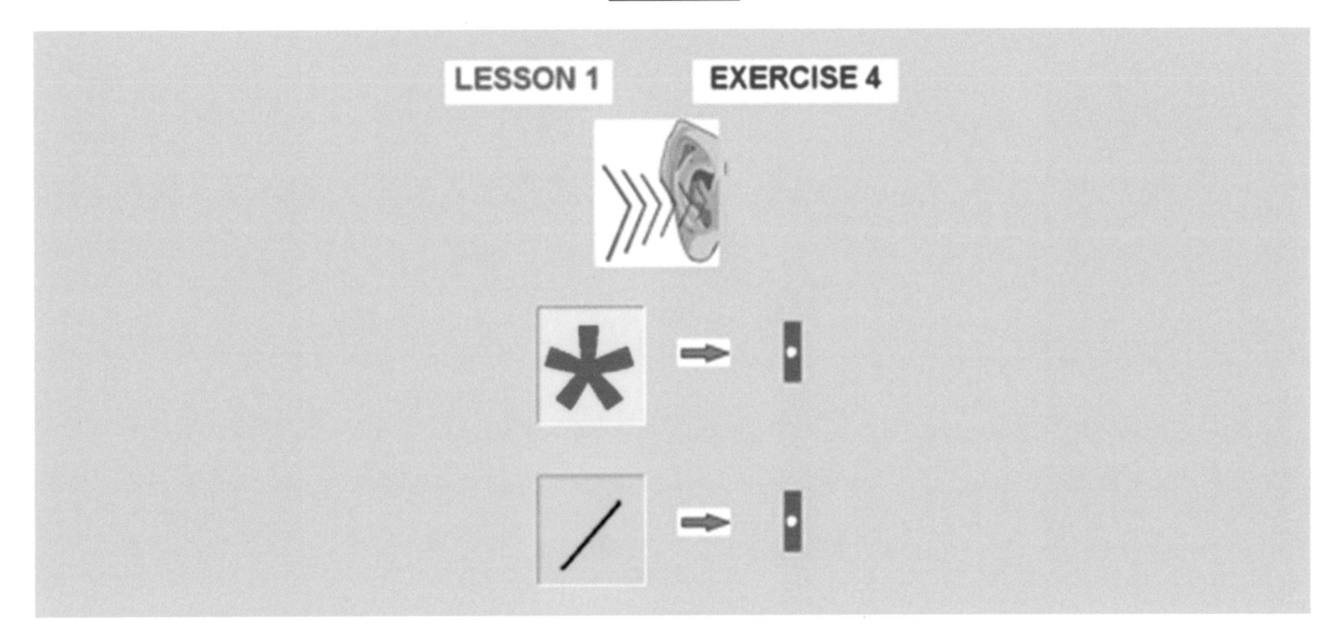

The multiplication sign is represented by one hand raised up. The division sign is represented by both hands extended up.

INTRODUCTION TO SIGNS

LESSON 1, EXERCISE 5

In Lesson 1 Ex. 5 there are two different signs below. The first picture is of an equal sign. The equal sign is represented by the sound of a **cling**. The second picture is of the addition sign. Addition is represented by the sound of a **tram**.

Exercise 5

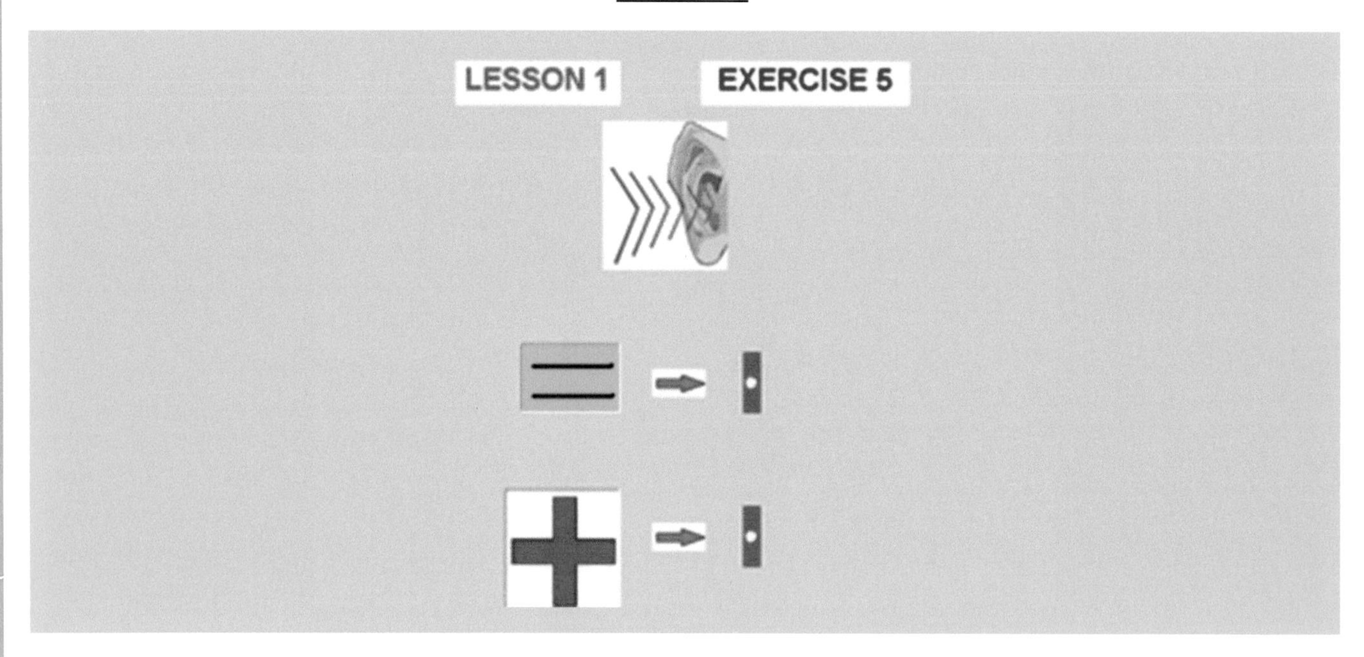

The equal sign is represented by crossing both hands. The addition sign is represented by one hand extended in front.

TRAINING, EXERCISE 1

INTRODUCTION TO SIGNS

In Exercise 1 Training, there are two sets of pictures below. The first picture contains an addition sign, and a multiplication sign. The sounds are represented by a **tram**, and then a **blick**. The second picture contains a division sign, and an equal sign. The sounds are represented by a **click**, and then a **cling**.

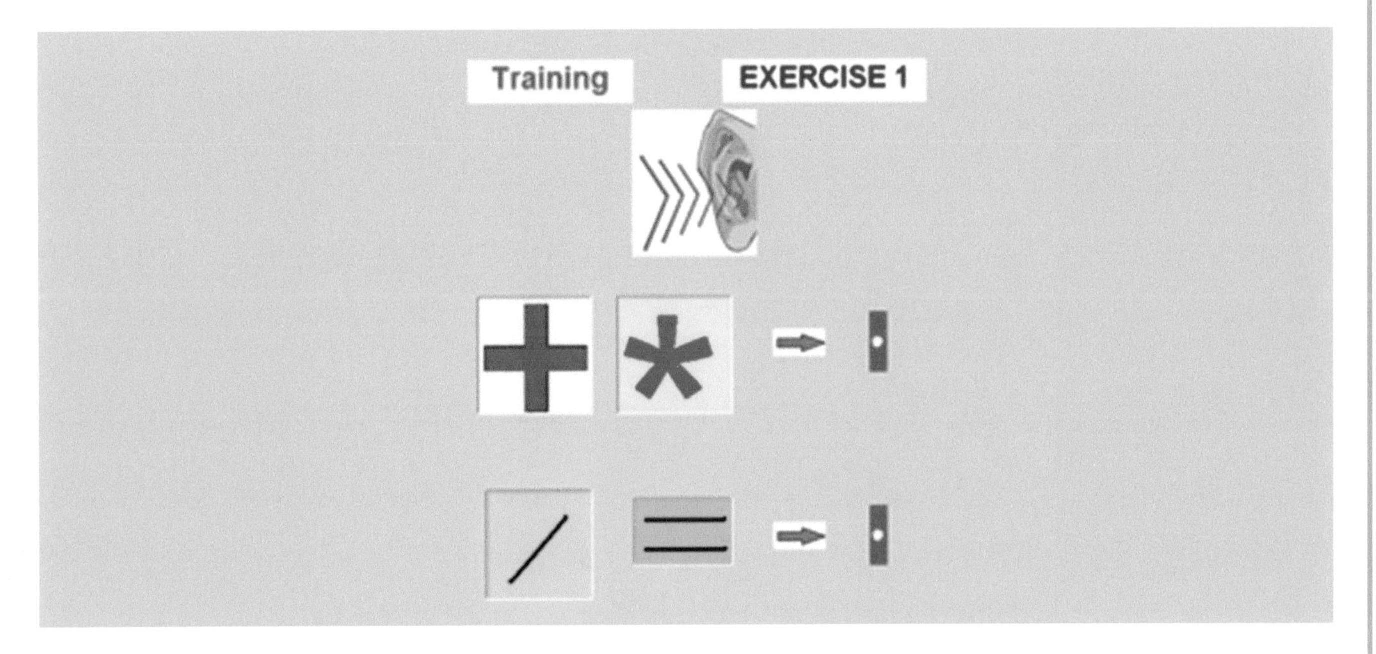

The first picture is represented by one hand extended in front, then one hand raised up. The second picture is represented by both hands extended up, then by crossing both hands.

INTRODUCTION TO SIGNS

TRAINING, EXERCISE 2

In Exercise 2 Training, there are two sets of pictures below. The first picture contains a division sign, and a multiplication sign. The sounds are represented by a **double click**, and then a **blick**. The second picture contains a subtraction sign, and an addition sign. The sounds are represented by a **double tram**, and then a **tram**.

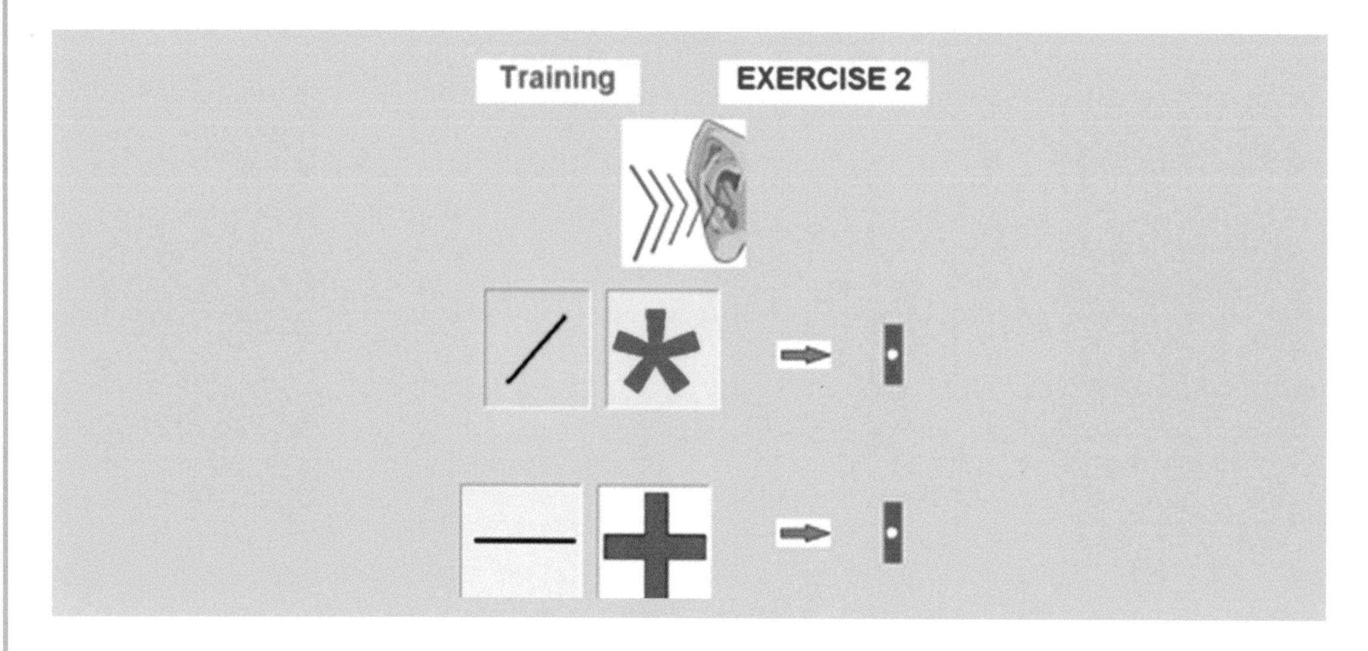

The first picture is represented by both hands extended up, then one hand raised up. The second picture is represented by both hands extended in front, then by one hand extended in front.

TRAINING, EXERCISE 3

INTRODUCTION TO SIGNS

In Exercise 3 Training, there are two sets of pictures below. The first picture contains a division sign, and an addition sign. The sounds are represented by a **double click**, and then a **tram**. The second picture contains a subtraction sign, and a multiplication sign. The sounds are represented by a **double tram**, and then a **blick**.

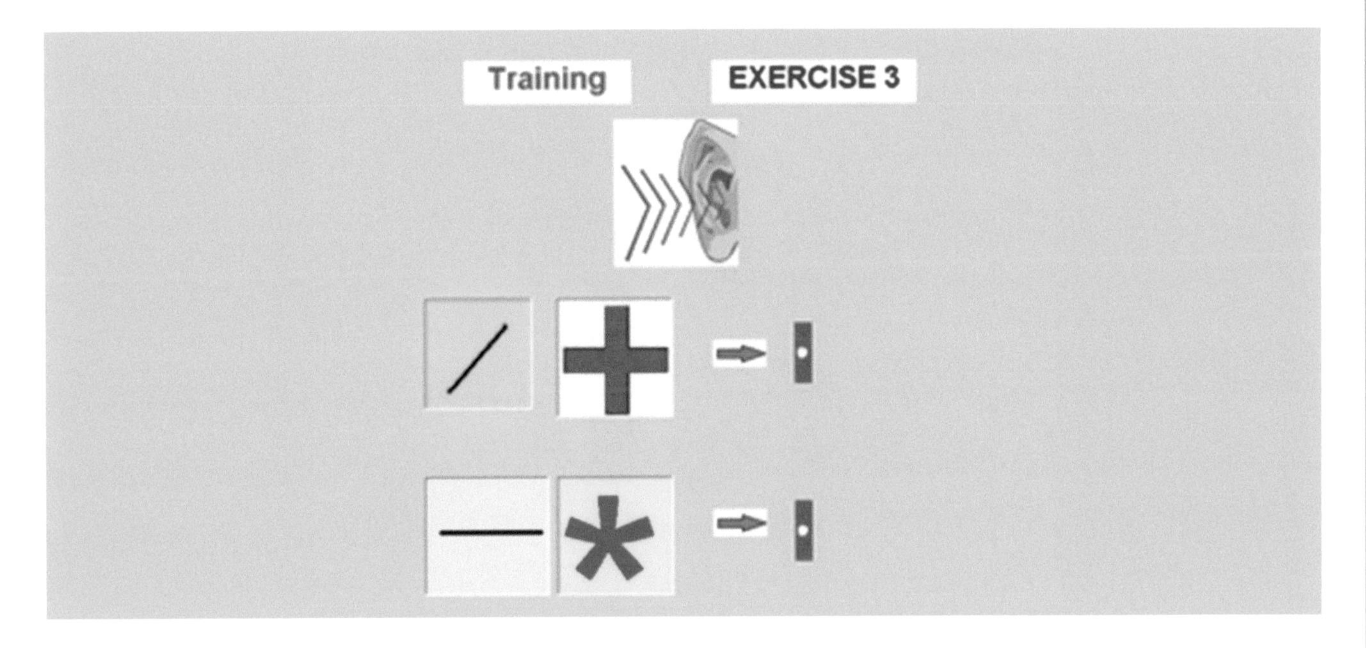

The first picture is represented by both hands extended up, then one hand extended in front. The second picture is represented by both hands extended in front, then by one hand raised up.

Dedicated to my nephew, David Gimelfarb,
Student of Adlerian School of Psychology in Chicago
Lost in Costa Rico in 2009.

CONTENTS

Litvin's Code uses the new approach for non-invasive brain stimulation to wake up the "sleeping" not-active brain. It provides an opportunity to translate letters from visual stimuli to audio and then to kinesthetic and vise versa. By translating the letters we stimulate the brain and make it more active. A stimulated brain is functioning on the more efficient levels and is increasing performance. We are using areas of the brain, which were previously inactive. After several lessons with translations of letters to different expressions we can observe the increase in memory, concentration and the ability to process information. Litvin's Code is a very simple and self-explanatory. Let's assume that to train our memory, we recall telephone numbers and other trivial information. However, instead of actual telephone numbers, we are encouraged to remember positions, where the numbers are located on the dial. Litvin's Code is recognizing the content of the positions, which have a particular sequence. The information is received by one type of stimuli and is easily translated to a different expression, which includes visual, audio, tactile, kinesthetic and olfactory types.

Litvin's Code provides the opportunity to stimulate the brain by translating letters from one mode of expression to another. Brain stimulation is an important part of learning. By disregarding the stimulation of a certain part of the brain, the learning process becomes only a process of the visual stimuli. This places a big emphasis on the visual faculty. The educational contribution of Litvin's Code includes the wide range of opportunities for brain stimulation. It is using a learning approach, which enhances learning by increasing the attention, concentration and the memory of the entire brain. The binary arithmetic allowed the processing of complex information by simple methods. The combinations of empty and filled positions stimulate the entire brain.

Litvin's Code is a new way of learning by translating the same meanings between different modes of expression and providing the opportunity to enhance learning by using alternative perceptual modes. It employs a simple approach, by using the letters and digits as the combination of empty and full positions. In Litvin's Code, the same logic is used through different perceptions. The result is very promising. Through the years, Litvin's Code has helped students increase their academic performance as well as their comprehension, attention, concentration, and the ability to focus and complete tasks. Depending on the sequenced number of the letters in alphabet the content of the position is either empty or filled. By using the recognizable combinations we are identifying the empty and filled positions as digits and letters.

To arrive at the desired digit or letter, we add together only contents of the filled positions, because the empty position has a content of zero. In Litvin's Code each letter has a corresponding

sequential number. This simplifies the spelling of the words because the letters have quantitative values. The word contains the letters, which are grouped in a unique pattern of filled and empty positions. The Litvin's Code can accommodate any existing alphabets by using the same logical structure. Consequently, children are able to memorize combinations of empty and full positions. The best way for children is using kinesthetic representations of the letters, because it makes the learning more fun and more productive. Previously, the youngest children who were able to understand Litvin's Code were about five years olds. However, it is perfectly possible for even younger children to comprehend letters by using Litvin's Code.

AUTHOR: Chester Litvin, Ph.D.

Chester Litvin, Ph.D., the author of Litvin's Code, is a Clinical Psychologist. He is licensed in the State of California and has a private practice in the city of Burbank in Southern California. Dr. Litvin moved to the United States from Saint Petersburg, Russia, where he was born during World War II. After Military service in the Red Army, he received teacher's credentials from the Hertzen Teacher's Academy. He was a teacher of science in a Saint Petersburg high school.
In 1984 Dr. Chester Litvin moved to Southern California, where he worked in San Diego under a contract with the Navy as a Senior Member of a Scientific Staff and took evening classes to study Psychology. He received a Master's degree as a Marriage Family Counselor and a School Psychologist. He also completed his Ph.D. program in Psychology in San Diego.

Dr. Litvin has been accredited as a School Psychologist. He is licensed as a Marriage, Family and Child Counselor and a Clinical Psychologist. He discovered new procedures of brain stimulation in 1991 and is used them successfully for many years. After using his discovery with patients of different ages, he decided to publish a book on several procedures of brain stimulation. Those procedures are very easy to understand, and mainly, are self-explanatory.

Illustrator: Dr. Niki Martirosyan, Ph. D

Dr. Niki Martirosyan, Ph. D. was born in Yerevan, Armenia. She studied literature, Liberal Arts, computer science and economics. She completed the State University in Yerevan and then worked on her dissertation. She received her Ph.D. in economics from the State University. She also has an excellent knowledge of computers.

Dr. Martirosyan is a very creative person. In Armenia she published a book of poetry and a separate piece on economics. She moved to Southern California in 2002 and was invited by Dr. Litvin to help illustrate Litvin's Code for publication. She has also designed different medical forms for a health group, worked on new developments for a home theater production company, written a second book of poetry and a script for a movie, which is now in its post-production stage. Presently Dr. Martirosyan works with a variety of Microsoft products in the city of Glendale, California.

FIGURE 1 INTRODUCTION TO ALPHABET

In figure 1, we have ***visual*** display of two pictures of boxes. The upper picture has the first position filled up with a symbol. The decimal numbers 1 and letter A correspond to the first picture.

The ***audio*** representation for the upper picture, number 1 and letter A, is **knock and double knock.**

The second picture in the bottom has the second position filled up with a symbol. The decimal number 2 and letter B correspond to the second picture.

The ***audio*** representation for the lower picture, number 2 and letter B, is **double knock and knock.**

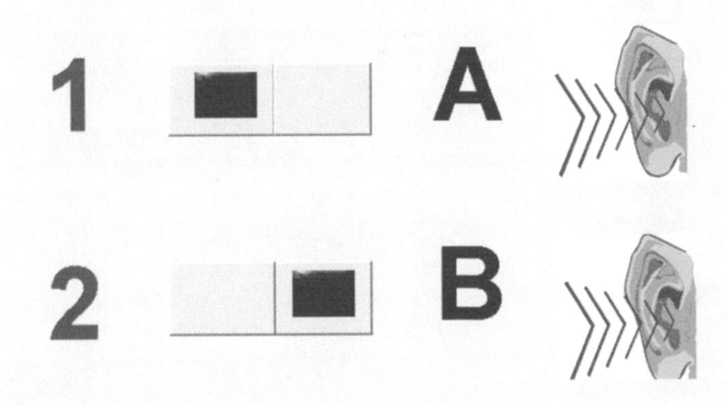

A knock represents a filled up position and a double knock represents an empty position. The clamping of the right hand represents a filled up position and the clamping of the left hand represents an empty position. The ***kinesthetic*** representation for the number 1 and letter A is clamping of the right hand and left hand. The ***kinesthetic*** representation for the number 2 and letter B is clamping of left hand and right hand.

INTRODUCTION TO ALPHABET

FIGURE 1.1

In figure 1.1, we have ***visual*** display of two pictures of boxes. We are given two choices of true or false; our objective is to observe whether the letters and pictures correspond to each other. The upper picture has the first position filled up with a symbol. The letter A is supposed to correspond to the first picture.

The ***audio*** representation for the upper picture and letter A is **knock and double knock.**

The second picture in the bottom has the second position filled up with a symbol. The letter B is supposed to correspond to the second picture.

The ***audio*** representation for the lower picture and letter B is **double knock and knock.**

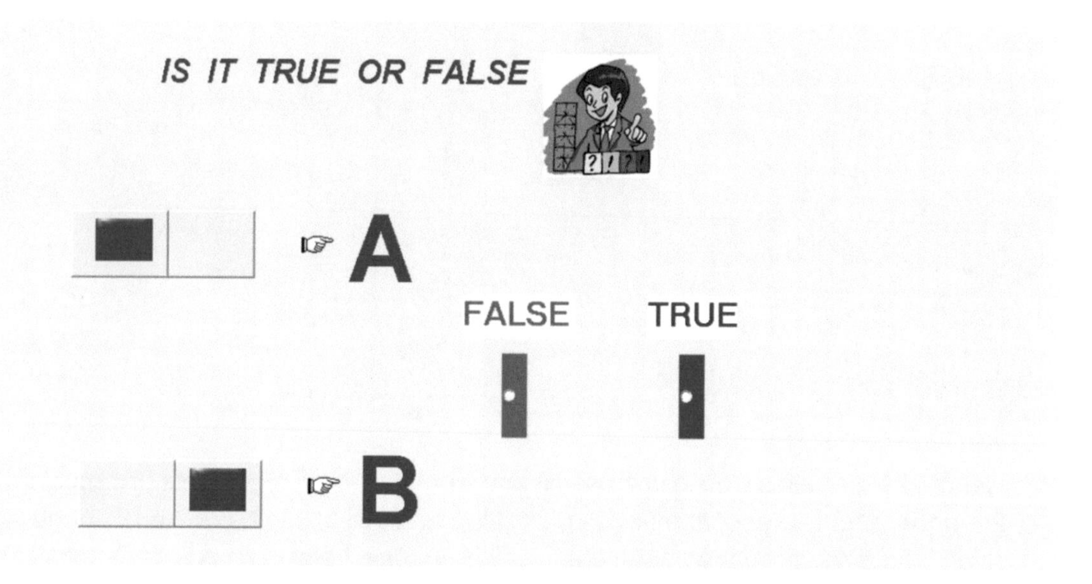

A knock represents a filled up position and a double knock represents an empty position. The clamping of the right hand represents a filled up position and the clamping of the left hand represents an empty position. The answer to this problem is true. The ***kinesthetic*** representation for the letter A is clamping of the right hand and left hand. The ***kinesthetic*** representation for the letter B is clamping of left hand and right hand.

FIGURE 2 INTRODUCTION TO ALPHABET

In figure 2, we have ***visual*** display of two pictures of boxes. The upper picture has both positions filled up with symbols. The decimal numbers 3 and letter C correspond to the first picture.

The ***audio*** representation for the upper picture, number 3 and letter C, is **knock and knock.**

The second picture in the bottom has the third position filled up with a symbol. The decimal number 4 and letter D correspond to the second picture.

The ***audio*** representation for the lower picture, number 4 and letter D, is **double knock, double knock, and knock.**

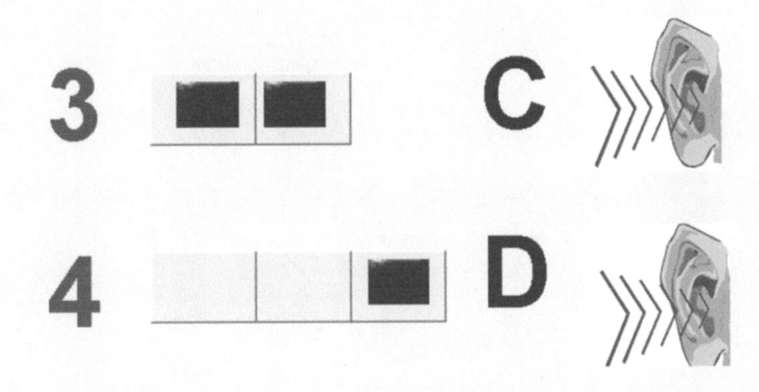

A knock represents a filled up position and a double knock represents an empty position. The clamping of the right hand represents a filled up position and the clamping of the left hand represents an empty position. The ***kinesthetic*** representation for the number 3 and letter C is clamping of the right hand twice. The ***kinesthetic*** representation for the number 4 and letter D is clamping of left hand twice and right hand.

INTRODUCTION TO ALPHABET

FIGURE 2.1

In figure 2.1, we have ***visual*** display of two pictures of boxes. We are given two choices of true or false; our objective is to observe whether the letters and pictures correspond to each other. The upper picture has both positions filled up with symbols. The letter D is supposed to correspond to the first picture.

The ***audio*** representation for the upper picture and letter D is **knock and knock.**

The second picture in the bottom has the third position filled up with a symbol. The letter C is supposed to correspond to the second picture.

The ***audio*** representation for the lower picture and letter C is **double knock, double knock, and knock.**

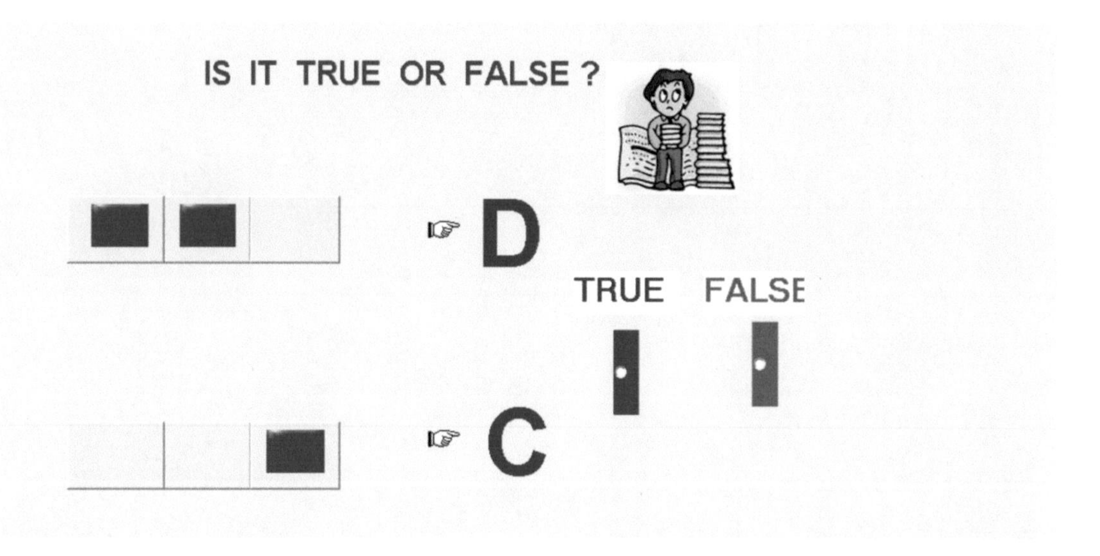

A knock represents a filled up position and a double knock represents an empty position. The clamping of the right hand represents a filled up position and the clamping of the left hand represents an empty position. The answer to the problem is false. The ***kinesthetic*** representation for the number 3 and letter C is clamping of the right hand twice. The ***kinesthetic*** representation for the number 4 and letter D is clamping of left hand twice and right hand.

FIGURE 2.2 INTRODUCTION TO ALPHABET

In figure 2.2, we have ***visual*** display of two pictures of boxes. The upper picture has both positions filled up with symbols. The decimal numbers 3 and letter C correspond to the first picture.

The ***audio*** representation for the upper picture, number 3 and letter C, is **knock and knock.**

The second picture in the bottom has the third position filled up with a symbol. The decimal number 4 and letter D correspond to the second picture.

The ***audio*** representation for the lower picture, number 4 and letter D, is **double knock, double knock, and knock.**

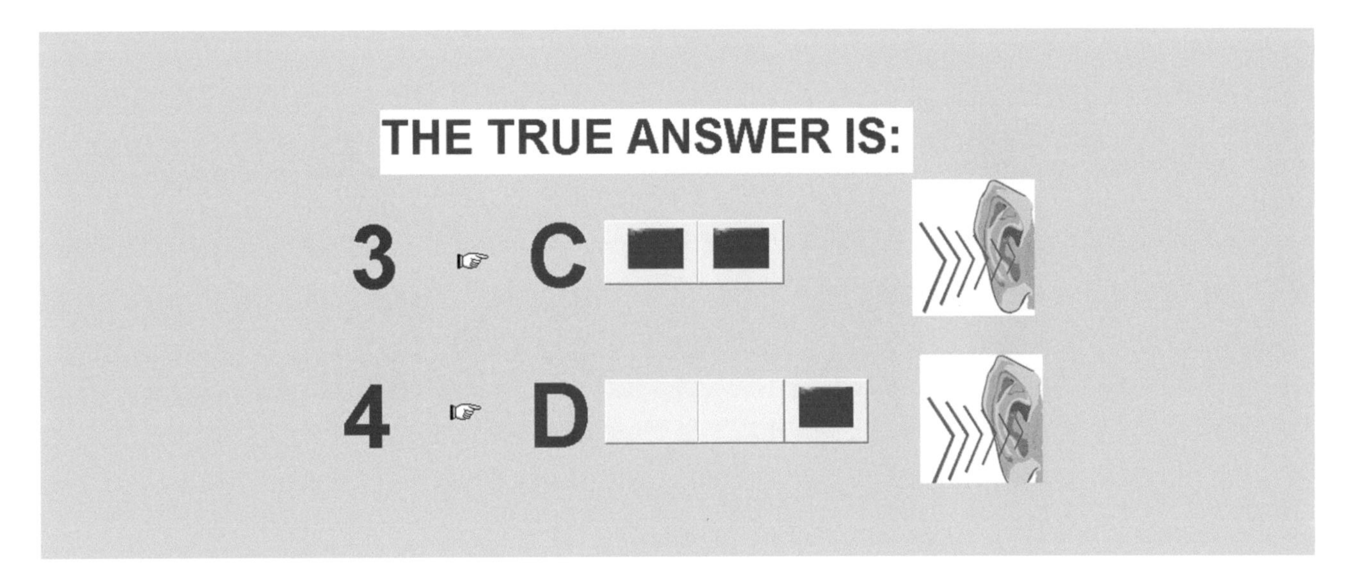

A knock represents a filled up position and a double knock represents an empty position. The clamping of the right hand represents a filled up position and the clamping of the left hand represents an empty position. The ***kinesthetic*** representation for the number 3 and letter C is clamping of the right hand twice. The ***kinesthetic*** representation for the number 4 and letter D is clamping of left hand twice and right hand.

INTRODUCTION TO ALPHABET

FIGURE 3

In figure 3, we have ***visual*** display of two pictures of boxes. The upper picture has the first and third positions filled up with symbols. The decimal numbers 5 and letter E correspond to the first picture.

The ***audio*** representation for the upper picture, number 5 and letter E, is **knock, double knock, and knock.**

The second picture in the bottom has the second and third position filled up with a symbol. The decimal number 6 and letter H correspond to the second picture.

The ***audio*** representation for the lower picture, number 6 and letter F, is **double knock, knock, and knock.**

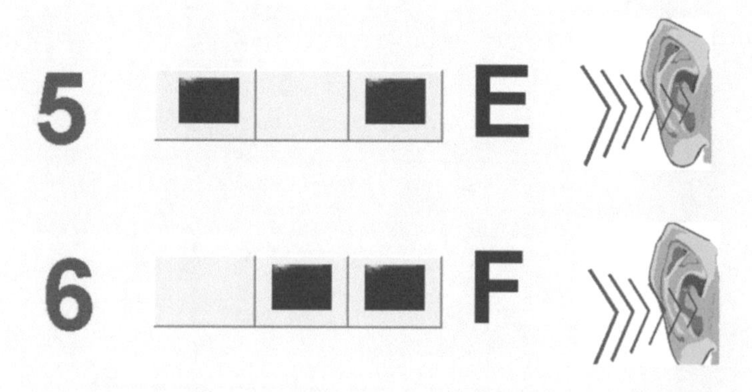

A knock represents a filled up position and a double knock represents an empty position. The clamping of the right hand represents a filled up position and the clamping of the left hand represents an empty position. The ***kinesthetic*** representation for the number 5 and letter E is clamping of the right hand, left hand, and right hand. The ***kinesthetic*** representation for the number 6 and letter F is clamping of left hand and right hand twice.

FIGURE 3.1 INTRODUCTION TO ALPHABET

In figure 3.1, we have ***visual*** display of two pictures of boxes. We are given two choices of true or false; our objective is to observe whether the letters and pictures correspond to each other. The upper picture has the first and third positions filled up with symbols. The letter E is supposed to correspond to the first picture.

The ***audio*** representation for the upper picture and letter E is **knock, double knock, and knock.**

The second picture in the bottom has the second and third position filled up with a symbol. The letter H is supposed to correspond to the second picture.

The ***audio*** representation for the lower picture and letter F is **double knock, knock, and knock.**

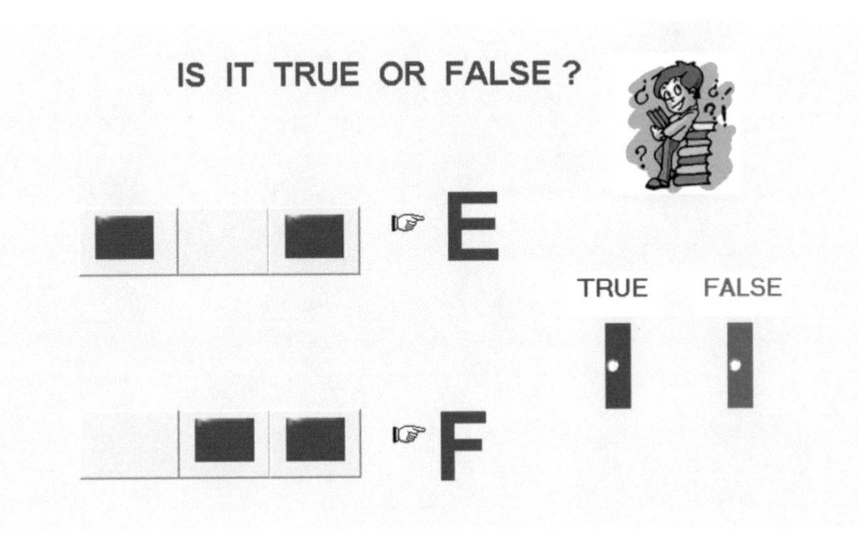

A knock represents a filled up position and a double knock represents an empty position. The clamping of the right hand represents a filled up position and the clamping of the left hand represents an empty position. The answer to the problem is true. The ***kinesthetic*** representation for the number 5 and letter E is clamping of the right hand, left hand, and right hand. The ***kinesthetic*** representation for the number 6 and letter F is clamping of left hand and right hand twice.

INTRODUCTION TO ALPHABET

FIGURE 4

In figure 4, we have ***visual*** display of two pictures of boxes. The upper picture has all three positions filled up with symbols. The decimal numbers 7 and letter G correspond to the first picture.

The ***audio*** representation for the upper picture, number 7 and letter G, is **knock, knock, and knock.**

The second picture in the bottom has the fourth position filled up with a symbol. The decimal number 8 and letter H correspond to the second picture.

The ***audio*** representation for the lower picture, number 8 and letter H, is **double knock, double knock, double knock, and knock.**

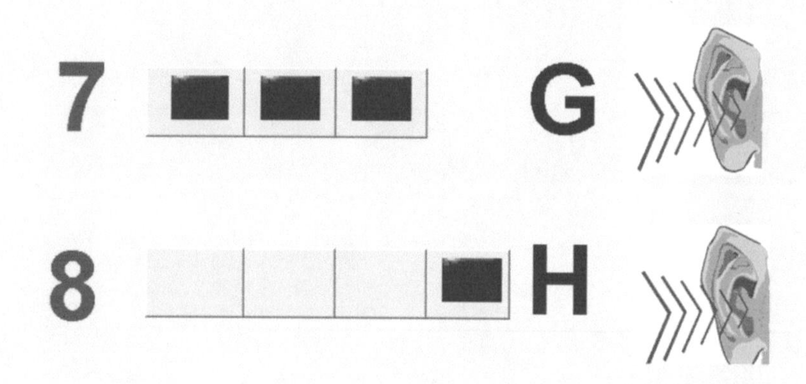

A knock represents a filled up position and a double knock represents an empty position. The clamping of the right hand represents a filled up position and the clamping of the left hand represents an empty position. The ***kinesthetic*** representation for the number 7 and letter G is clamping of the right hand thrice. The ***kinesthetic*** representation for the number 8 and letter H is clamping of left hand thrice and right hand.

FIGURE 4.1 INTRODUCTION TO ALPHABET

In figure 4.1, we have ***visual*** display of two pictures of boxes. We are given two choices of true or false; our objective is to observe whether the letters and pictures correspond to each other. The upper picture has all three positions filled up with symbols. The letter H is supposed to correspond to the first picture.

The ***audio*** representation for the upper picture and letter H is **knock, knock, and knock.**

The second picture in the bottom has the fourth position filled up with a symbol. The decimal and letter G is supposed to correspond to the second picture.

The ***audio*** representation for the lower picture and letter G is **double knock, double knock, double knock, and knock.**

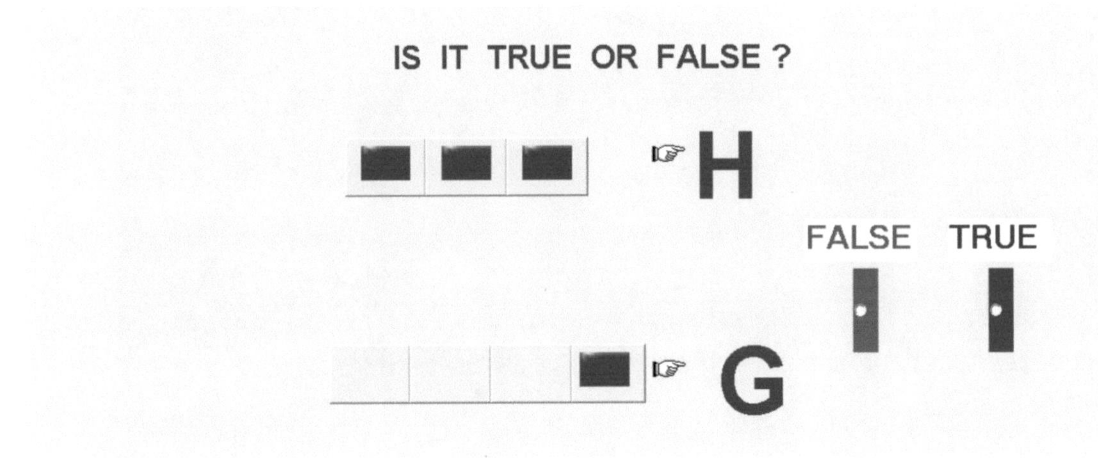

A knock represents a filled up position and a double knock represents an empty position. The clamping of the right hand represents a filled up position and the clamping of the left hand represents an empty position. The answer to the problem is false. The ***kinesthetic*** representation for the number 7 and letter G is clamping of the right hand thrice. The ***kinesthetic*** representation for the number 8 and letter H is clamping of left hand thrice and right hand.

INTRODUCTION TO ALPHABET

FIGURE 4.2

In figure 4.2, we have ***visual*** display of two pictures of boxes. The upper picture has all three positions filled up with symbols. The decimal numbers 7 and letter G correspond to the first picture.

The ***audio*** representation for the upper picture, number 7 and letter G, is **knock, knock, and knock.**

The second picture in the bottom has the fourth position filled up with a symbol. The decimal number 8 and letter H correspond to the second picture.

The ***audio*** representation for the lower picture, number 8 and letter H, is **double knock, double knock, double knock, and knock.**

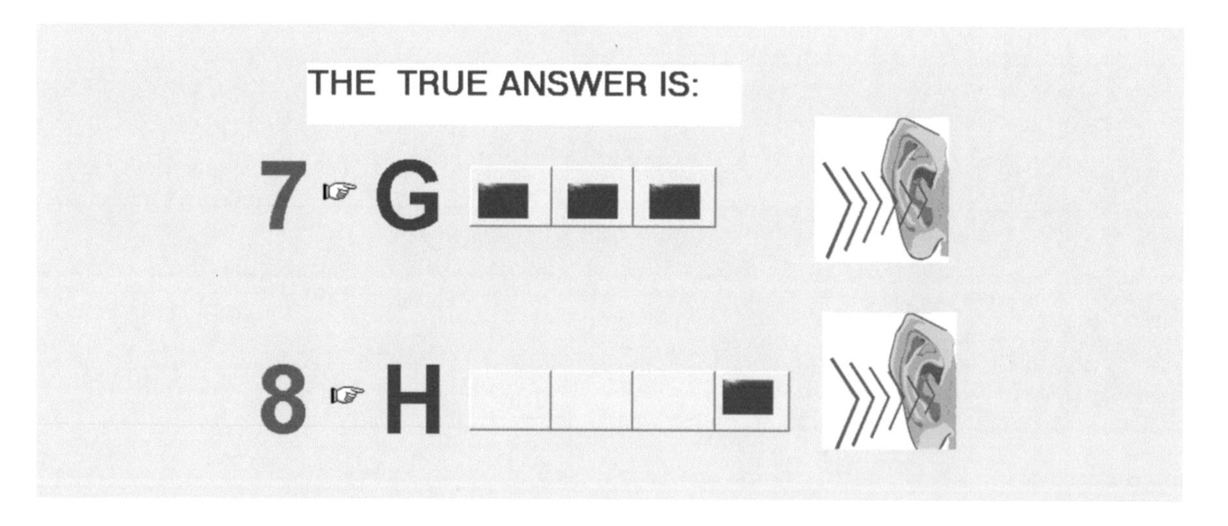

A knock represents a filled up position and a double knock represents an empty position. The clamping of the right hand represents a filled up position and the clamping of the left hand represents an empty position. The ***kinesthetic*** representation for the number 7 and letter G is clamping of the right hand thrice. The ***kinesthetic*** representation for the number 8 and letter H is clamping of left hand thrice and right hand.

FIGURE 5 INTRODUCTION TO ALPHABET

In figure 5, we have ***visual*** display of two pictures of boxes. The upper picture has the first and fourth positions filled up with symbols. The decimal numbers 9 and letter I correspond to the first picture.

The ***audio*** representation for the upper picture, number 9 and letter I, is **knock, double knock, double knock, and knock.**

The second picture in the bottom has the second and fourth positions filled up with symbols. The decimal number 10 and letter J correspond to the second picture.

The ***audio*** representation for the lower picture, number 10 and letter J, is **double knock, knock, double knock, and knock.**

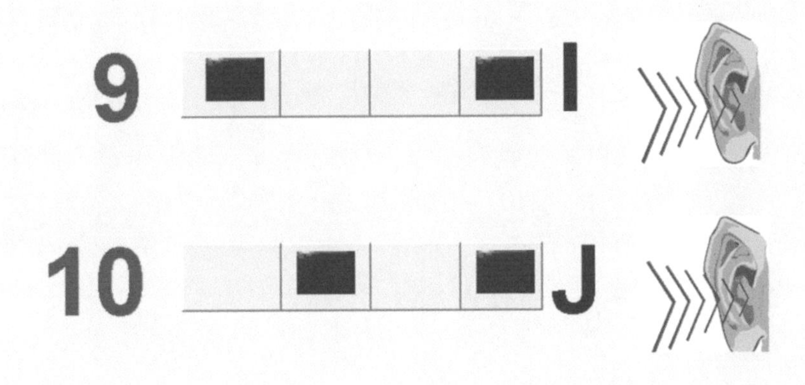

A knock represents a filled up position and a double knock represents an empty position. The clamping of the right hand represents a filled up position and the clamping of the left hand represents an empty position. The ***kinesthetic*** representation for the number 9 and letter I is clamping of the right hand, left hand twice, and right hand. The ***kinesthetic*** representation for the number 10 and letter J is clamping of left hand, right hand, left hand, and right hand.

INTRODUCTION TO ALPHABET

FIGURE 5.1

In figure 5.1, we have ***visual*** display of two pictures of boxes. We are given two choices of true or false; our objective is to observe whether the letters and pictures correspond to each other. The upper picture has the first and fourth positions filled up with symbols. The letter I is supposed to correspond to the first picture.

The ***audio*** representation for the upper picture letter I is **knock, double knock, double knock, and knock.**

The second picture in the bottom has the second and fourth positions filled up with symbols. The decimal letter J is supposed to correspond to the second picture.

The ***audio*** representation for the lower picture letter J is **double knock, knock, double knock, and knock.**

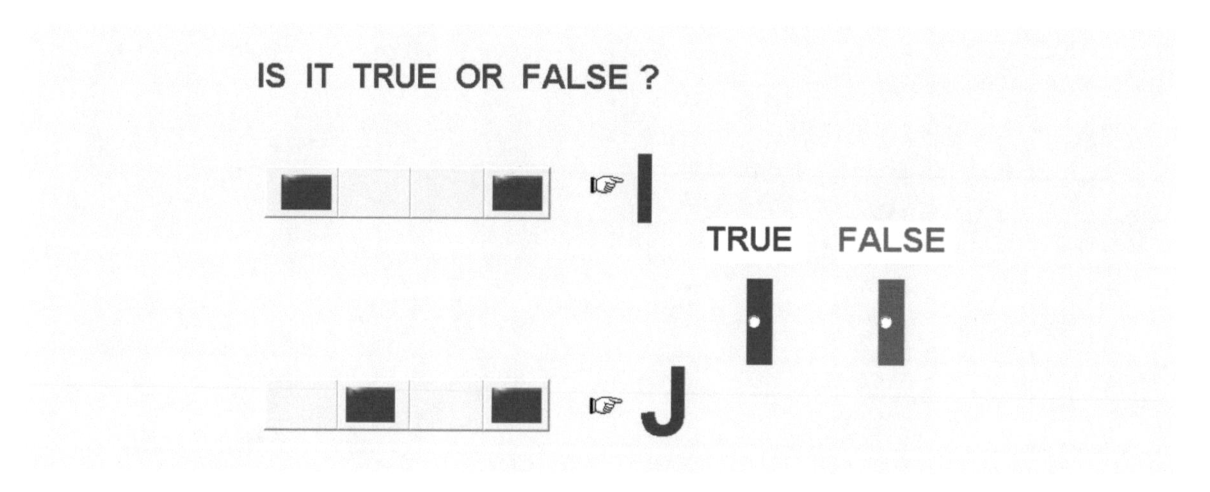

A knock represents a filled up position and a double knock represents an empty position. The clamping of the right hand represents a filled up position and the clamping of the left hand represents an empty position. The answer to the problem is true. The ***kinesthetic*** representation for the number 9 and letter I is clamping of the right hand, left hand twice, and right hand. The ***kinesthetic*** representation for the number 10 and letter J is clamping of left hand, right hand, left hand, and right hand.

FIGURE 6 INTRODUCTION TO ALPHABET

In figure 6, we have ***visual*** display of two pictures of boxes. The upper picture has the first, second, and fourth positions filled up with symbols. The decimal numbers 11 and letter K correspond to the first picture.

The ***audio*** representation for the upper picture, number 11 and letter K, is **knock, knock, double knock, and knock.**

The second picture in the bottom has the third and fourth positions filled up with symbols. The decimal number 12 and letter L correspond to the second picture.

The ***audio*** representation for the upper picture, number 12 and letter L, is **double knock, knock, double knock, and knock.**

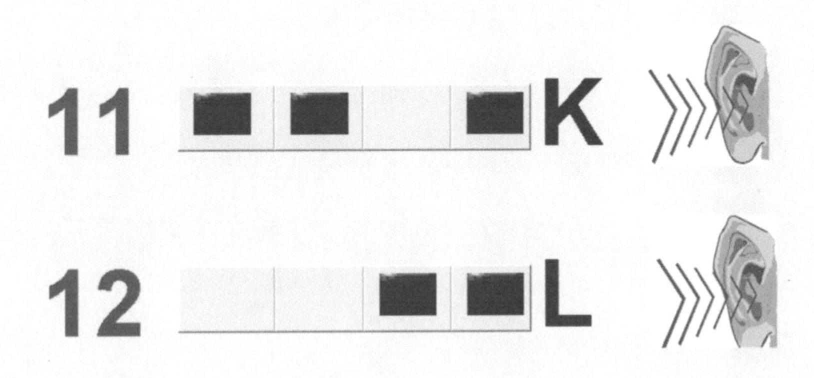

A knock represents a filled up position and a double knock represents an empty position. The clamping of the right hand represents a filled up position and the clamping of the left hand represents an empty position. The ***kinesthetic*** representation for the number 11 and letter K is clamping of the right hand twice, left hand, and right hand. The ***kinesthetic*** representation for the number 12 and letter L is clamping of left hand twice and right hand twice.

INTRODUCTION TO ALPHABET

FIGURE 6.1

In figure 6.1, we have ***visual*** display of two pictures of boxes. We are given two choices of true or false; our objective is to observe whether the letters and pictures correspond to each other. The upper picture has the first, second, and fourth positions filled up with symbols. The letter L is supposed to correspond to the first picture.

The ***audio*** representation for the upper picture and letter L is **knock, knock, double knock, and knock.**

The second picture in the bottom has the third and fourth positions filled up with symbols. The letter K is supposed to correspond to the second picture.

The ***audio*** representation for the upper picture and letter K is **double knock, knock, double knock, and knock.**

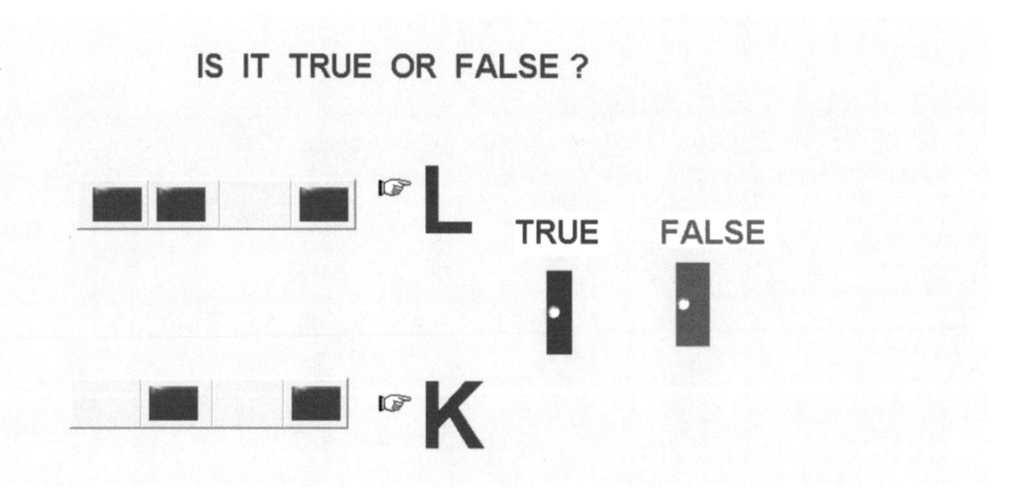

A knock represents a filled up position and a double knock represents an empty position. The clamping of the right hand represents a filled up position and the clamping of the left hand represents an empty position. The answer to the problem is false. The ***kinesthetic*** representation for the number 11 and letter K is clamping of the right hand twice, left hand, and right hand. The ***kinesthetic*** representation for the number 12 and letter L is clamping of left hand twice and right hand twice.

FIGURE 6.2 INTRODUCTION TO ALPHABET

In figure 6.2, we have ***visual*** display of two pictures of boxes. The upper picture has the first, second, and fourth positions filled up with symbols. The decimal numbers 11 and letter K correspond to the first picture.

The ***audio*** representation for the upper picture, number 11 and letter K, is **knock, knock, double knock, and knock.**

The second picture in the bottom has the third and fourth positions filled up with symbols. The decimal number 12 and letter L correspond to the second picture.

The ***audio*** representation for the upper picture, number 12 and letter L, is **double knock, knock, double knock, and knock.**

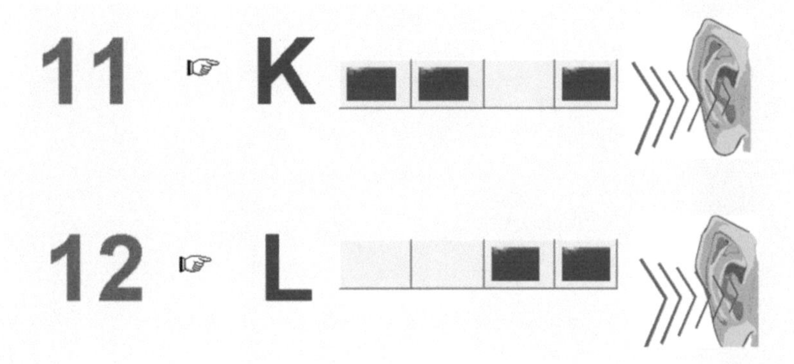

A knock represents a filled up position and a double knock represents an empty position. The clamping of the right hand represents a filled up position and the clamping of the left hand represents an empty position. The ***kinesthetic*** representation for the number 11 and letter K is clamping of the right hand twice, left hand, and right hand. The ***kinesthetic*** representation for the number 12 and letter L is clamping of left hand twice and right hand twice.

INTRODUCTION TO ALPHABET

FIGURE 7

In figure 7, we have ***visual*** display of two pictures of boxes. The upper picture has the first, third, and fourth positions filled up with symbols. The decimal numbers 13 and letter M correspond to the first picture.

The ***audio*** representation for the upper picture, number 13 and letter M, is **knock, double knock, knock, and knock.**

The second picture in the bottom has the second, third, and fourth positions filled up with symbols. The decimal number 14 and letter N correspond to the second picture.

The ***audio*** representation for the lower picture, number 14 and letter N, is **double knock, knock, knock, and knock.**

A knock represents a filled up position and a double knock represents an empty position. The clamping of the right hand represents a filled up position and the clamping of the left hand represents an empty position. The ***kinesthetic*** representation for the number 13 and letter M is clamping of the right hand, left hand, and right hand twice. The ***kinesthetic*** representation for the number 14 and letter N is clamping of left hand and right hand thrice.

FIGURE 7.1 INTRODUCTION TO ALPHABET

In figure 7.1, we have ***visual*** display of two pictures of boxes. We are given two choices of true or false; our objective is to observe whether the letters and pictures correspond to each other. The upper picture has the first, third, and fourth positions filled up with symbols. The letter N is supposed to correspond to the first picture.

The ***audio*** representation for the upper picture and letter N is **knock, double knock, knock, and knock.**

The second picture in the bottom has the second, third, and fourth positions filled up with symbols. The letter M is supposed to correspond to the second picture.

The ***audio*** representation for the lower picture letter M is **double knock, knock, knock, and knock.**

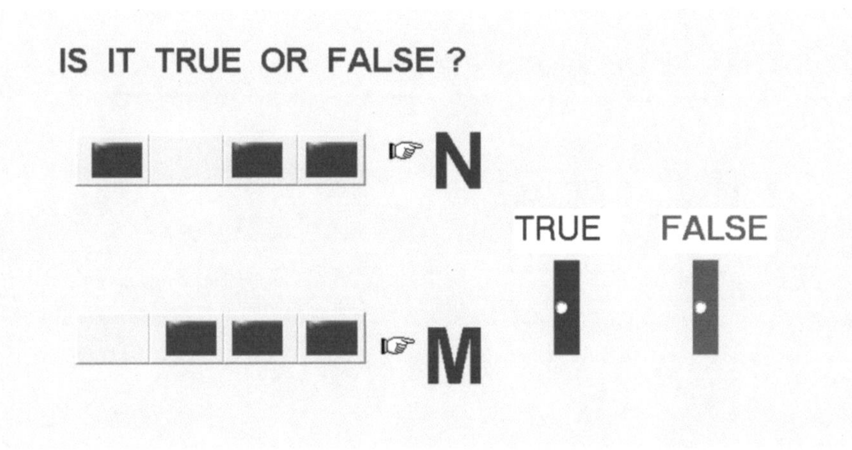

A knock represents a filled up position and a double knock represents an empty position. The clamping of the right hand represents a filled up position and the clamping of the left hand represents an empty position. The answer to the problem is false. The ***kinesthetic*** representation for the number 13 and letter M is clamping of the right hand, left hand, and right hand twice. The ***kinesthetic*** representation for the number 14 and letter N is clamping of left hand and right hand thrice.

INTRODUCTION TO ALPHABET

FIGURE 7.2

In figure 7.2, we have ***visual*** display of two pictures of boxes. The upper picture has the first, third, and fourth positions filled up with symbols. The decimal numbers 13 and letter M correspond to the first picture.

The ***audio*** representation for the upper picture, number 13 and letter M, is **knock, double knock, knock, and knock.**

The second picture in the bottom has the second, third, and fourth positions filled up with symbols. The decimal number 14 and letter N correspond to the second picture.

The ***audio*** representation for the lower picture, number 14 and letter N, is **double knock, knock, knock, and knock.**

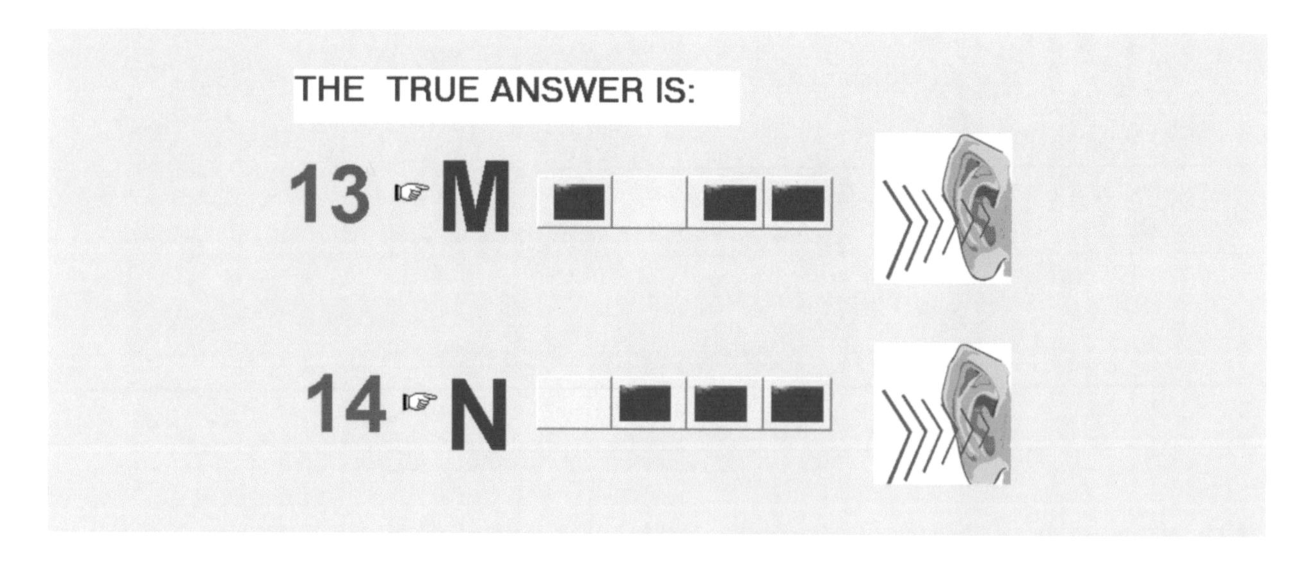

A knock represents a filled up position and a double knock represents an empty position. The clamping of the right hand represents a filled up position and the clamping of the left hand represents an empty position. The ***kinesthetic*** representation for the number 13 and letter M is clamping of the right hand, left hand, and right hand twice. The ***kinesthetic*** representation for the number 14 and letter N is clamping of left hand and right hand thrice.

FIGURE 8 INTRODUCTION TO ALPHABET

In figure 8, we have ***visual*** display of two pictures of boxes. The upper picture has the all four positions filled up with symbols. The decimal numbers 15 and letter O correspond to the first picture.

The ***audio*** representation for the upper picture, number 15 and letter M, is **knock, knock, knock, and knock.**

The second picture in the bottom has the fifth position filled up with a symbol. The decimal number 16 and letter P correspond to the second picture.

The ***audio*** representation for the lower picture, number 16 and letter P, is **double knock, double knock, double knock, double knock, and knock.**

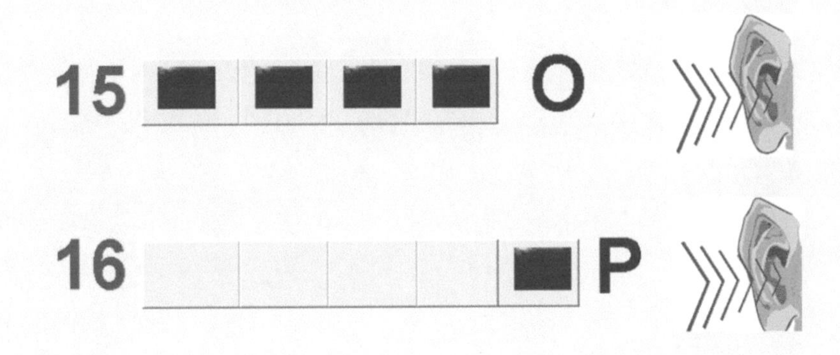

A knock represents a filled up position and a double knock represents an empty position. The clamping of the right hand represents a filled up position and the clamping of the left hand represents an empty position. The ***kinesthetic*** representation for the number 15 and letter O is clamping of the right hand four times. The ***kinesthetic*** representation for the number 16 and letter P is clamping of left hand four times and right hand.

INTRODUCTION TO ALPHABET

FIGURE 8.1

In figure 8.1, we have ***visual*** display of two pictures of boxes. We are given two choices of true or false; our objective is to observe whether the letters and pictures correspond to each other. The upper picture has the all four positions filled up with symbols. The letter O is supposed to correspond to the first picture.

The ***audio*** representation for the upper picture letter M is **knock, knock, knock, and knock.**

The second picture in the bottom has the fifth position filled up with a symbol. The letter P is supposed correspond to the second picture.

The ***audio*** representation for the lower picture and letter P is **double knock, double knock, double knock, double knock, and knock.**

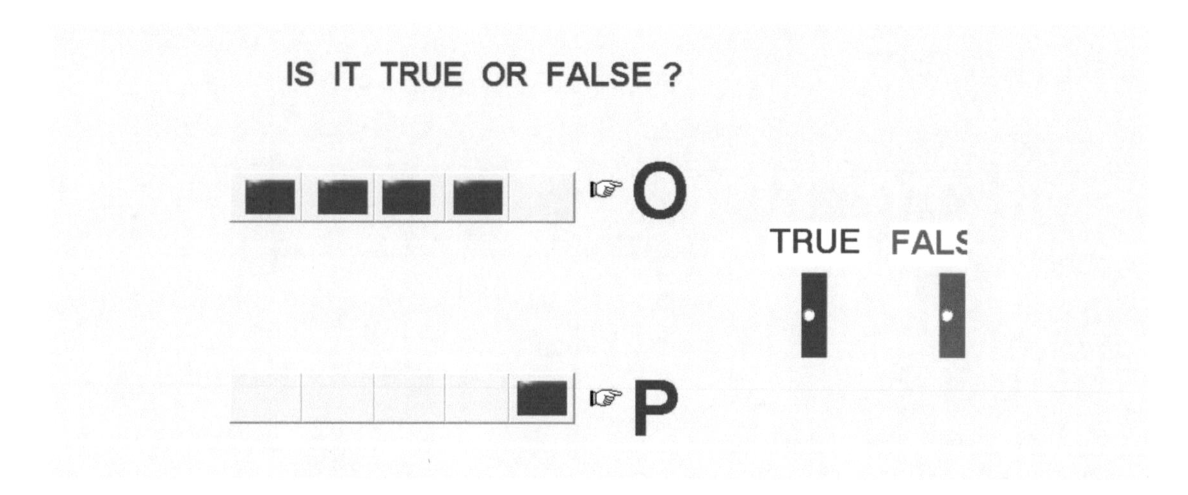

A knock represents a filled up position and a double knock represents an empty position. The clamping of the right hand represents a filled up position and the clamping of the left hand represents an empty position. The answer to the problem is true. The ***kinesthetic*** representation for the number 15 and letter O is clamping of the right hand four times. The ***kinesthetic*** representation for the number 16 and letter P is clamping of left hand four times and right hand.

FIGURE 9 INTRODUCTION TO ALPHABET

In figure 9, we have ***visual*** display of two pictures of boxes. The upper picture has the first and fifth positions filled up with symbols. The decimal numbers 17 and letter Q correspond to the first picture.

The ***audio*** representation for the upper picture, number 17 and letter Q, is **knock, double knock, double knock, double knock, and knock.**

The second picture in the bottom has the second and fifth positions filled up with symbols. The decimal number 18 and letter R correspond to the second picture.

The ***audio*** representation for the lower picture, number 18 and letter R, is **double knock, knock, double knock, double knock, and knock.**

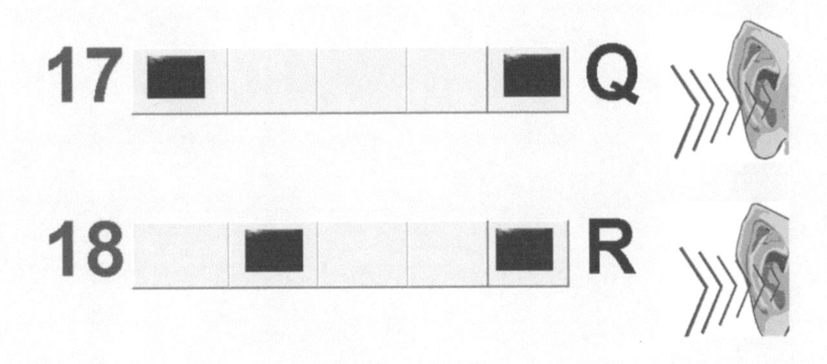

A knock represents a filled up position and a double knock represents an empty position. The clamping of the right hand represents a filled up position and the clamping of the left hand represents an empty position. The ***kinesthetic*** representation for the number 17 and letter Q is clamping of the right hand, left hand thrice, and right hand. The ***kinesthetic*** representation for the number 18and letter R is clamping of left hand, right hand, left hand twice, and right hand.

INTRODUCTION TO ALPHABET

FIGURE 9.1

In figure 9.1, we have ***visual*** display of two pictures of boxes. We are given two choices of true or false; our objective is to observe whether the letters and pictures correspond to each other. The upper picture has the first and fifth positions filled up with symbols. The letter R is supposed to correspond to the first picture.

The ***audio*** representation for the upper picture and letter R is **knock, double knock, double knock, double knock, and knock.**

The second picture in the bottom has the second and fifth positions filled up with symbols. The letter Q is supposed to correspond to the second picture.

The ***audio*** representation for the lower picture and letter Q is **double knock, knock, double knock, double knock, and knock.**

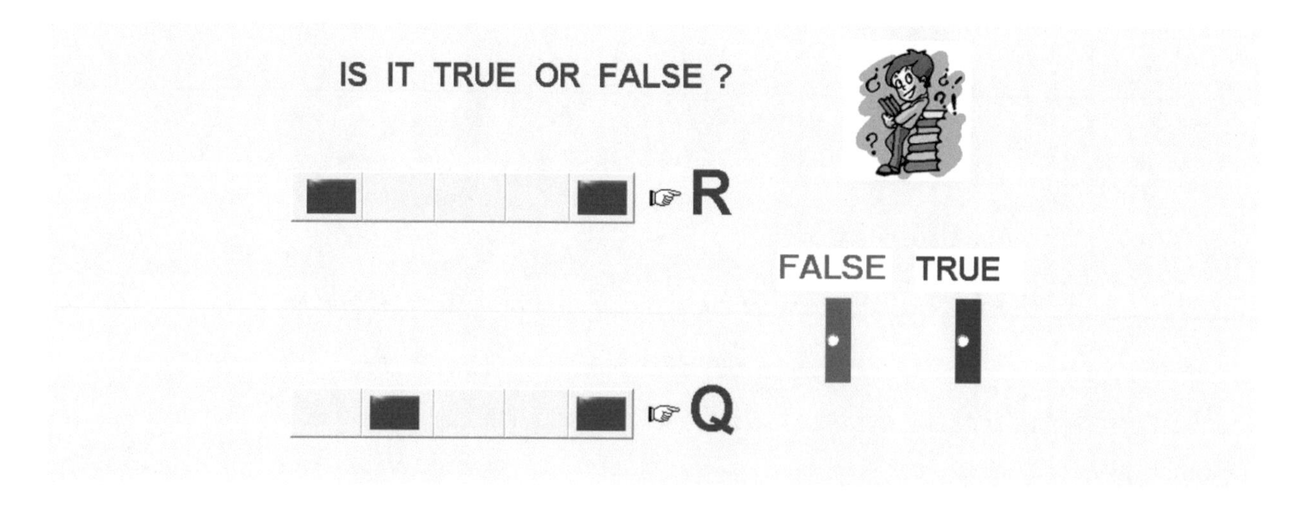

A knock represents a filled up position and a double knock represents an empty position. The clamping of the right hand represents a filled up position and the clamping of the left hand represents an empty position. The answer to the problem is false. The ***kinesthetic*** representation for the number 17 and letter Q is clamping of the right hand, left hand thrice, and right hand. The ***kinesthetic*** representation for the number 18and letter R is clamping of left hand, right hand, left hand twice, and right hand.

FIGURE 9.2 INTRODUCTION TO ALPHABET

In figure 9.2, we have ***visual*** display of two pictures of boxes. The upper picture has the first and fifth positions filled up with symbols. The decimal numbers 17 and letter Q correspond to the first picture.

The ***audio*** representation for the upper picture, number 17 and letter Q, is **knock, double knock, double knock, double knock, and knock.**

The second picture in the bottom has the second and fifth positions filled up with symbols. The decimal number 18 and letter R correspond to the second picture.

The ***audio*** representation for the lower picture, number 18 and letter R, is **double knock, knock, double knock, double knock, and knock.**

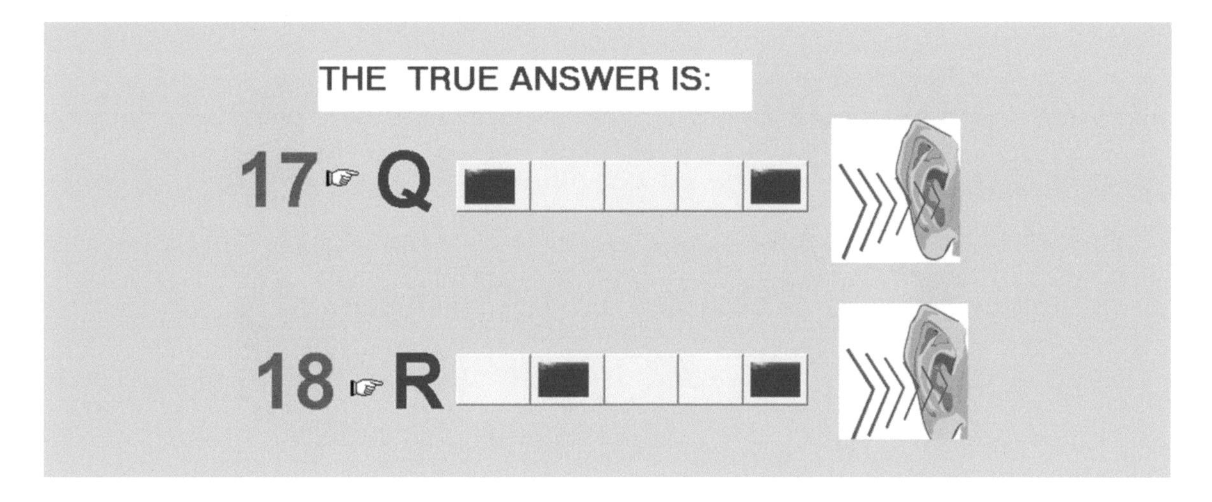

A knock represents a filled up position and a double knock represents an empty position. The clamping of the right hand represents a filled up position and the clamping of the left hand represents an empty position. The ***kinesthetic*** representation for the number 17 and letter Q is clamping of the right hand, left hand thrice, and right hand. The ***kinesthetic*** representation for the number 18and letter R is clamping of left hand, right hand, left hand twice, and right hand.

INTRODUCTION TO ALPHABET

FIGURE 10

In figure 10, we have ***visual*** display of two pictures of boxes. The upper picture has the first, second, and fifth positions filled up with symbols. The decimal numbers 19 and letter S correspond to the first picture.

The ***audio*** representation for the upper picture, number 19 and letter S, is **knock, knock, double knock, double knock, and knock.**

The second picture in the bottom has the third and fifth positions filled up with symbols. The decimal number 20 and letter T correspond to the second picture.

The ***audio*** representation for the lower picture, number 20 and letter T, is **double knock, double knock, knock, double knock, and knock.**

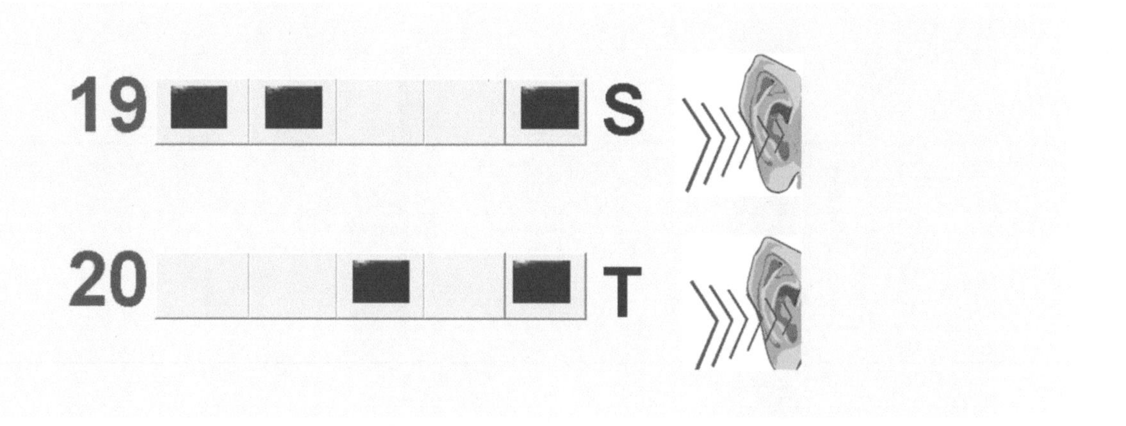

A knock represents a filled up position and a double knock represents an empty position. The clamping of the right hand represents a filled up position and the clamping of the left hand represents an empty position. The ***kinesthetic*** representation for the number 19 and letter S is clamping of the right hand twice, left hand twice, and right hand. The ***kinesthetic*** representation for the number 20 and letter T is clamping of left hand twice, right hand, left hand, and right hand.

FIGURE 10.1 INTRODUCTION TO ALPHABET

In figure 10.1, we have ***visual*** display of two pictures of boxes. We are given two choices of true or false; our objective is to observe whether the letters and pictures correspond to each other. The upper picture has the first, second, and fifth positions filled up with symbols. The letter S is supposed to correspond to the first picture.

The ***audio*** representation for the upper picture and letter S is **knock, knock, double knock, double knock, and knock.**

The second picture in the bottom has the third and fifth positions filled up with symbols. The letter T is supposed to correspond to the second picture.

The ***audio*** representation for the lower picture and letter T is **double knock, double knock, knock, double knock, and knock.**

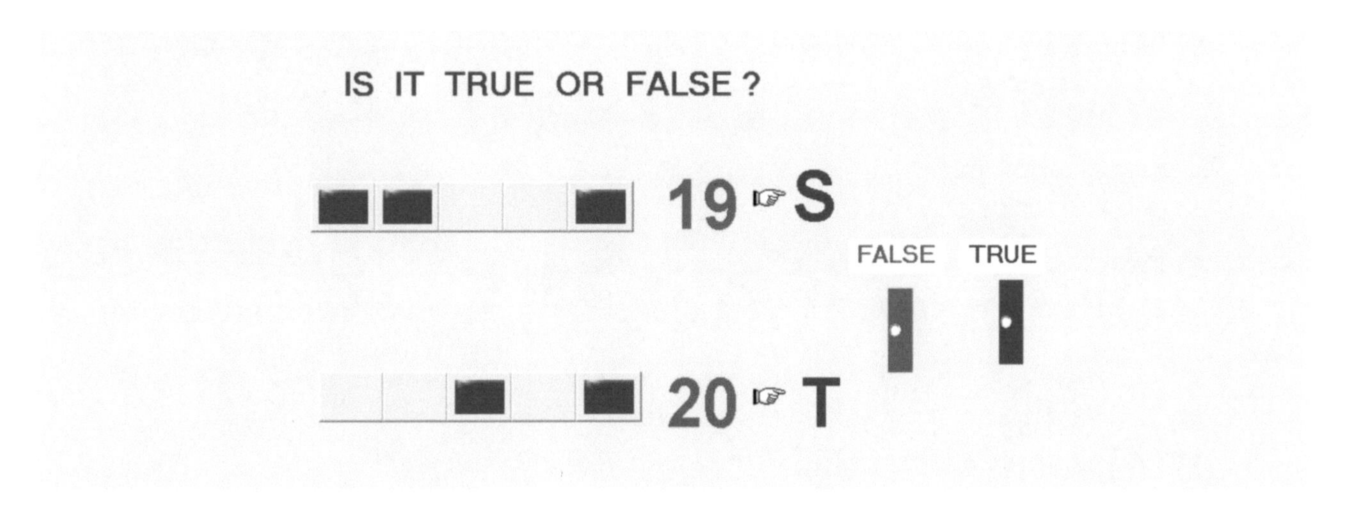

A knock represents a filled up position and a double knock represents an empty position. The clamping of the right hand represents a filled up position and the clamping of the left hand represents an empty position. The answer to the problem is true. The ***kinesthetic*** representation for the number 19 and letter S is clamping of the right hand twice, left hand twice, and right hand. The ***kinesthetic*** representation for the number 20 and letter T is clamping of left hand twice, right hand, left hand, and right hand.

INTRODUCTION TO ALPHABET

FIGURE 11

In figure 11, we have ***visual*** display of two pictures of boxes. The upper picture has the first, third, and fifth positions filled up with symbols. The decimal numbers 21 and letter U correspond to the first picture.

The ***audio*** representation for the upper picture, number 21 and letter U, is **knock, double knock, knock, double knock, and knock.**

The second picture in the bottom has the second, third, and fifth positions filled up with symbols. The decimal number 22 and letter V correspond to the second picture.

The ***audio*** representation for the lower picture, number 22 and letter V, is **double knock, knock, knock, double knock, and knock.**

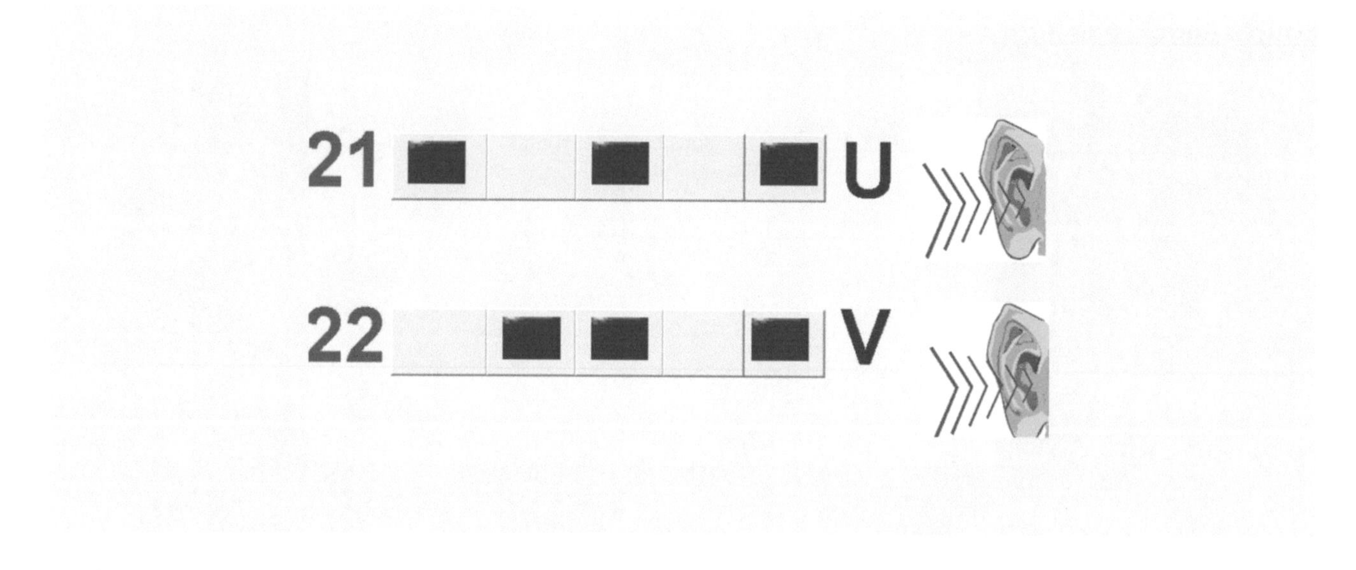

A knock represents a filled up position and a double knock represents an empty position. The clamping of the right hand represents a filled up position and the clamping of the left hand represents an empty position. The ***kinesthetic*** representation for the number 21 and letter U is clamping of the right hand, left hand, right hand, left hand, and right hand. The ***kinesthetic*** representation for the number 22 and letter V is clamping of left hand, right hand twice, left hand, and right hand.

FIGURE 11.1 INTRODUCTION TO ALPHABET

In figure 11.1, we have ***visual*** display of two pictures of boxes. We are given two choices of true or false; our objective is to observe whether the letters and pictures correspond to each other. The upper picture has the first, third, and fifth positions filled up with symbols. The letter V is supposed to correspond to the first picture.

The ***audio*** representation for the upper picture and letter V is **knock, double knock, knock, double knock, and knock.**

The second picture in the bottom has the second, third, and fifth positions filled up with symbols. The letter U is supposed to correspond to the second picture.

The ***audio*** representation for the lower picture, number 22 and letter U is **double knock, knock, knock, double knock, and knock.**

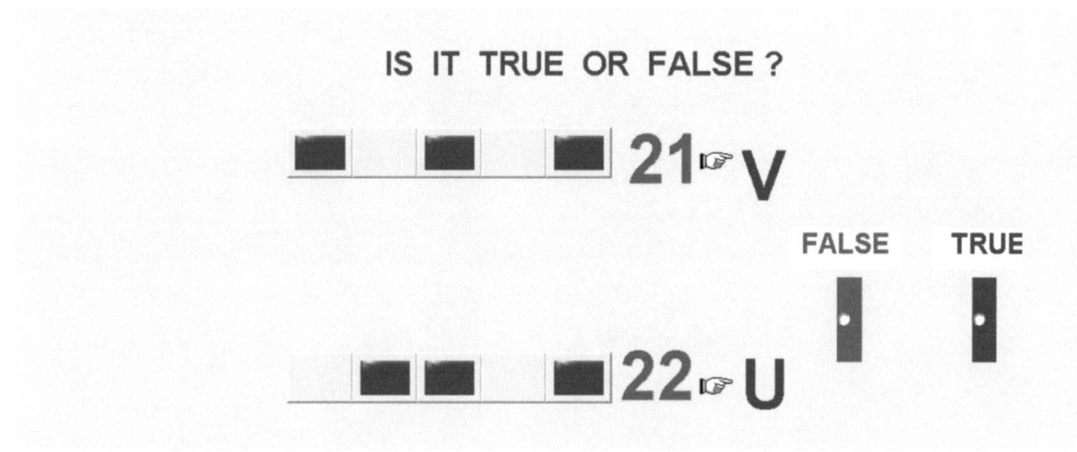

A knock represents a filled up position and a double knock represents an empty position. The clamping of the right hand represents a filled up position and the clamping of the left hand represents an empty position. The answer to the problem is false. The ***kinesthetic*** representation for the number 21 and letter U is clamping of the right hand, left hand, right hand, left hand, and right hand. The ***kinesthetic*** representation for the number 22 and letter V is clamping of left hand, right hand twice, left hand, and right hand.

INTRODUCTION TO ALPHABET

FIGURE 11.2

In figure 11.2, we have ***visual*** display of two pictures of boxes. The upper picture has the first, third, and fifth positions filled up with symbols. The decimal numbers 21 and letter U correspond to the first picture.

The ***audio*** representation for the upper picture, number 21 and letter U, is **knock, double knock, knock, double knock, and knock.**

The second picture in the bottom has the second, third, and fifth positions filled up with symbols. The decimal number 22 and letter V correspond to the second picture.

The ***audio*** representation for the lower picture, number 22 and letter V, is **double knock, knock, knock, double knock, and knock.**

A knock represents a filled up position and a double knock represents an empty position. The clamping of the right hand represents a filled up position and the clamping of the left hand represents an empty position. The ***kinesthetic*** representation for the number 21 and letter U is clamping of the right hand, left hand, right hand, left hand, and right hand. The ***kinesthetic*** representation for the number 22 and letter V is clamping of left hand, right hand twice, left hand, and right hand.

FIGURE 12 INTRODUCTION TO ALPHABET

In figure 12, we have ***visual*** display of two pictures of boxes. The upper picture has the first, second, third, and fifth positions filled up with symbols. The decimal numbers 23 and letter W correspond to the first picture.

The ***audio*** representation for the upper picture, number 23 and letter W, is **knock, knock, knock, double knock, and knock.**

The second picture in the bottom has the fourth and fifth positions filled up with symbols. The decimal number 24 and letter X correspond to the second picture.

The ***audio*** representation for the lower picture, number 24 and letter X, is **double knock, double knock, double knock, knock, and knock.**

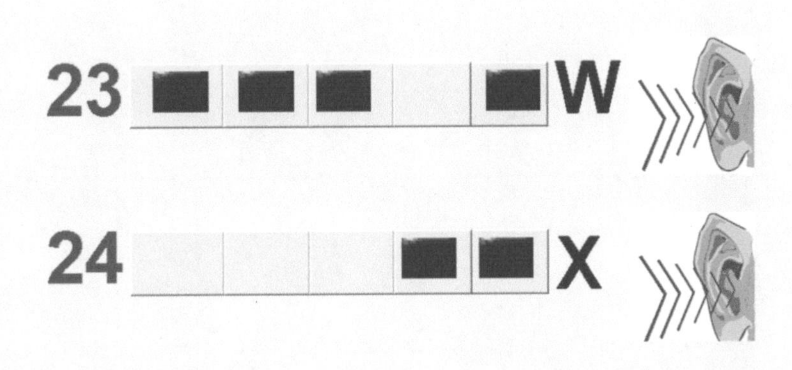

A knock represents a filled up position and a double knock represents an empty position. The clamping of the right hand represents a filled up position and the clamping of the left hand represents an empty position. The ***kinesthetic*** representation for the number 23 and letter W is clamping of the right hand thrice, left hand, and right hand. The ***kinesthetic*** representation for the number 24 and letter X is clamping of left hand thrice and right hand twice.

INTRODUCTION TO ALPHABET

FIGURE 12.1

In figure 12.1, we have ***visual*** display of two pictures of boxes. We are given two choices of true or false; our objective is to observe whether the letters and pictures correspond to each other. The upper picture has the first, second, third, and fifth positions filled up with symbols. The letter W is supposed to correspond to the first picture.

The ***audio*** representation for the upper picture and letter W is **knock, knock, knock, double knock, and knock.**

The second picture in the bottom has the fourth and fifth positions filled up with symbols. The letter X is supposed to correspond to the second picture.

The ***audio*** representation for the lower picture and letter X is **double knock, double knock, double knock, knock, and knock.**

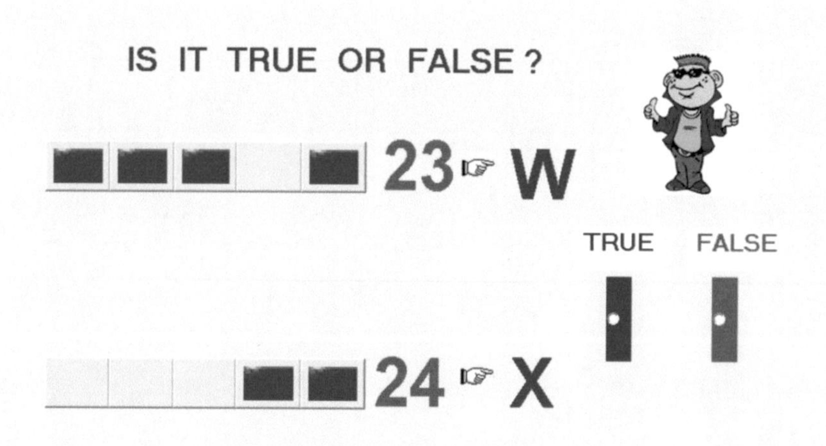

A knock represents a filled up position and a double knock represents an empty position. The clamping of the right hand represents a filled up position and the clamping of the left hand represents an empty position. The answer to the problem is true. The ***kinesthetic*** representation for the number 23 and letter W is clamping of the right hand thrice, left hand, and right hand. The ***kinesthetic*** representation for the number 24 and letter X is clamping of left hand thrice and right hand twice.

FIGURE 13 INTRODUCTION TO ALPHABET

In figure 13, we have ***visual*** display of two pictures of boxes. The upper picture has the first, fourth, and fifth positions filled up with symbols. The decimal numbers 25 and letter Y correspond to the first picture.

The ***audio*** representation for the upper picture, number 25 and letter Y, is **knock, double knock, double knock, knock, and knock.**

The second picture in the bottom has the second, fourth, and fifth positions filled up with symbols. The decimal number 26 and letter Z correspond to the second picture.

The ***audio*** representation for the lower picture, number 26 and letter Z, is **double knock, knock, double knock, knock, and knock.**

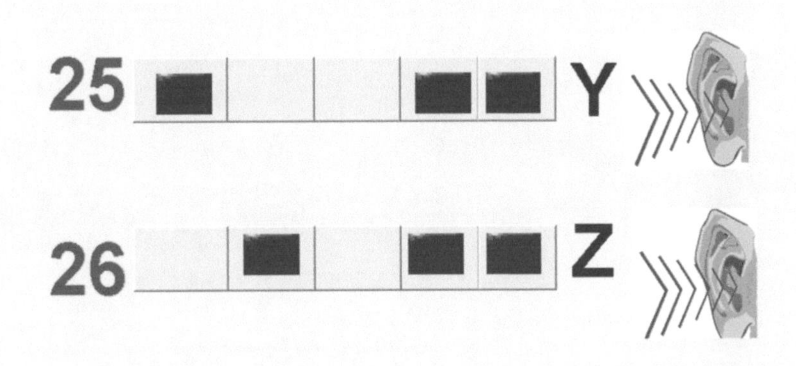

A knock represents a filled up position and a double knock represents an empty position. The clamping of the right hand represents a filled up position and the clamping of the left hand represents an empty position. The ***kinesthetic*** representation for the number 25 and letter Y is clamping of the right hand, left hand twice, and right hand. The ***kinesthetic*** representation for the number 26 and letter Z is clamping of left hand, right hand, left hand, and right hand twice.

INTRODUCTION TO ALPHABET

FIGURE 13.1

In figure 13.1, we have ***visual*** display of two pictures of boxes. We are given two choices of true or false; our objective is to observe whether the letters and pictures correspond to each other. The upper picture has the first, fourth, and fifth positions filled up with symbols. The letter Z is supposed to correspond to the first picture.

The ***audio*** representation for the upper picture and letter Z is **knock, double knock, double knock, knock, and knock.**

The second picture in the bottom has the second, fourth, and fifth positions filled up with symbols. The letter Y is supposed to correspond to the second picture.

The ***audio*** representation for the lower picture and letter Y is **double knock, knock, double knock, knock, and knock.**

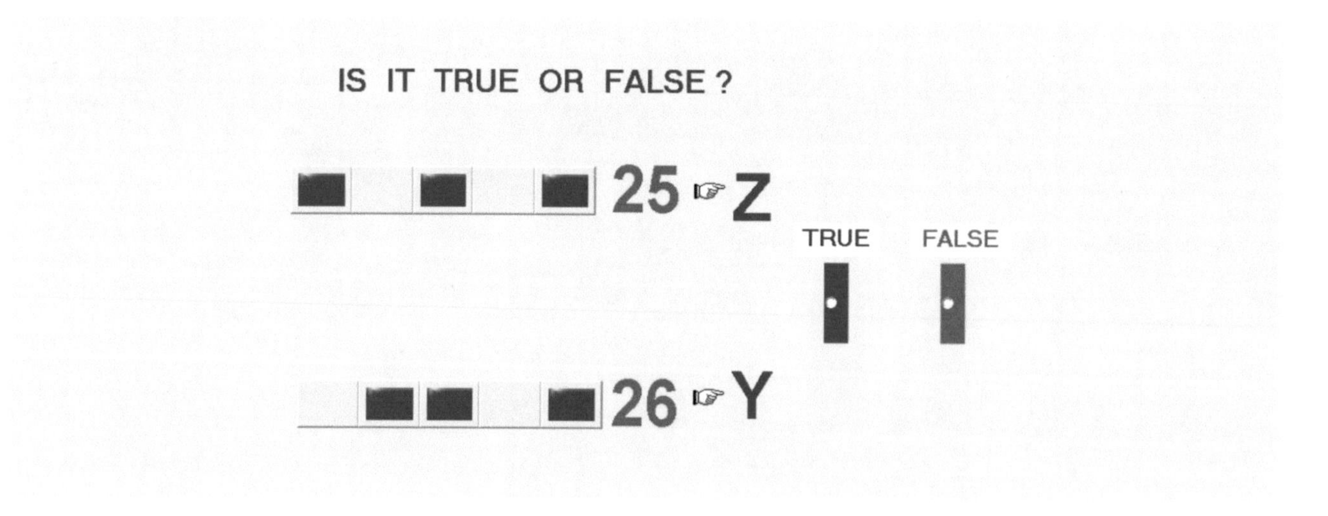

A knock represents a filled up position and a double knock represents an empty position. The clamping of the right hand represents a filled up position and the clamping of the left hand represents an empty position. The answer to the problem is false. The ***kinesthetic*** representation for the number 25 and letter Y is clamping of the right hand, left hand twice, and right hand. The ***kinesthetic*** representation for the number 26 and letter Z is clamping of left hand, right hand, left hand, and right hand twice.

FIGURE 13.2 INTRODUCTION TO ALPHABET

In figure 13.2, we have ***visual*** display of two pictures of boxes. The upper picture has the first, fourth, and fifth positions filled up with symbols. The decimal numbers 25 and letter Y correspond to the first picture.

The ***audio*** representation for the upper picture, number 25 and letter Y, is **knock, double knock, double knock, knock, and knock.**

The second picture in the bottom has the second, fourth, and fifth positions filled up with symbols. The decimal number 26 and letter Z correspond to the second picture.

The ***audio*** representation for the lower picture, number 26 and letter Z, is **double knock, knock, double knock, knock, and knock.**

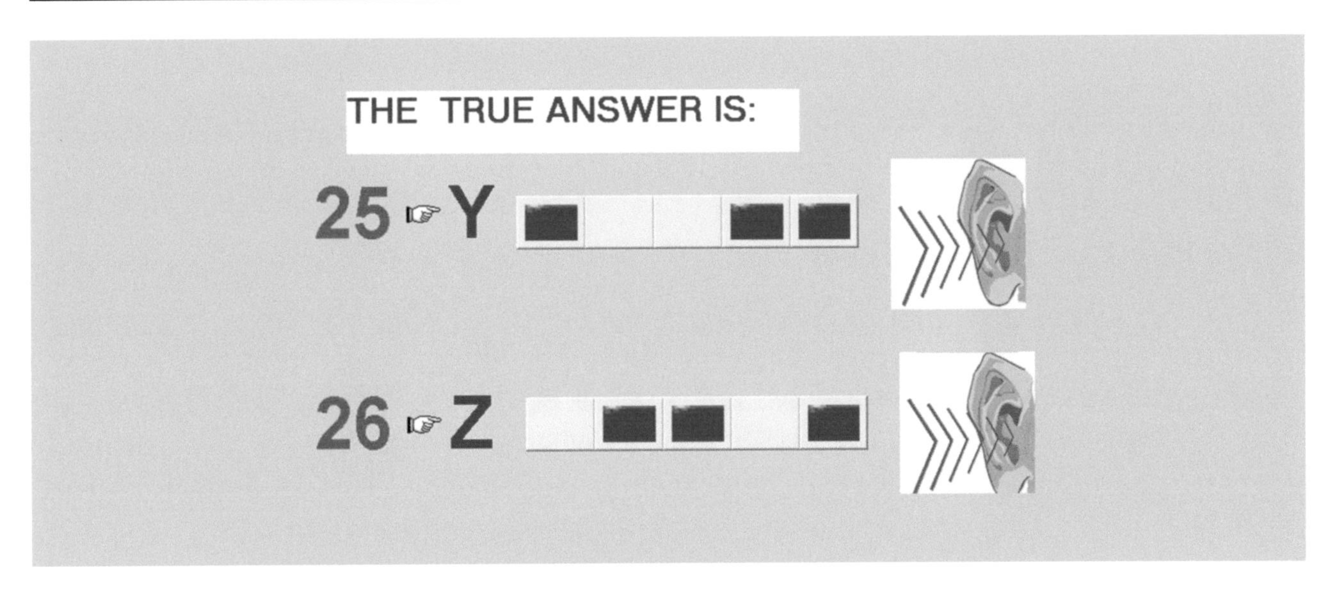

A knock represents a filled up position and a double knock represents an empty position. The clamping of the right hand represents a filled up position and the clamping of the left hand represents an empty position. The ***kinesthetic*** representation for the number 25 and letter Y is clamping of the right hand, left hand twice, and right hand. The ***kinesthetic*** representation for the number 26 and letter Z is clamping of left hand, right hand, left hand, and right hand twice.

INTRODUCTION TO ALPHABET

FIGURE 1

In figure 1, we have ***visual*** display of a picture of boxes. The picture has the first position filled up with a symbol. The decimal numbers 1 and letter A correspond to the first picture. The letter A has the first position in the English alphabet.

The ***audio*** representation for the picture, number 1 and letter A, is **knock and double knock.**

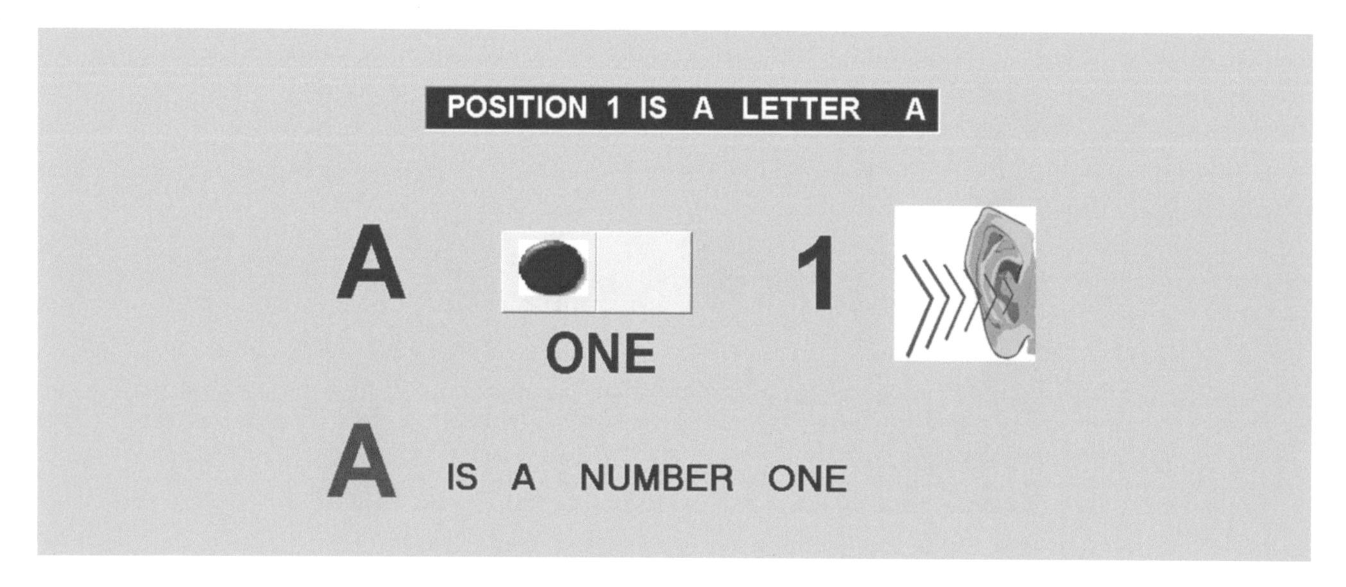

A knock represents a filled up position and a double knock represents an empty position. The clamping of the right hand represents a filled up position and the clamping of the left hand represents an empty position. The ***kinesthetic*** representation for the number 1 and letter A is clamping of the right hand and left hand.

EXERCISE 1.1 WHAT LETTER IS IT?

In Exercise 1.1, we have ***visual*** display of a picture of boxes and the letter A. We have to select whether the audio representation is corresponding to the picture of boxes or the letter A.

The ***audio*** representation is **knock and double knock.**

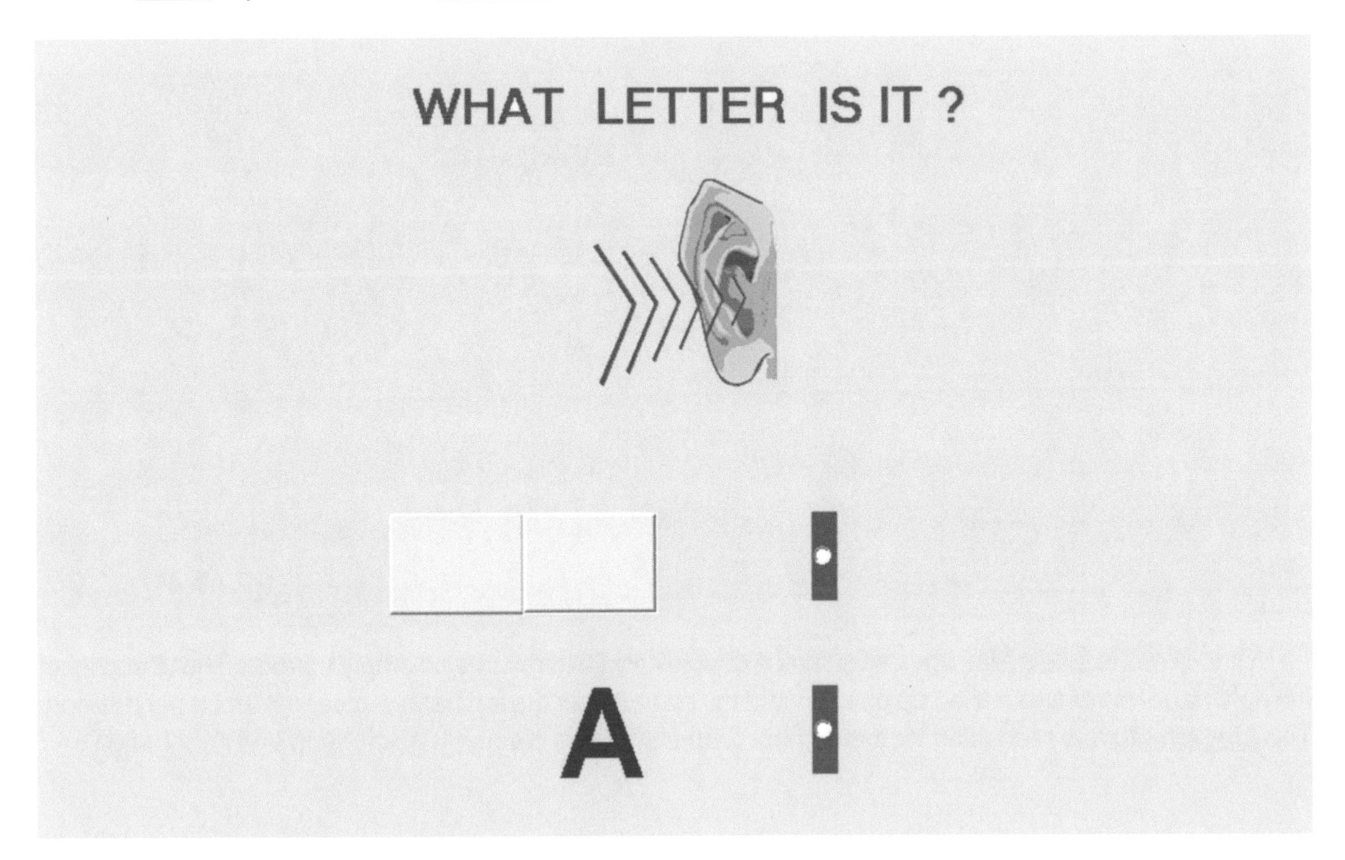

A knock represents a filled up position and a double knock represents an empty position. The clamping of the right hand represents a filled up position and the clamping of the left hand represents an empty position. The audio representation is corresponding to letter A. The ***kinesthetic*** representation for letter A is clamping of the right hand and left hand.

INTRODUCTION TO ALPHABET

FIGURE 2

In figure 2, we have ***visual*** display of a picture of boxes. The picture has the second position filled up with a symbol. The decimal number 2 and letter B correspond to the picture. The letter B has the second position in the English alphabet.

The ***audio*** representation for the picture, number 2 and letter B, is **double knock and knock.**

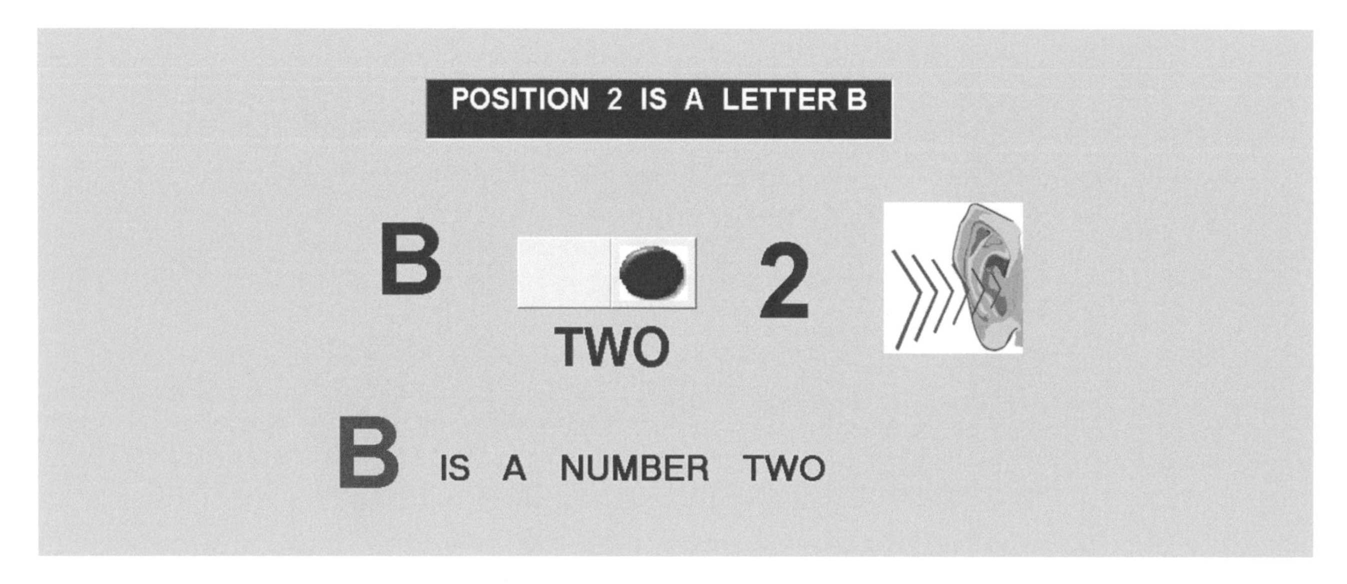

A knock represents a filled up position and a double knock represents an empty position. The clamping of the right hand represents a filled up position and the clamping of the left hand represents an empty position. The ***kinesthetic*** representation for the number 2 and letter B is clamping of left hand and right hand.

EXERCISE 2.1 WHAT LETTER IS IT?

In exercise 2.1, we have ***visual*** display of a picture of boxes, the letter A, and the letter B. We have to select whether the audio representation is corresponding to the picture of boxes, the letter B, or the letter A.

The ***audio*** representation is **double knock and knock.**

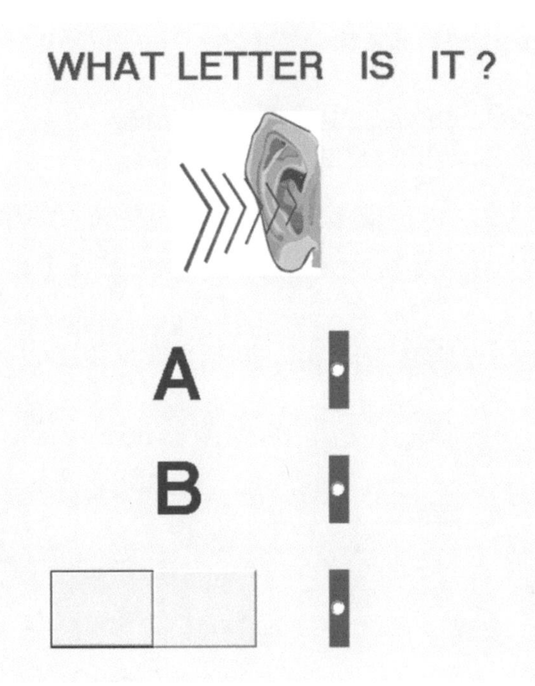

A knock represents a filled up position and a double knock represents an empty position. The clamping of the right hand represents a filled up position and the clamping of the left hand represents an empty position. The audio representation is corresponding to letter B. The ***kinesthetic*** representation for the letter B is clamping of left hand and right hand.

WHAT LETTERS ARE THESE? EXERCISE 2.2

In exercise 2.2, in ***visual*** display of alphabet we see two pictures of boxes. In the bottom we see letters which are probable answers. The first picture has the first position filled with a symbol and the second picture has the second position filled with a symbol. The **audio** representation for the first picture is **knock and double knock.** The **audio** representation for the second picture is **double knock and knock.**

From the probable answers we need to find the right one. The audio signals are:

1.) **double knock, knock, chick, knock, and double knock.**
2.) **knock, double knock, chick, double knock, and knock.**

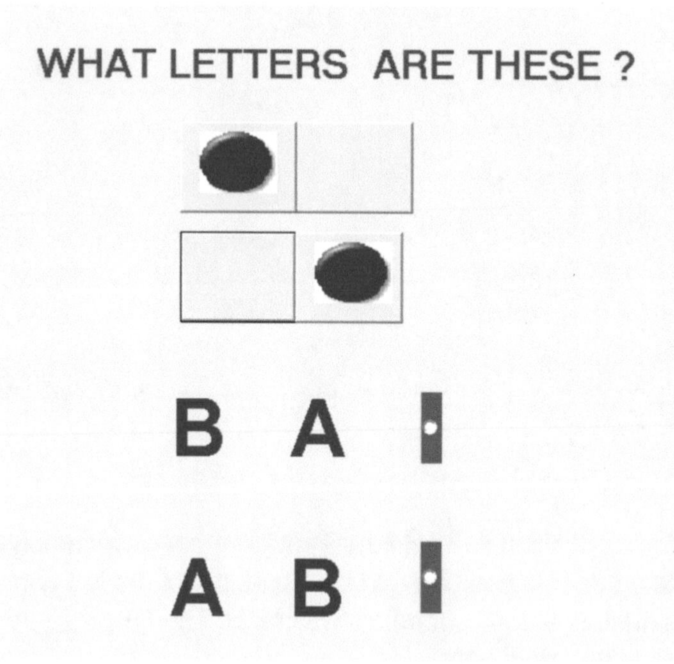

A knock represents a filled up position and a double knock represents an empty position. The clamping of the right hand represents a filled up position and the clamping of the left hand represents an empty position. The audio representation is corresponding to the letter A and letter B. The ***kinesthetic*** representation is clamping of the right hand, left hand, one hand shaking, left hand, and right hand.

FIGURE 3 INTRODUCTION TO ALPHABET

In figure 3, we have ***visual*** display of a picture of boxes. The picture has both positions filled up with symbols. The decimal numbers 3 and letter C correspond to the picture. The letter C has the third position in the English alphabet.

The ***audio*** representation for the picture, number 3 and letter C, is **knock and knock.**

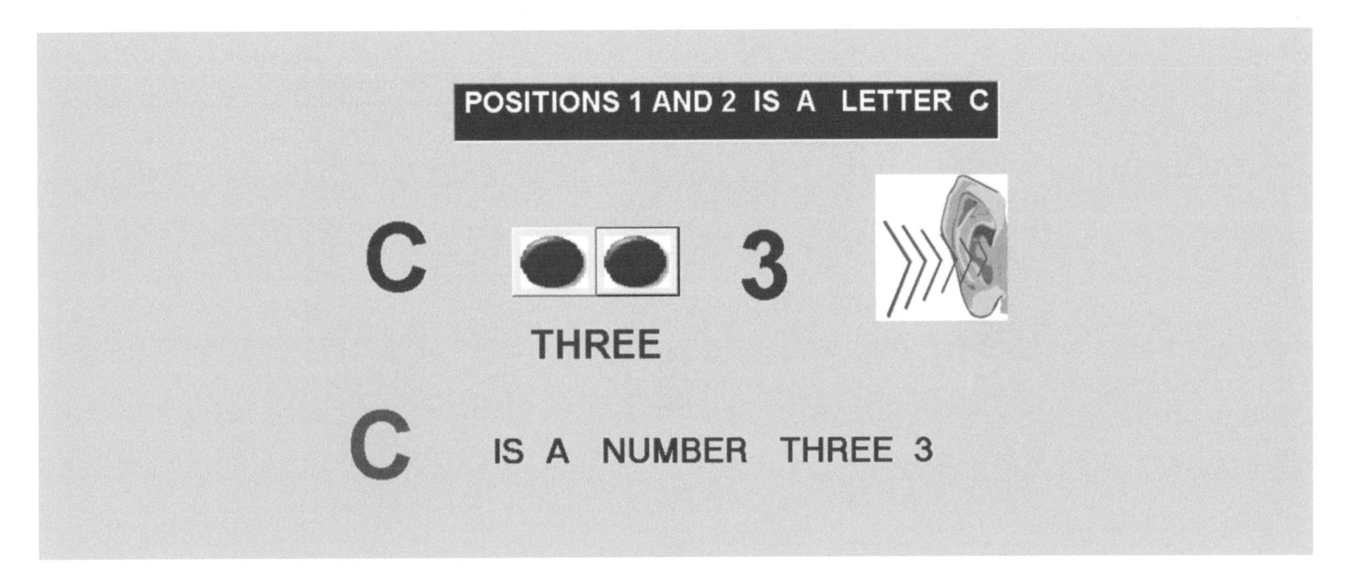

A knock represents a filled up position and a double knock represents an empty position. The clamping of the right hand represents a filled up position and the clamping of the left hand represents an empty position. The ***kinesthetic*** representation for the number 3 and letter C is clamping of the right hand twice.

WHAT LETTER IS IT? EXERCISE 3.1

In exercise 3.1, we have ***visual*** display of the letters C, B, and A. We have to select whether the audio representation is corresponding to the letter C, B, or A.

The ***audio*** representation is **knock and knock.**

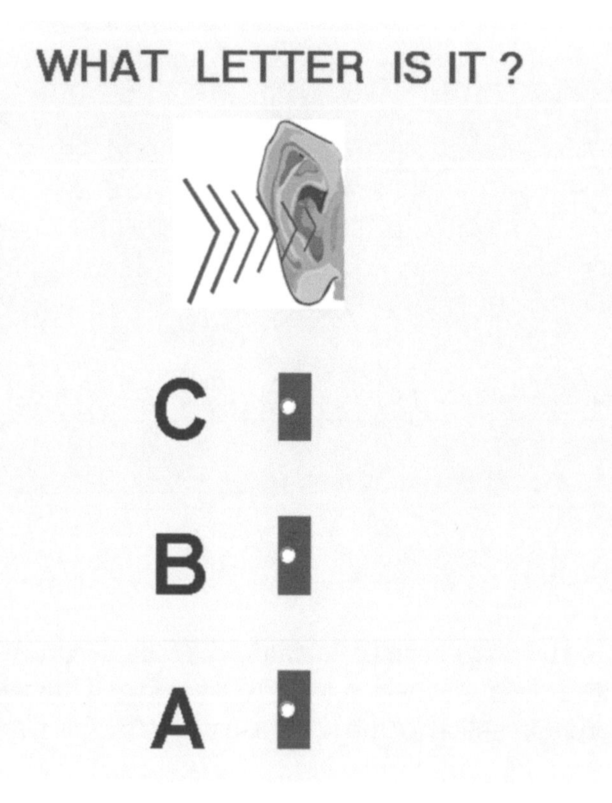

A knock represents a filled up position and a double knock represents an empty position. The clamping of the right hand represents a filled up position and the clamping of the left hand represents an empty position. The audio representation is corresponding to letter C. The ***kinesthetic*** representation for the letter C is clamping of right hand twice.

FIGURE 4 INTRODUCTION TO ALPHABET

In figure 4, we have ***visual*** display of two pictures of boxes. The picture has the third position filled up with a symbol. The decimal number 4 and letter D correspond to the picture. The letter D has the fourth position in the English alphabet.

The ***audio*** representation for the picture, number 4 and letter D, is **double knock, double knock, and knock.**

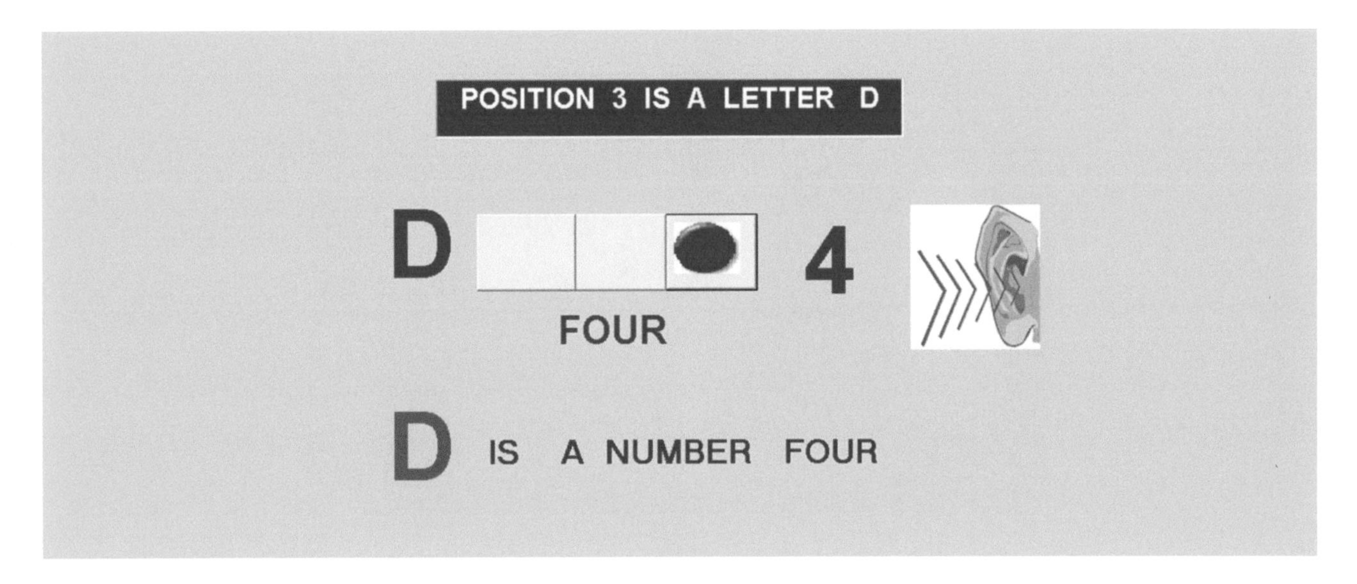

A knock represents a filled up position and a double knock represents an empty position. The clamping of the right hand represents a filled up position and the clamping of the left hand represents an empty position. The ***kinesthetic*** representation for the number 4 and letter D is clamping of left hand twice and right hand.

WHAT LETTER IS IT? EXERCISE 4.1

In exercise 4.1, we have ***visual*** display of the letters B, D, and A. We have to select whether the audio representation is corresponding to the letter B, D, or A.

The ***audio*** representation is **double knock, double knock, and knock.**

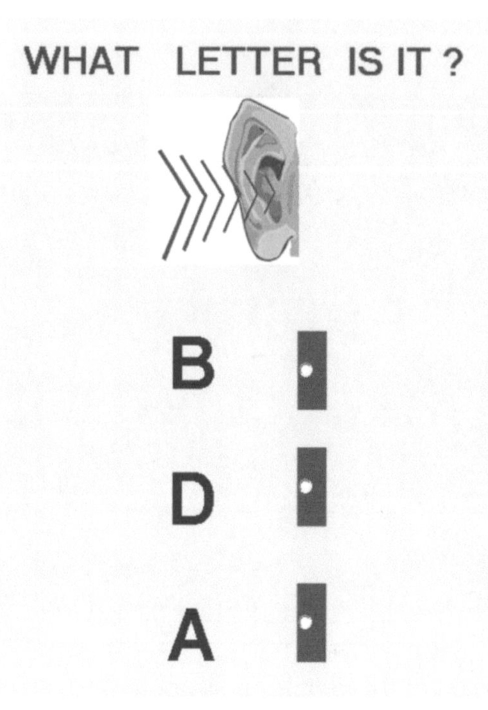

A knock represents a filled up position and a double knock represents an empty position. The clamping of the right hand represents a filled up position and the clamping of the left hand represents an empty position. The audio representation is corresponding to letter D. The ***kinesthetic*** representation for the D is clamping of left hand twice and right hand.

EXERCISE 4.2 WHAT LETTERS ARE THESE?

In exercise 4.2, in ***visual*** display of alphabet we see two pictures of boxes. In the bottom we see letters which are probable answers. The first picture has the third position filled with a symbol and the second picture has the first and second positions filled with symbols. The **audio** representation for the first picture is **double knock, double knock, and knock.** The **audio** representation for the second picture is **knock, knock, and double knock.**

From the probable answers we need to find the right one. The audio signals are:

1.) **knock, knock, double knock, chick, double knock, double knock, and knock.**
2.) **double knock, double knock, knock, chick, knock, knock, and double knock.**

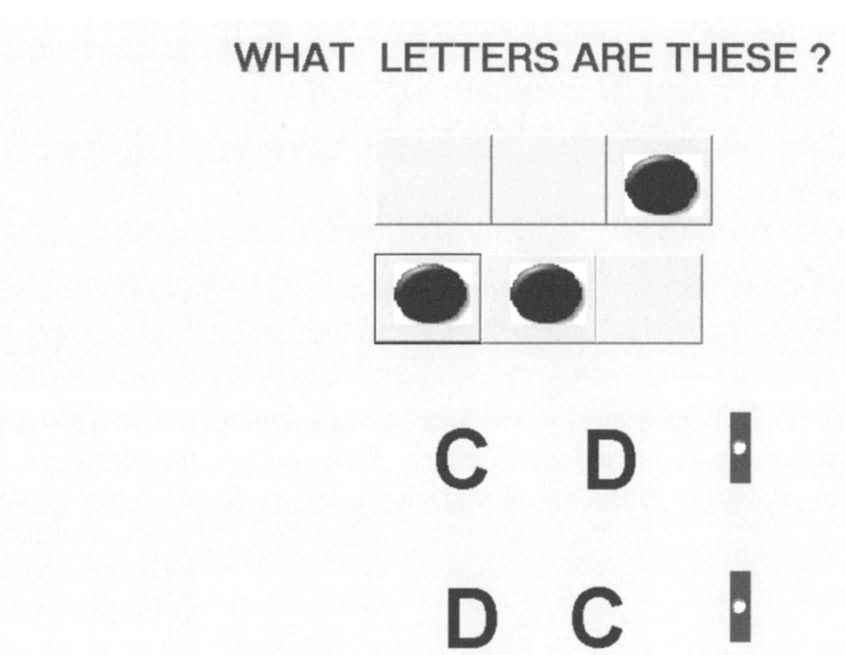

A knock represents a filled up position and a double knock represents an empty position. The clamping of the right hand represents a filled up position and the clamping of the left hand represents an empty position. The audio representation is corresponding to the letter D and letter C. The ***kinesthetic*** representation is clamping of the left hand twice, right hand, one hand shaking, right hand twice, and left hand.

INTRODUCTION TO ALPHABET

FIGURE 5

In figure 5, we have ***visual*** display of a picture of boxes. The picture has the first and third positions filled up with symbols. The decimal numbers 5 and letter E correspond to the picture. The letter E has the fifth position in the English alphabet.

The ***audio*** representation for the left picture, number 5 and letter E, is **knock, double knock, and knock.**

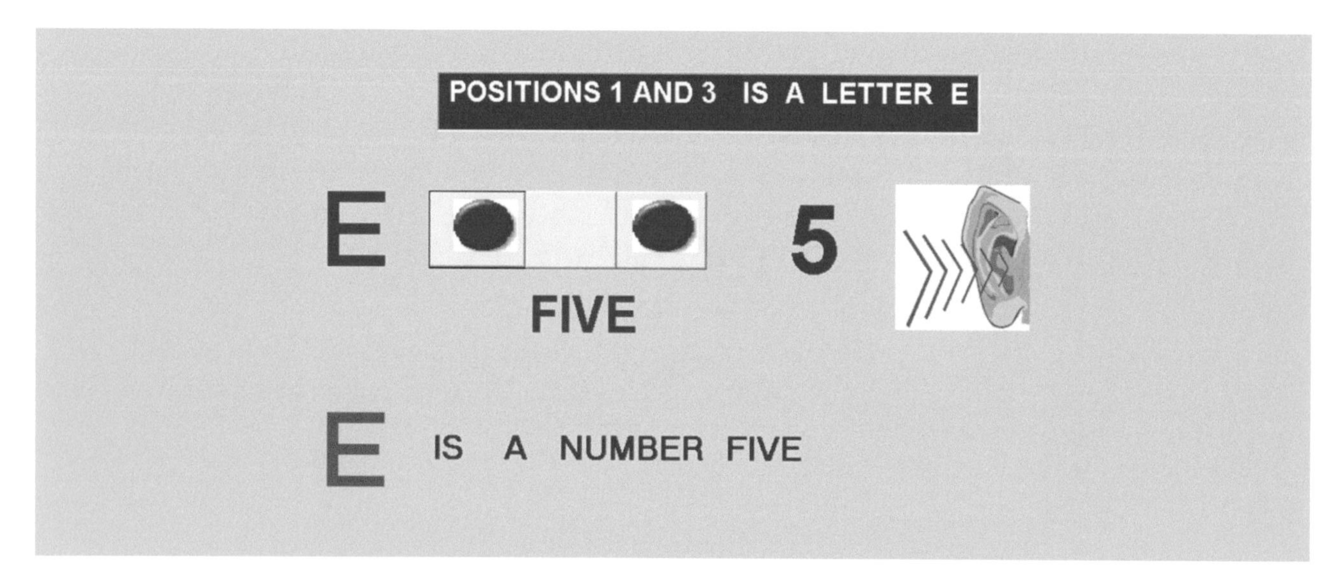

A knock represents a filled up position and a double knock represents an empty position. The clamping of the right hand represents a filled up position and the clamping of the left hand represents an empty position. The ***kinesthetic*** representation for the number 5 and letter E is clamping of the right hand, left hand, and right hand.

EXERCISE 5.1 WHAT LETTER IS IT?

In exercise 5.1, we have ***visual*** display of the letters C, D, and E. We have to select whether the audio representation is corresponding to the letter C, D, or E.

The ***audio*** representation is **knock, double knock, and knock.**

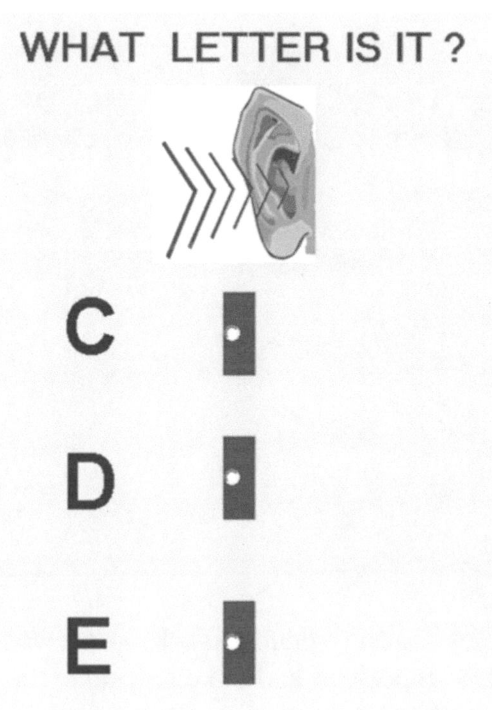

A knock represents a filled up position and a double knock represents an empty position. The clamping of the right hand represents a filled up position and the clamping of the left hand represents an empty position. The audio representation is corresponding to letter E. The ***kinesthetic*** representation for the letter E is clamping of the right hand, left hand, and right hand.

INTRODUCTION TO ALPHABET FIGURE 6

In figure 6, we have ***visual*** display of a picture of boxes. The picture has the second and third position filled up with a symbol. The decimal number 6 and letter F correspond to the picture. The letter F has the 6th position in the English alphabet.

The ***audio*** representation for the picture, number 6 and letter F, is **double knock, knock, and knock.**

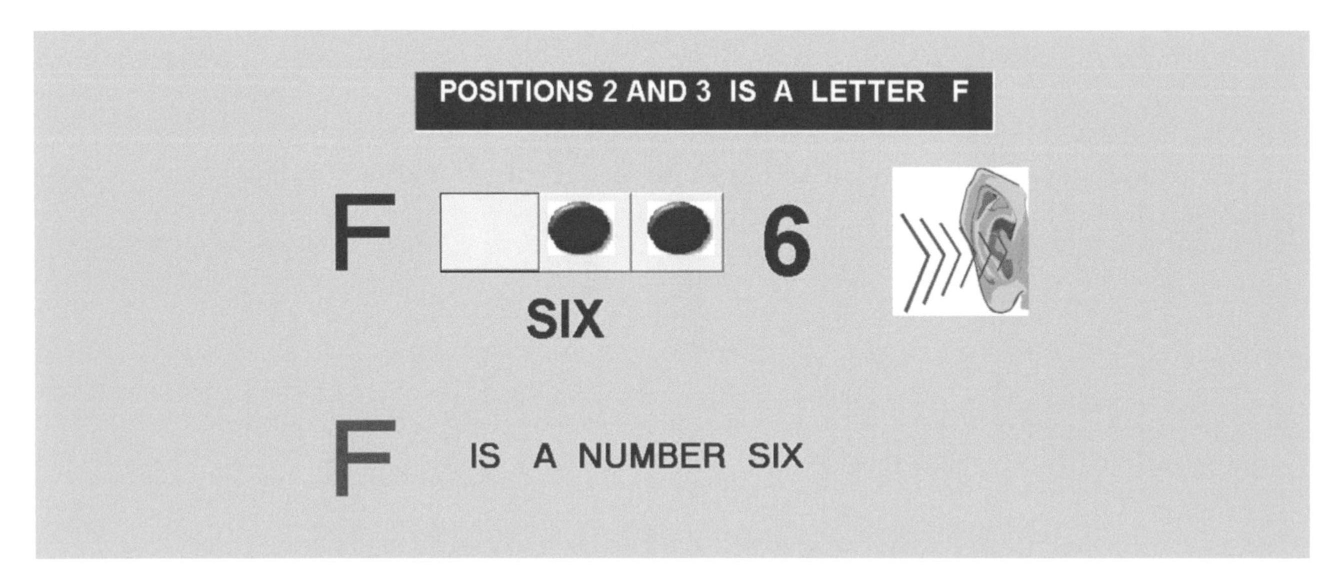

A knock represents a filled up position and a double knock represents an empty position. The clamping of the right hand represents a filled up position and the clamping of the left hand represents an empty position. The ***kinesthetic*** representation for the number 6 and letter F is clamping of left hand and right hand twice.

EXERCISE 6.1 WHAT LETTER IS IT?

In exercise 6.1, we have ***visual*** display of the letters F, B, and C. We have to select whether the audio representation is corresponding to the letter F, B, or C.

The ***audio*** representation is **knock, double knock, and knock.**

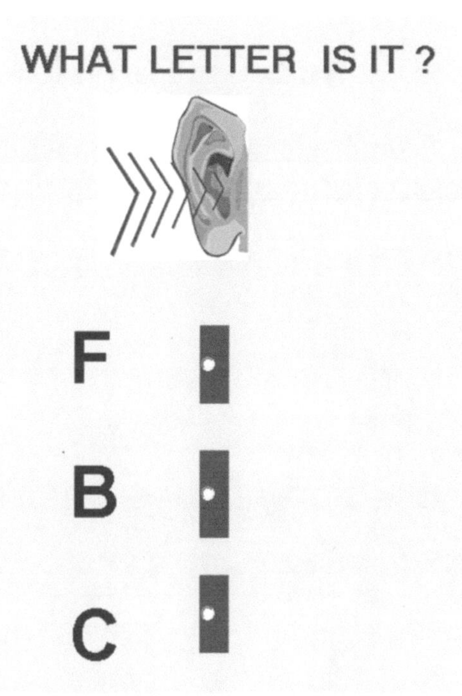

A knock represents a filled up position and a double knock represents an empty position. The clamping of the right hand represents a filled up position and the clamping of the left hand represents an empty position. The audio representation is corresponding to letter F. The ***kinesthetic*** representation for the letter F is clamping of left hand and right hand twice.

WHAT LETTERS ARE THESE? EXERCISE 6.2

In exercise 6.2, in ***visual*** display of alphabet we see two pictures of boxes. In the bottom we see letters which are probable answers. The first picture has the first and third positions filled with symbols and the second picture has the thrid and second positions filled with symbols. The **audio** representation for the first picture is **knock, double knock, and knock.** The **audio** representation for the second picture is **double knock, knock, and knock.**

From the probable answers we need to find the right one. The audio signals are:

1.) **double knock, knock, chick, double knock, double knock, and knock.**
2.) **knock, double knock, knock, chick, double knock, knock, and knock.**

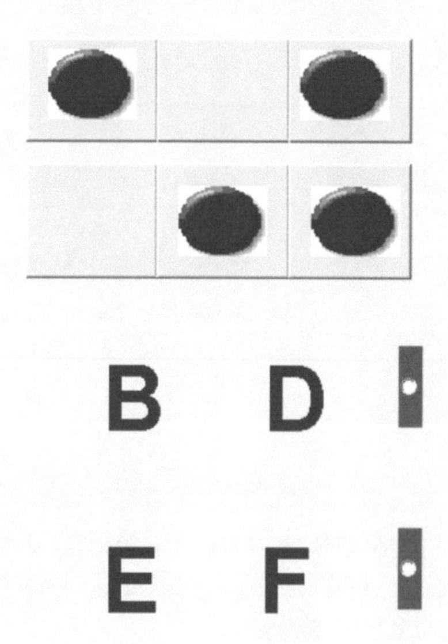

A knock represents a filled up position and a double knock represents an empty position. The clamping of the right hand represents a filled up position and the clamping of the left hand represents an empty position. The audio representation is corresponding to the letter E and letter F. The ***kinesthetic*** representation is clamping of the right hand, left hand, right hand, one hand shaking, left hand and right hand twice.

FIGURE 7 INTRODUCTION TO ALPHABET

In figure 7, we have ***visual*** display of a picture of boxes. The picture has all three positions filled up with symbols. The decimal numbers 7 and letter G correspond to the picture. The letter G has the 7th position in the English alphabet.

The ***audio*** representation for the picture, number 7 and letter G, is **knock, knock, and knock.**

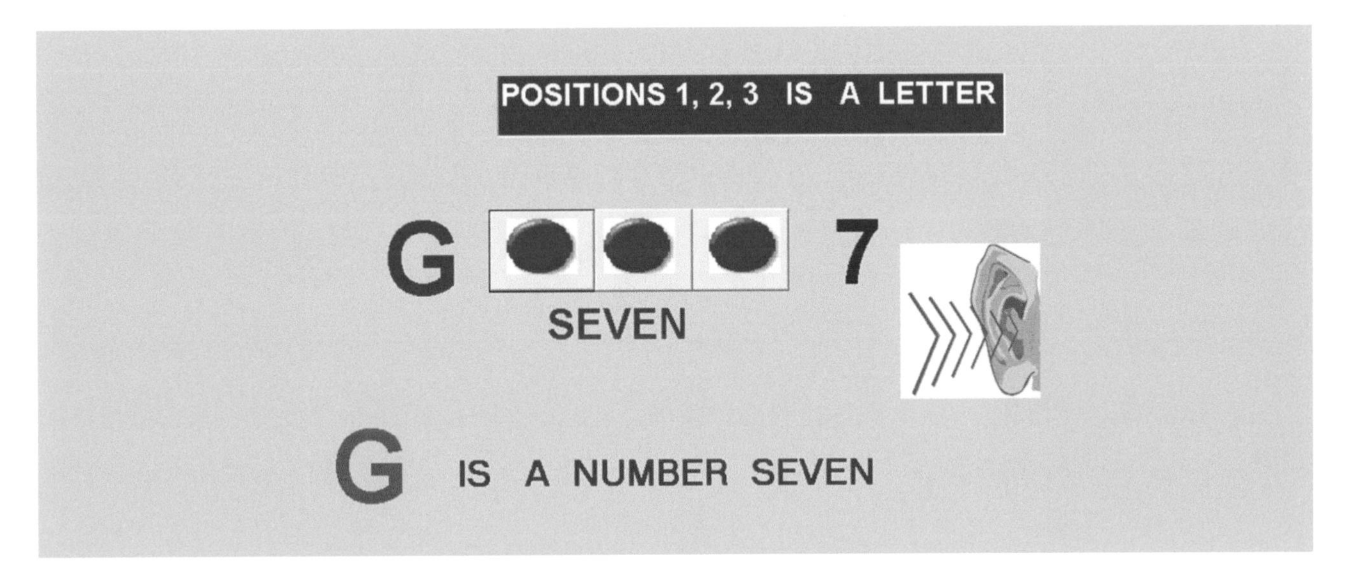

A knock represents a filled up position and a double knock represents an empty position. The clamping of the right hand represents a filled up position and the clamping of the left hand represents an empty position. The ***kinesthetic*** representation for the number 7 and letter G is clamping of the right hand thrice.

WHAT LETTER IS IT? EXERCISE 7.1

In exercise 7.1, we have ***visual*** display of the letters B, C, and G. We have to select whether the audio representation is corresponding to the letter B, C, or G.

The ***audio*** representation is **knock, knock, and knock.**

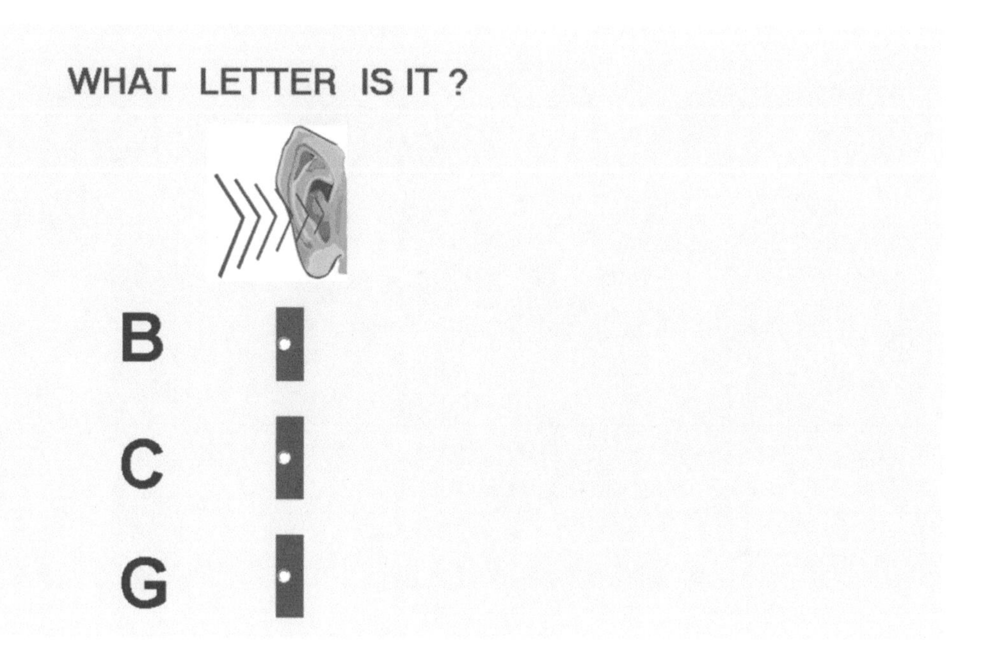

A knock represents a filled up position and a double knock represents an empty position. The clamping of the right hand represents a filled up position and the clamping of the left hand represents an empty position. The audio representation is corresponding to letter G. The ***kinesthetic*** representation for the letter G is clamping of right hand thrice.

FIGURE 8 INTRODUCTION TO ALPHABET

In figure 8, we have ***visual*** display of a picture of boxes. The picture has the fourth position filled up with a symbol. The decimal number 8 and letter H correspond to the picture. The letter H has the 8th position in the English alphabet.

The ***audio*** representation for the right picture, number 8 and letter H, is **double knock, double knock, double knock, and knock.**

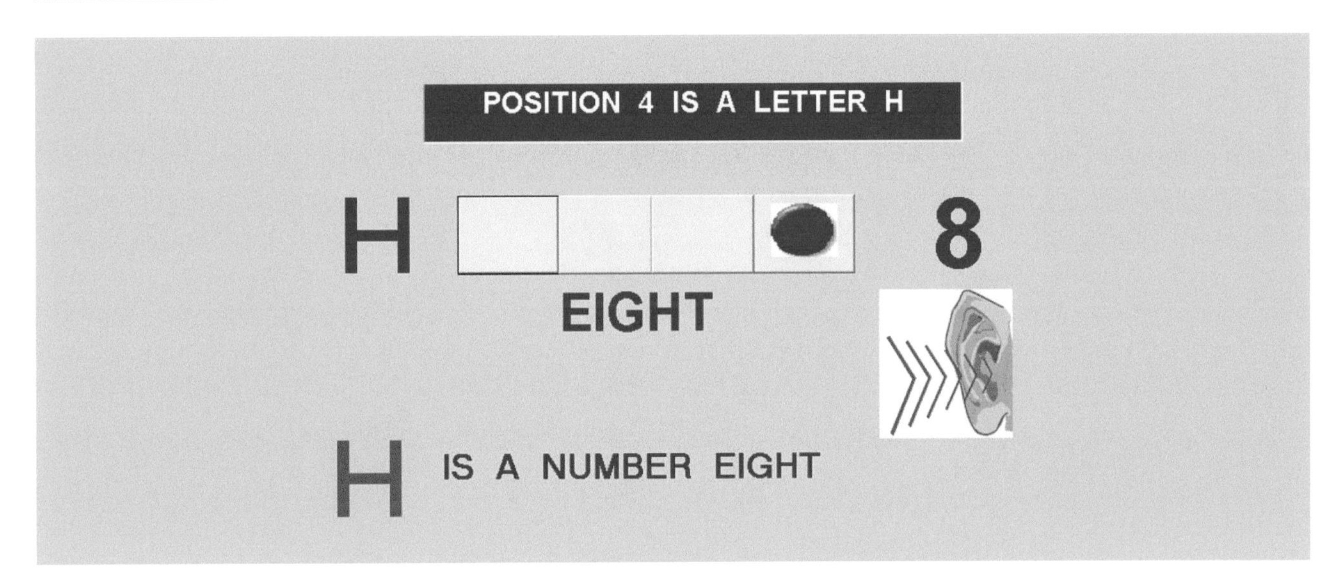

A knock represents a filled up position and a double knock represents an empty position. The clamping of the right hand represents a filled up position and the clamping of the left hand represents an empty position. The ***kinesthetic*** representation for the number 8 and letter H is clamping of left hand thrice and right hand.

WHAT LETTER IS IT? EXERCISE 8.1

In exercise 8.1, we have ***visual*** display of the letters H, C, and E. We have to select whether the audio representation is corresponding to the letter H, C, or E.

The ***audio*** representation is **double knock, double knock, double knock, and knock.**

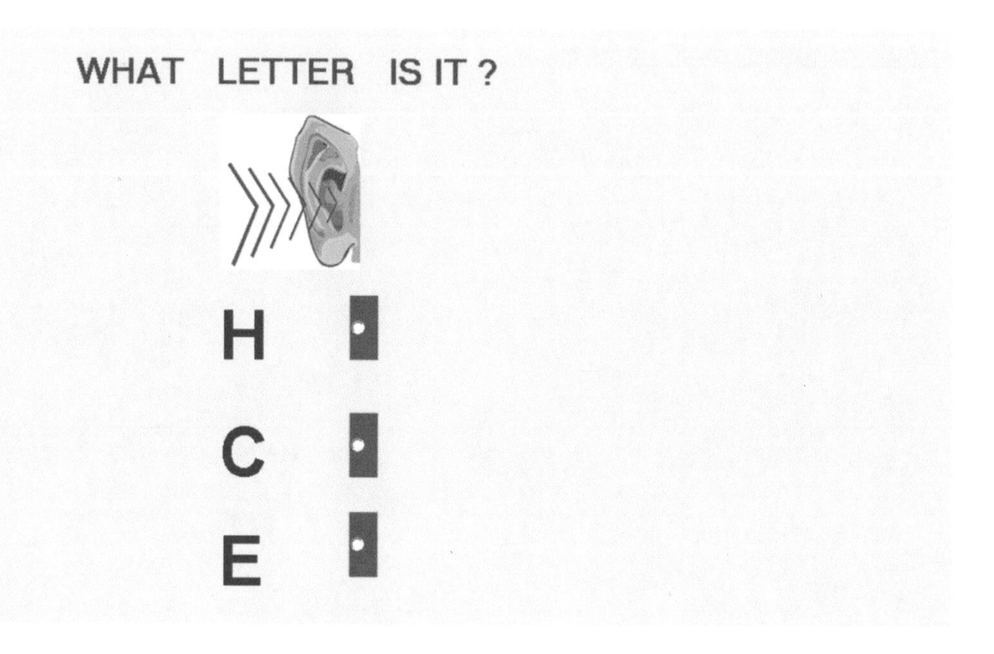

A knock represents a filled up position and a double knock represents an empty position. The clamping of the right hand represents a filled up position and the clamping of the left hand represents an empty position. The audio representation is corresponding to letter H. The ***kinesthetic*** representation for the letter H is clamping of left hand thrice and right hand.

EXERCISE 8.2 WHAT LETTERS ARE THESE?

In exercise 8.2, in ***visual*** display of alphabet we see two pictures of boxes. In the bottom we see letters which are probable answers. The first picture has the first, second, and third positions filled with symbols and the second picture has the fourth position filled with a symbol. The **audio** representation for the first picture is **knock, knock, knock, and double knock.** The **audio** representation for the second picture is **double knock, double knock, double knock, and knock.**

From the probable answers we need to find the right one. The audio signals are:

1.) **double knock, double knock, double knock, knock, chick, knock, knock, and knock.**
2.) **knock, knock, knock, chick, double knock, double knock, double knock, and knock.**

A knock represents a filled up position and a double knock represents an empty position. The clamping of the right hand represents a filled up position and the clamping of the left hand represents an empty position. The audio representation is corresponding to the letter G and letter H. The ***kinesthetic*** representation is clamping of the right hand thrice, one hand shaking, left hand thrice, and right hand.

INTRODUCTION TO ALPHABET **FIGURE 9**

In figure 9, we have ***visual*** display of a picture of boxes. The picture has the first and fourth positions filled up with symbols. The decimal numbers 9 and letter I correspond to the picture. The letter I has the 9th position in the English alphabet.

The ***audio*** representation for the picture, number 9 and letter I, is **knock, double knock, double knock, and knock.**

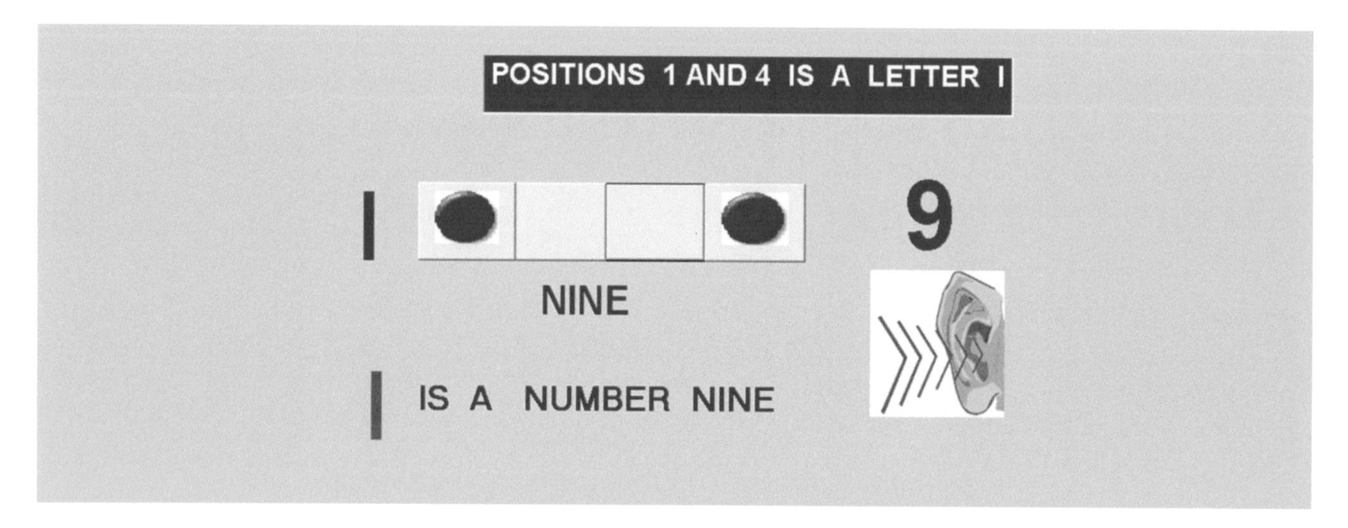

A knock represents a filled up position and a double knock represents an empty position. The clamping of the right hand represents a filled up position and the clamping of the left hand represents an empty position. The ***kinesthetic*** representation for the number 9 and letter I is clamping of the right hand, left hand twice, and right hand.

EXERCISE 9.1 WHAT LETTER IS IT?

In exercise 9.1, we have ***visual*** display of the letters C, D, and I. We have to select whether the audio representation is corresponding to the letter C, D, or I.

The ***audio*** representation is **knock, double knock, double knock, and knock.**

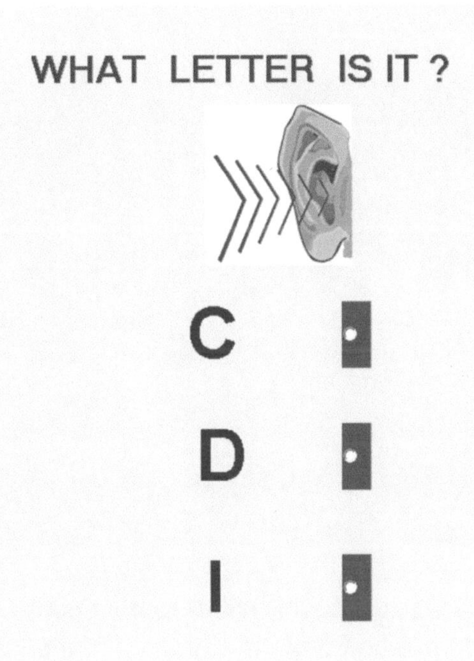

A knock represents a filled up position and a double knock represents an empty position. The clamping of the right hand represents a filled up position and the clamping of the left hand represents an empty position. The audio representation is corresponding to letter I. The ***kinesthetic*** representation for the letter I is clamping of the right hand, left hand twice, and right hand.

INTRODUCTION TO ALPHABET

FIGURE 10

In figure 10, we have ***visual*** display of a picture of boxes. The picture has the second and fourth positions filled up with symbols. The decimal number 10 and letter J correspond to the picture. The letter J has the 10th position in the English alphabet.

The ***audio*** representation for the picture, number 10 and letter J, is **double knock, knock, double knock, and knock.**

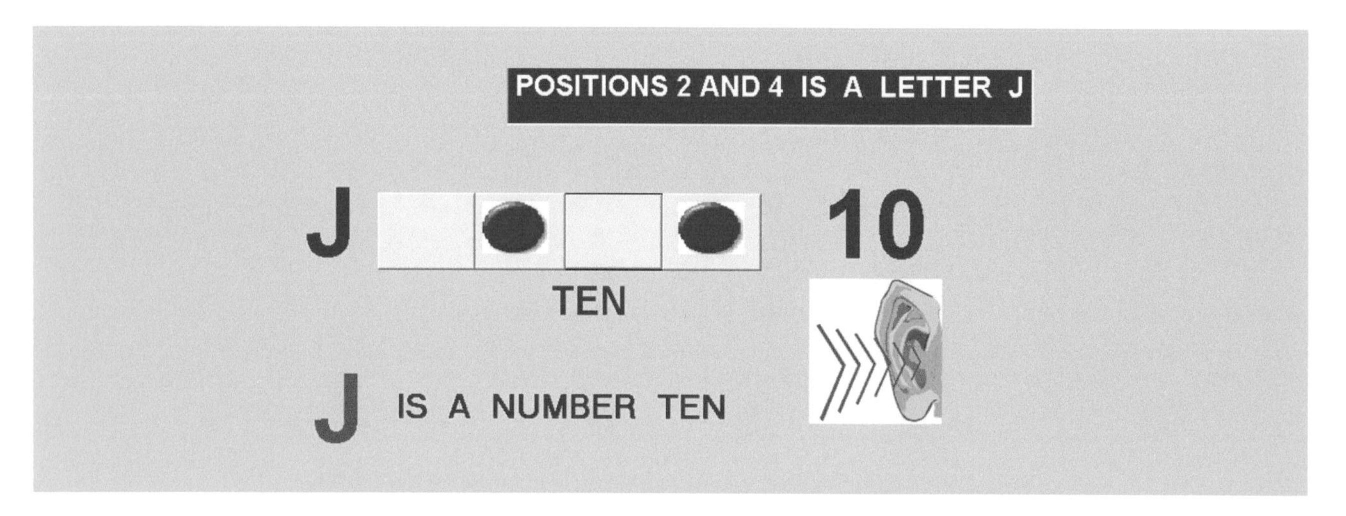

A knock represents a filled up position and a double knock represents an empty position. The clamping of the right hand represents a filled up position and the clamping of the left hand represents an empty position. The ***kinesthetic*** representation for the number 10 and letter J is clamping of left hand, right hand, left hand, and right hand.

EXERCISE 10.1 WHAT LETTER IS IT?

In exercise 10.1, we have ***visual*** display of the letters D, F, and J. We have to select whether the audio representation is corresponding to the letter D, F, or J.

The ***audio*** representation is **double knock, knock, double knock, and knock.**

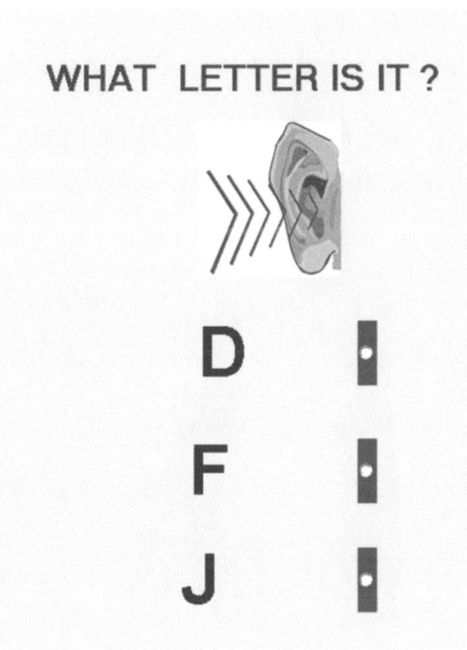

A knock represents a filled up position and a double knock represents an empty position. The clamping of the right hand represents a filled up position and the clamping of the left hand represents an empty position. The audio representation is corresponding to letter J. The ***kinesthetic*** representation for the letter J is clamping of left hand, right hand, left hand, and right hand.

WHAT LETTERS ARE THESE? EXERCISE 10.2

In exercise 10.2, in ***visual*** display of alphabet we see two pictures of boxes. In the bottom we see letters which are probable answers. The first picture has the first and fourth positions filled with symbols and the second picture has the second and fourth positions filled with symbols. The **audio** representation for the first picture is **knock, double knock, double knock, and knock.** The **audio** representation for the second picture is **double knock, knock, double knock, and knock.**

From the probable answers we need to find the right one. The audio signals are:

1.) **knock, knock, chick, double knock, double knock, and knock.**
2.) **knock, double knock, double knock, knock, chick, double knock, knock, double knock, and knock.**

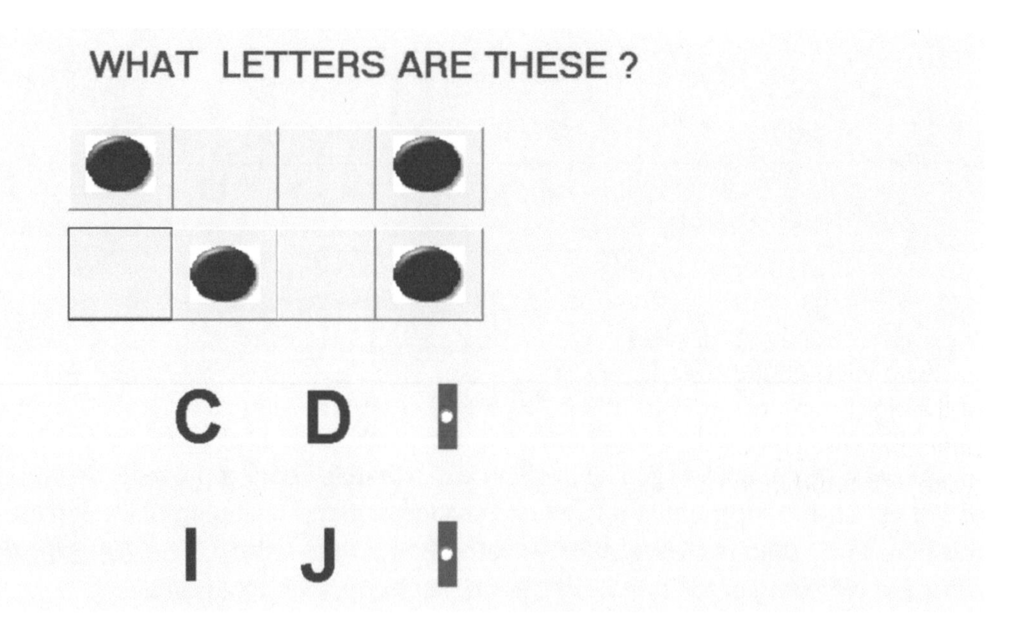

A knock represents a filled up position and a double knock represents an empty position. The clamping of the right hand represents a filled up position and the clamping of the left hand represents an empty position. The audio representation is corresponding to the letter I and letter J. The ***kinesthetic*** representation is clamping of the right hand, left hand twice, right hand, one hand shaking, left hand, right hand, left hand, and right hand.

FIGURE 11 INTRODUCTION TO ALPHABET

In figure 11, we have ***visual*** display of a picture of boxes. The picture has the first, second, and fourth positions filled up with symbols. The decimal numbers 11 and letter K correspond to the picture. The letter K has the 11th position in the English alphabet.

The ***audio*** representation for the picture, number 11 and letter K, is **knock, knock, double knock, and knock.**

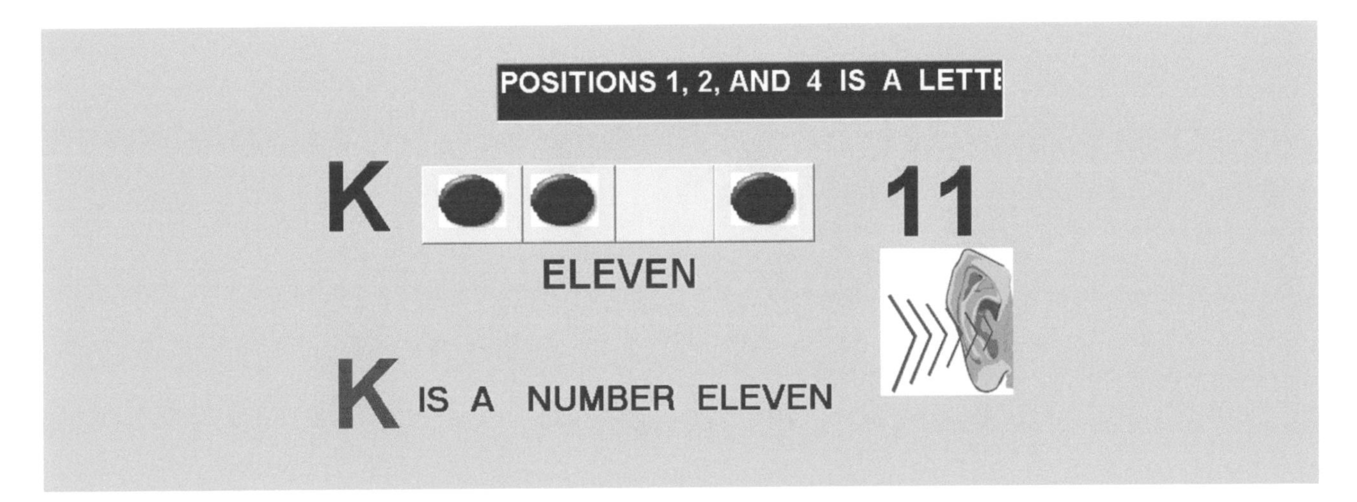

A knock represents a filled up position and a double knock represents an empty position. The clamping of the right hand represents a filled up position and the clamping of the left hand represents an empty position. The ***kinesthetic*** representation for the number 11 and letter K is clamping of the right hand twice, left hand, and right hand.

WHAT LETTER IS IT? EXERCISE 11.1

In exercise 11.1, we have ***visual*** display of the letters K, E, and C. We have to select whether the audio representation is corresponding to the letter K, E, or C.

The ***audio*** representation is **knock, knock, double knock, and knock.**

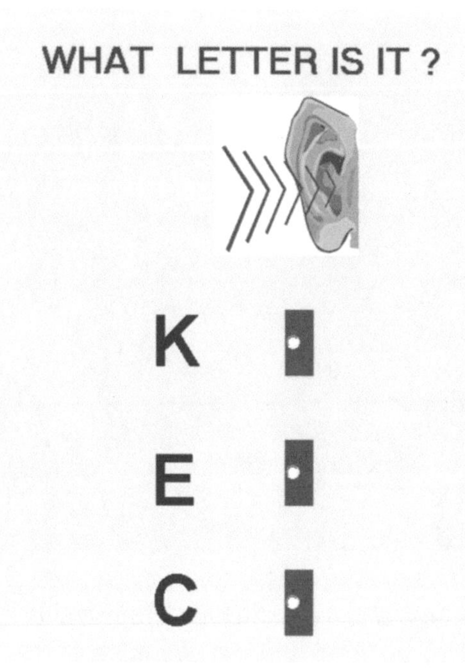

A knock represents a filled up position and a double knock represents an empty position. The clamping of the right hand represents a filled up position and the clamping of the left hand represents an empty position. The audio representation is corresponding to letter K. The ***kinesthetic*** representation for the letter K is clamping of the right hand twice, left hand, and right hand.

FIGURE 12 INTRODUCTION TO ALPHABET

In figure 12, we have ***visual*** display of a picture of boxes. The picture has the third and fourth positions filled up with symbols. The decimal number 12 and letter L correspond to the picture. The letter L has the 12th position in the English alphabet.

The ***audio*** representation for the picture, number 12 and letter L, is **double knock, knock, double knock, and knock.**

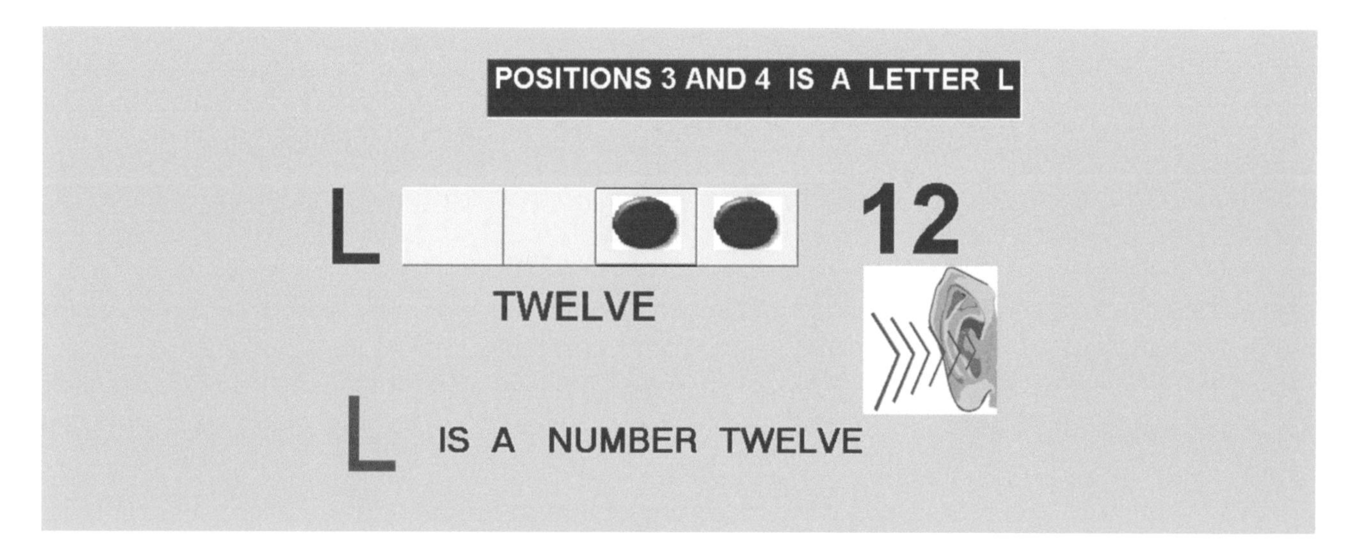

A knock represents a filled up position and a double knock represents an empty position. The clamping of the right hand represents a filled up position and the clamping of the left hand represents an empty position. The ***kinesthetic*** representation for the number 12 and letter L is clamping of left hand twice and right hand twice.

WHAT LETTER IS IT? EXERCISE 12.1

In exercise 12.1, we have ***visual*** display of the letters D, L, and B. We have to select whether the audio representation is corresponding to the letter D, L, or B.

The ***audio*** representation is **double knock, double knock, knock, and knock.**

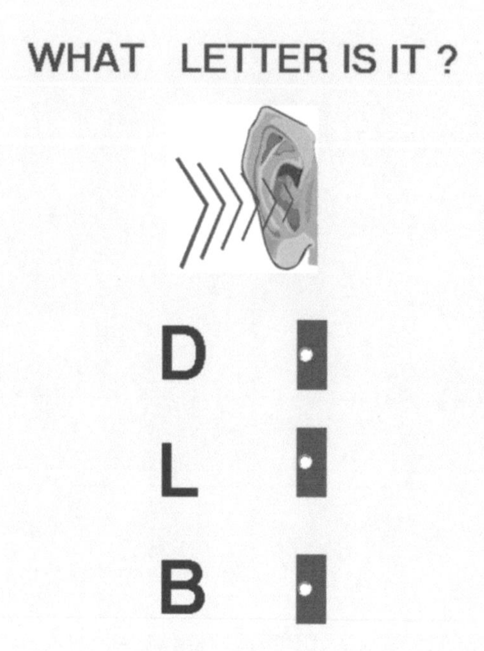

A knock represents a filled up position and a double knock represents an empty position. The clamping of the right hand represents a filled up position and the clamping of the left hand represents an empty position. The audio representation is corresponding to letter L. The ***kinesthetic*** representation for the letter L is clamping of left hand twice and right hand twice.

EXERCISE 12.2 WHAT LETTERS ARE THESE?

In exercise 12.2, in ***visual*** display of alphabet we see two pictures of boxes. In the bottom we see letters which are probable answers. The first picture has the first, second, and fourth positions filled with symbols and the second picture has the third and fourth positions filled with symbols. The **audio** representation for the first picture is **knock, knock, double knock, and knock.** The **audio** representation for the second picture is **double knock, double knock, knock, and knock.**

From the probable answers we need to find the right one. The audio signals are:

1.) **knock, knock, double knock, knock, chick, double knock, double knock, knock, and knock.**
2.) **knock, chick, double knock, and knock.**

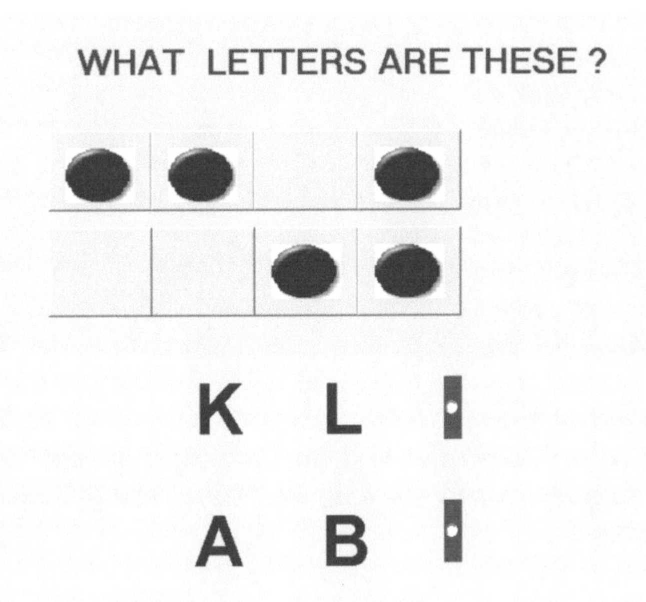

A knock represents a filled up position and a double knock represents an empty position. The clamping of the right hand represents a filled up position and the clamping of the left hand represents an empty position. The audio representation is corresponding to the letter K and letter L. The ***kinesthetic*** representation is clamping of the right hand twice, left hand, right hand, one hand shaking, left hand twice, and right hand twice.

INTRODUCTION TO ALPHABET

FIGURE 13

In figure 13, we have ***visual*** display of a picture of boxes. The picture has the first, third, and fourth positions filled up with symbols. The decimal numbers 13 and letter M correspond to the picture. The letter M has the 13th position in the English alphabet.

The ***audio*** representation for the picture, number 13 and letter M, is **knock, double knock, knock, and knock.**

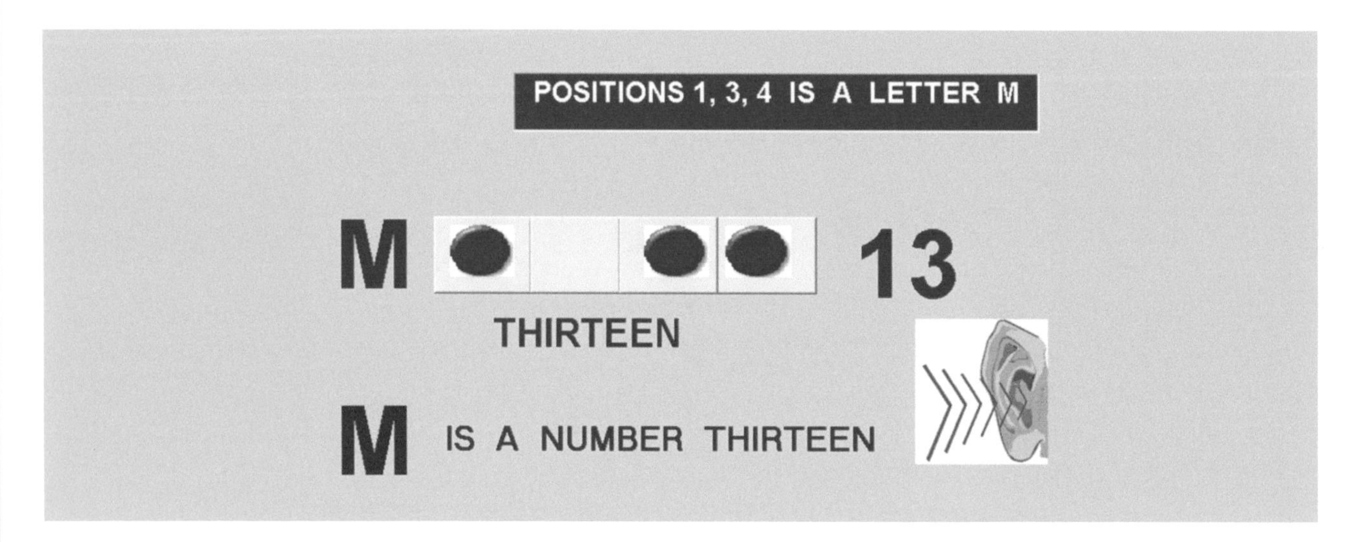

A knock represents a filled up position and a double knock represents an empty position. The clamping of the right hand represents a filled up position and the clamping of the left hand represents an empty position. The ***kinesthetic*** representation for the number 13 and letter M is clamping of the right hand, left hand, and right hand twice.

EXERCISE 13.1 WHAT LETTER IS IT?

In exercise 13.1, we have ***visual*** display of the letters M, C, and G. We have to select whether the audio representation is corresponding to the letter M, C, or G.

The ***audio*** representation is **knock, double knock, knock, and knock.**

A knock represents a filled up position and a double knock represents an empty position. The clamping of the right hand represents a filled up position and the clamping of the left hand represents an empty position. The audio representation is corresponding to letter M. The ***kinesthetic*** representation for the letter M is clamping of the right hand, left hand, and right hand twice.

INTRODUCTION TO ALPHABET

FIGURE 14

In figure 14, we have ***visual*** display of a picture of boxes. The picture has the second, third, and fourth positions filled up with symbols. The decimal number 14 and letter N correspond to the picture. The letter N has the 14th position in the English alphabet.

The ***audio*** representation for the picture, number 14 and letter N, is **double knock, knock, knock, and knock.**

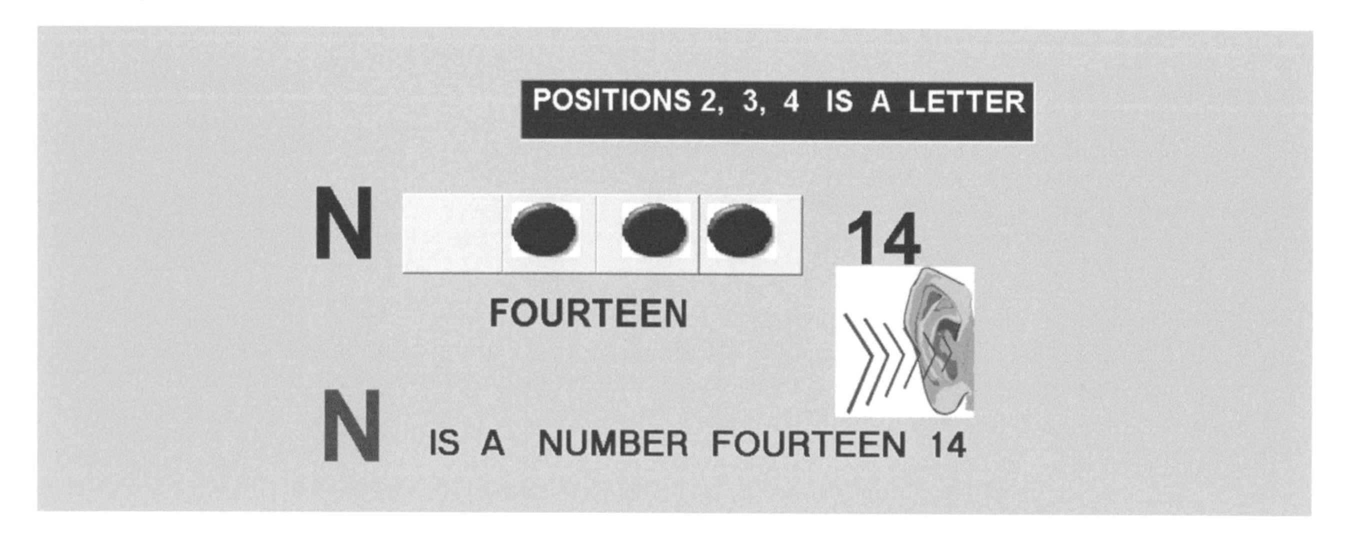

A knock represents a filled up position and a double knock represents an empty position. The clamping of the right hand represents a filled up position and the clamping of the left hand represents an empty position. The ***kinesthetic*** representation for the number 14 and letter N is clamping of left hand and right hand thrice.

EXERCISE 14.1 WHAT LETTER IS IT?

In exercise 14.1, we have ***visual*** display of the letters E, N, and C. We have to select whether the audio representation is corresponding to the letter E, N, or C.

The ***audio*** representation is **double knock, knock, knock, and knock.**

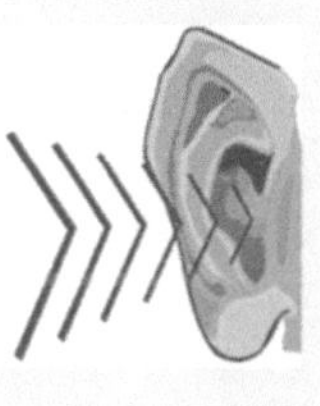

E

N

C

A knock represents a filled up position and a double knock represents an empty position. The clamping of the right hand represents a filled up position and the clamping of the left hand represents an empty position. The audio representation is corresponding to letter N. The ***kinesthetic*** representation for the letter N is clamping of left hand and right hand thrice.

WHAT LETTERS ARE THESE? EXERCISE 14.2

In exercise 14.2, in ***visual*** display of alphabet we see two pictures of boxes. In the bottom we see letters which are probable answers. The first picture has the first, third, and fourth positions filled with symbols and the second picture has the second, third, and fourth positions filled with symbols. The **audio** representation for the first picture is **knock, double knock, knock, and knock.** The **audio** representation for the second picture is **double knock, knock, knock, and knock.**

From the probable answers we need to find the right one. The audio signals are:

1.) **knock, knock, double knock, knock, chick, double knock, double knock, knock, and knock.**
2.) **knock, double knock, knock, knock, chick, double knock, knock, knock, and knock.**

A knock represents a filled up position and a double knock represents an empty position. The clamping of the right hand represents a filled up position and the clamping of the left hand represents an empty position. The audio representation is corresponding to the letter M and letter N. The ***kinesthetic*** representation is clamping of the right hand, left hand, right hand twice, one hand shaking, left hand, and right hand thrice.

FIGURE 15 INTRODUCTION TO ALPHABET

In figure 15, we have ***visual*** display of a picture of boxes. The picture has the all four positions filled up with symbols. The decimal numbers 15 and letter O correspond to the picture. The letter O has the 15th position in the English alphabet.

The ***audio*** representation for the picture, number 15 and letter M, is **knock, knock, knock, and knock.**

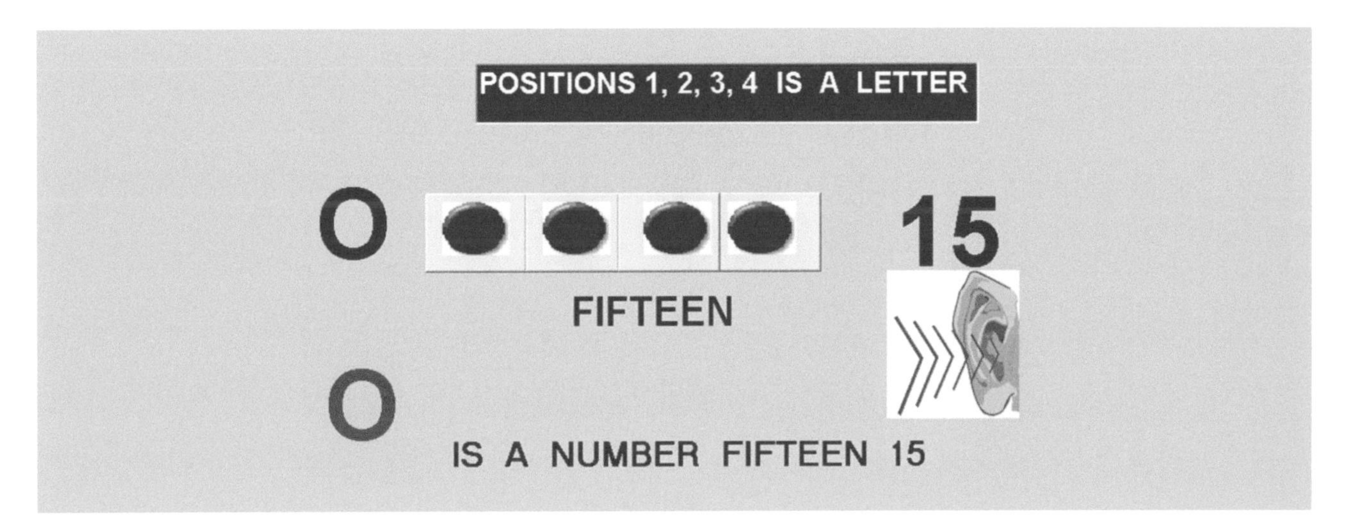

A knock represents a filled up position and a double knock represents an empty position. The clamping of the right hand represents a filled up position and the clamping of the left hand represents an empty position. The ***kinesthetic*** representation for the number 15 and letter O is clamping of the right hand four times.

WHAT LETTER IS IT? EXERCISE 15.1

In exercise 15.1, we have ***visual*** display of the letters O, D, and B. We have to select whether the audio representation is corresponding to the letter O, D, or B.

The ***audio*** representation is **knock, knock, knock, and knock.**

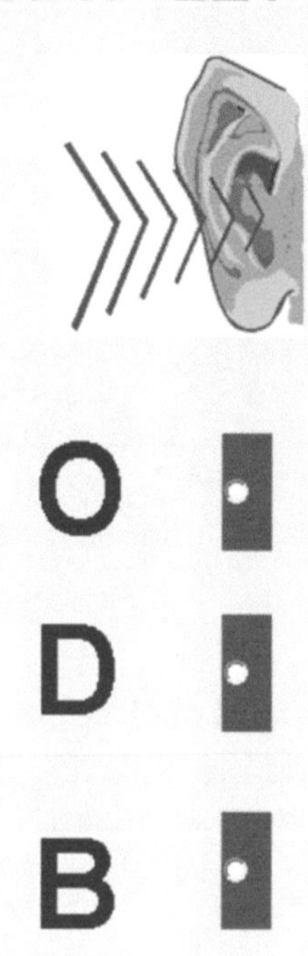

A knock represents a filled up position and a double knock represents an empty position. The clamping of the right hand represents a filled up position and the clamping of the left hand represents an empty position. The audio representation is corresponding to letter O. The ***kinesthetic*** representation for the number 15 and letter O is clamping of the right hand four times.

FIGURE 16 INTRODUCTION TO ALPHABET

In figure 16, we have ***visual*** display of a picture of boxes. The picture has the fifth position filled up with a symbol. The decimal number 16 and letter P correspond to the picture. The letter N has the 16th position in the English alphabet.

The ***audio*** representation for the picture, number 16 and letter P, is **double knock, double knock, double knock, double knock, and knock.**

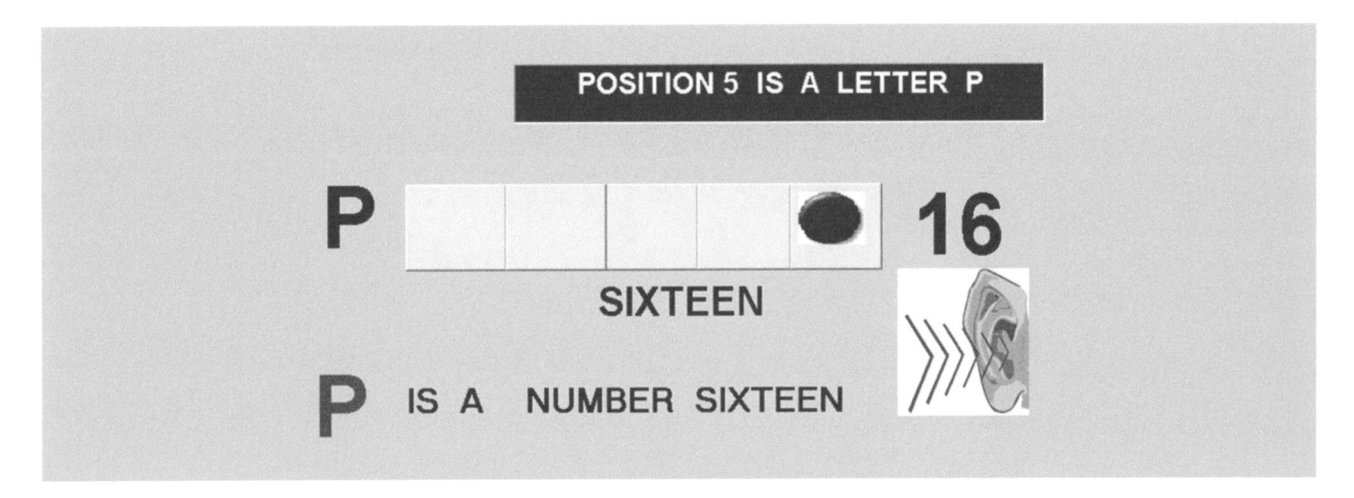

A knock represents a filled up position and a double knock represents an empty position. The clamping of the right hand represents a filled up position and the clamping of the left hand represents an empty position. The ***kinesthetic*** representation for the number 16 and letter P is clamping of left hand four times and right hand.

WHAT LETTER IS IT? EXERCISE 16.1

In exercise 16.1, we have ***visual*** display of the letters P, F, and B. We have to select whether the audio representation is corresponding to the letter P, F, or B.

The ***audio*** representation is **double knock, double knock, double knock, double knock, and knock.**

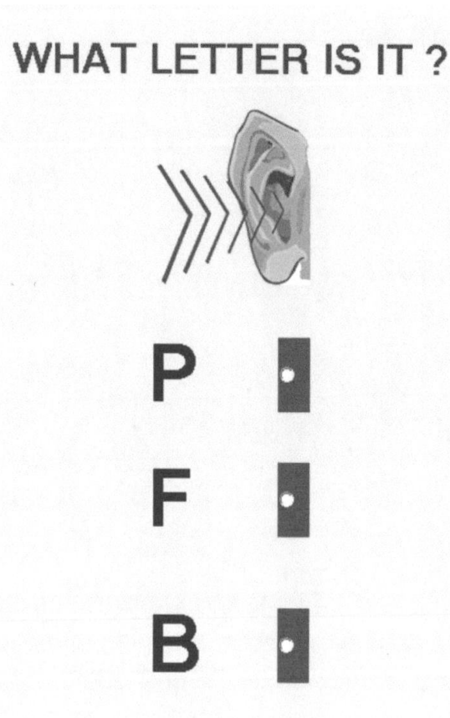

A knock represents a filled up position and a double knock represents an empty position. The clamping of the right hand represents a filled up position and the clamping of the left hand represents an empty position. The audio representation is corresponding to letter P. The ***kinesthetic*** representation for the letter P is clamping of left hand four times and right hand.

EXERCISE 16.2 WHAT LETTERS ARE THESE?

In exercise 16.2, in ***visual*** display of alphabet we see two pictures of boxes. In the bottom we see letters which are probable answers. The first picture has the first, second, third, and fourth positions filled with symbols and the second picture has the fifth position filled with a symbol. The **audio** representation for the first picture is **knock, knock, knock, and knock.** The **audio** representation for the second picture is **double knock, double knock, double knock, double knock, and knock.**

From the probable answers we need to find the right one. The audio signals are:

1.) **knock, knock, knock, knock, chick, double knock, double knock, double knock, double knock, and knock.**
2.) **knock, chick, double knock, and knock.**

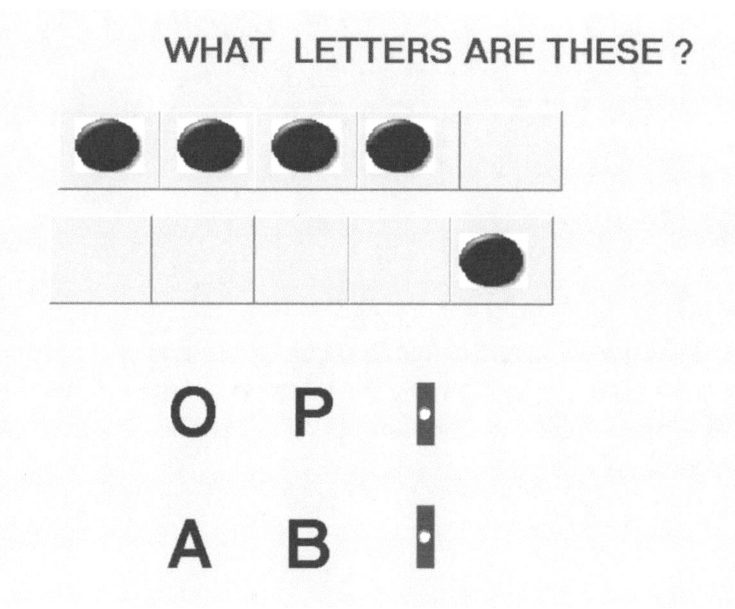

A knock represents a filled up position and a double knock represents an empty position. The clamping of the right hand represents a filled up position and the clamping of the left hand represents an empty position. The audio representation is corresponding to the letter O and letter P. The ***kinesthetic*** representation is clamping of the right hand four times, one hand shaking, left hand four times, and right hand.

INTRODUCTION TO ALPHABET

FIGURE 17

In figure 9, we have ***visual*** display of a picture of boxes. The picture has the first and fifth positions filled up with symbols. The decimal numbers 17 and letter Q correspond to the picture. The letter Q has the 17th position in the English alphabet.

The ***audio*** representation for the picture, number 17 and letter Q, is **knock, double knock, double knock, double knock, and knock.**

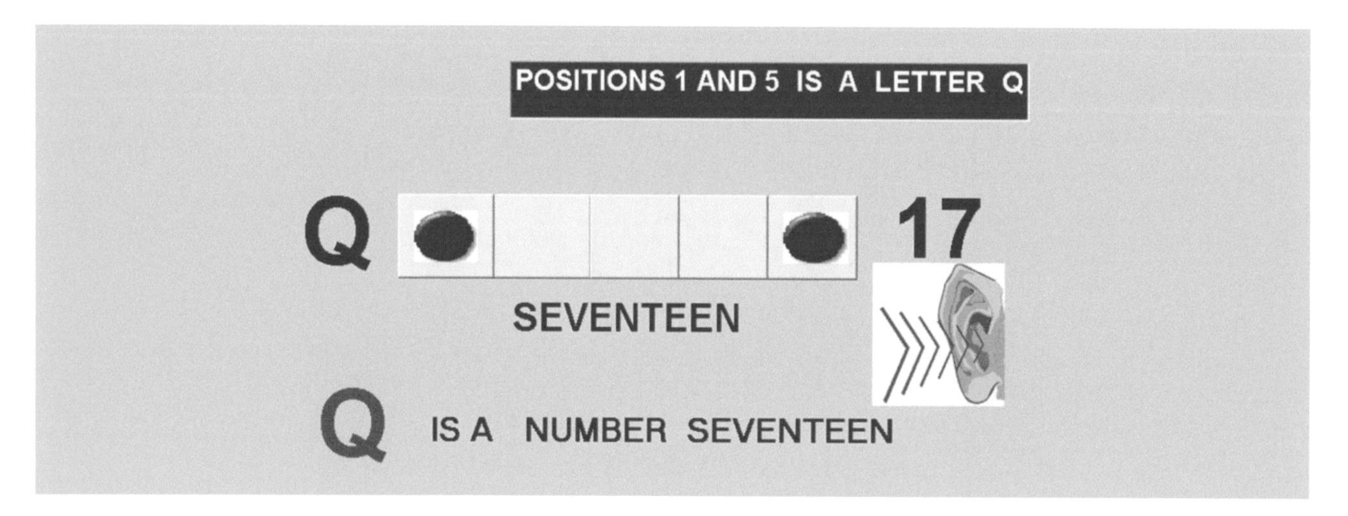

A knock represents a filled up position and a double knock represents an empty position. The clamping of the right hand represents a filled up position and the clamping of the left hand represents an empty position. The ***kinesthetic*** representation for the number 17 and letter Q is clamping of the right hand, left hand thrice, and right hand.

EXERCISE 17.1 WHAT LETTER IS IT?

In exercise 17.1, we have ***visual*** display of the letters F, C, and Q. We have to select whether the audio representation is corresponding to the letter F, C, or Q.

The ***audio*** representation is **knock, double knock, double knock, double knock, and knock.**

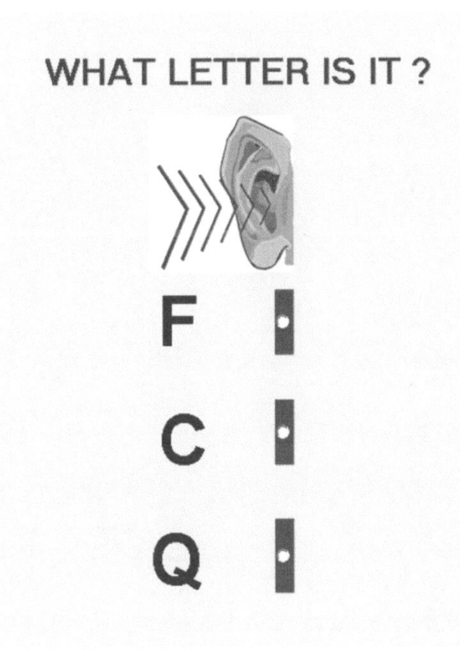

A knock represents a filled up position and a double knock represents an empty position. The clamping of the right hand represents a filled up position and the clamping of the left hand represents an empty position. The audio representation is corresponding to letter Q. The ***kinesthetic*** representation for the letter P is clamping of right hand, left hand thrice, and right hand.

INTRODUCTION TO ALPHABET

FIGURE 18

In figure 17, we have ***visual*** display of a picture of boxes. The picture has the second and fifth positions filled up with symbols. The decimal number 18 and letter R correspond to the picture. The letter R has the 18th position in the English alphabet.

The ***audio*** representation for the picture, number 18 and letter R, is **double knock, knock, double knock, double knock, and knock.**

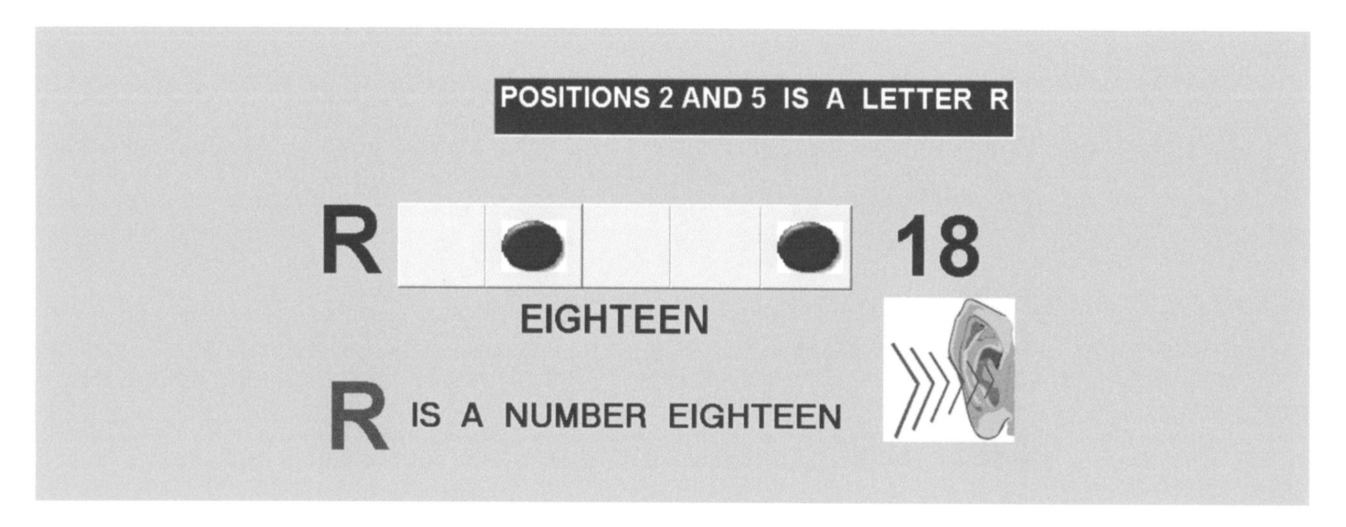

A knock represents a filled up position and a double knock represents an empty position. The clamping of the right hand represents a filled up position and the clamping of the left hand represents an empty position. The ***kinesthetic*** representation for the number 18and letter R is clamping of left hand, right hand, left hand twice, and right hand.

EXERCISE 18.1 WHAT LETTER IS IT?

In exercise 18.1, we have ***visual*** display of the letters R, D, and B. We have to select whether the audio representation is corresponding to the letter R, D, or B.

The ***audio*** representation is **double knock, knock, double knock, double knock, and knock.**

WHAT LETTER IS IT ?

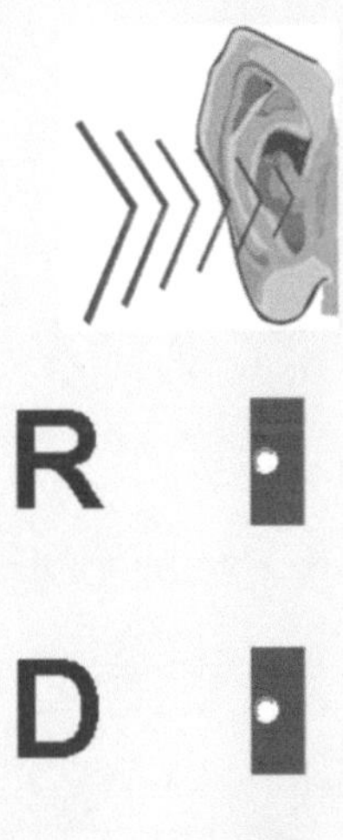

A knock represents a filled up position and a double knock represents an empty position. The clamping of the right hand represents a filled up position and the clamping of the left hand represents an empty position. The audio representation is corresponding to letter R. The ***kinesthetic*** representation for the letter R is clamping of left hand, right hand, left hand twice, and right hand.

WHAT LETTERS ARE THESE? EXERCISE 18.2

In exercise 18.2, in ***visual*** display of alphabet we see two pictures of boxes. In the bottom we see letters which are probable answers. The first picture has the first and fifth positions filled with symbols and the second picture has the second and fifth positions filled with symbols. The **audio** representation for the first picture is **knock, double knock, double knock, double knock, and knock.** The **audio** representation for the second picture is **double knock, knock, double knock, double knock, and knock.**

From the probable answers we need to find the right one. The audio signals are:

1.) **knock, knock, knock, knock, chick, double knock, double knock, double knock, double knock, and knock.**
2.) **knock, chick, double knock, and knock.**

A knock represents a filled up position and a double knock represents an empty position. The clamping of the right hand represents a filled up position and the clamping of the left hand represents an empty position. The audio representation is corresponding to the letter O and letter P. The ***kinesthetic*** representation is clamping of the right hand, left hand thrice, right hand, one hand shaking, left hand, right hand, left hand twice, and right hand.

FIGURE 19 INTRODUCTION TO ALPHABET

In figure 19, we have ***visual*** display of a picture of boxes. The picture has the first, second, and fifth positions filled up with symbols. The decimal numbers 19 and letter S correspond to the picture. The letter S has the 19th position in the English alphabet.

The ***audio*** representation for the picture, number 19 and letter S, is **knock, knock, double knock, double knock, and knock.**

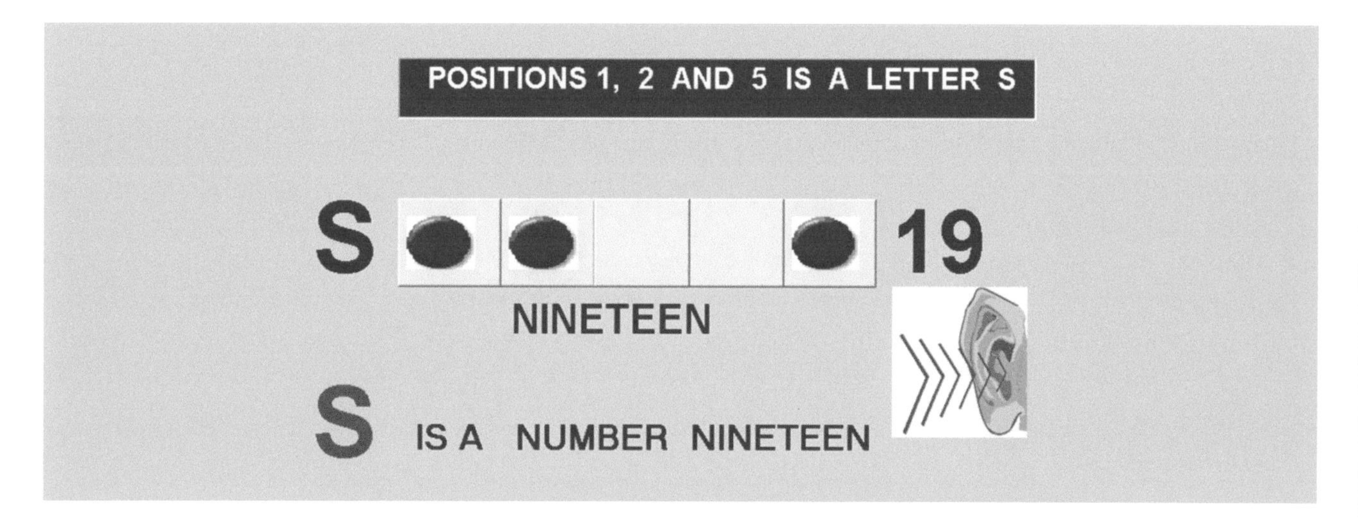

A knock represents a filled up position and a double knock represents an empty position. The clamping of the right hand represents a filled up position and the clamping of the left hand represents an empty position. The ***kinesthetic*** representation for the number 19 and letter S is clamping of the right hand twice, left hand twice, and right hand.

WHAT LETTER IS IT? EXERCISE 19.1

In exercise 19.1, we have ***visual*** display of the letters S, F, and D. We have to select whether the audio representation is corresponding to the letter S, F, or D.

The ***audio*** representation is **knock, knock, double knock, double knock, and knock.**

A knock represents a filled up position and a double knock represents an empty position. The clamping of the right hand represents a filled up position and the clamping of the left hand represents an empty position. The audio representation is corresponding to letter S. The ***kinesthetic*** representation for the letter S is clamping of the right hand twice, left hand twice, and right hand.

FIGURE 20 INTRODUCTION TO ALPHABET

In figure 20, we have ***visual*** display of two pictures of boxes. The picture has the third and fifth positions filled up with symbols. The decimal number 20 and letter T correspond to the picture.

The ***audio*** representation for the picture, number 20 and letter T, is **double knock, double knock, knock, double knock, and knock.**

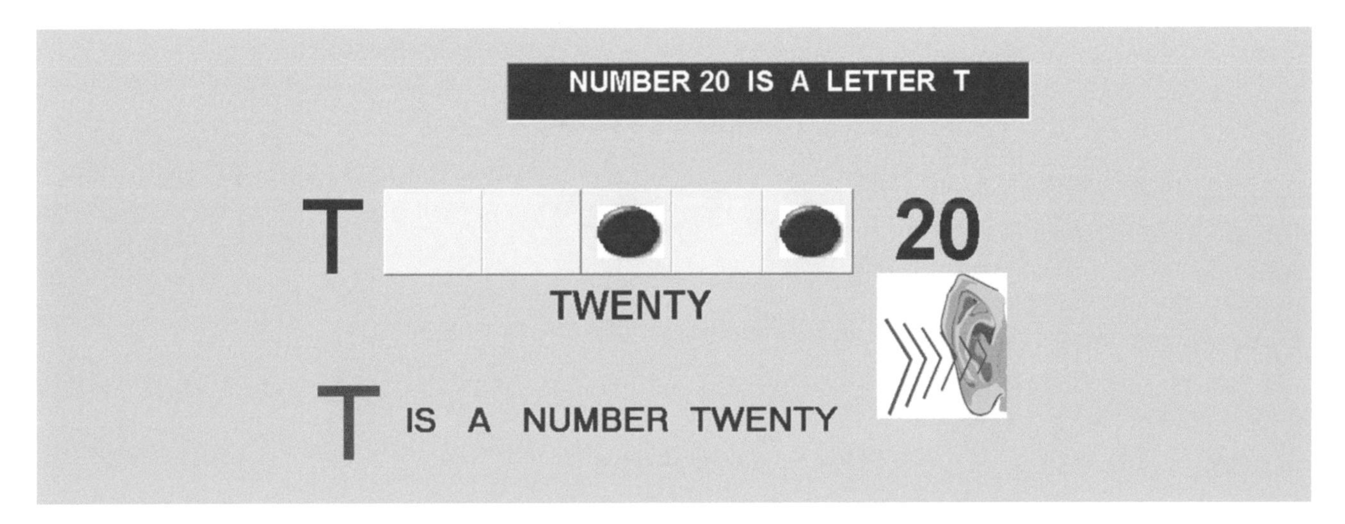

A knock represents a filled up position and a double knock represents an empty position. The clamping of the right hand represents a filled up position and the clamping of the left hand represents an empty position. The ***kinesthetic*** representation for the number 20 and letter T is clamping of left hand twice, right hand, left hand, and right hand.

WHAT LETTER IS IT? EXERCISE 20.1

In exercise 20.1, we have ***visual*** display of the letters L, T, and E. We have to select whether the audio representation is corresponding to the letter L, T, or E.

The ***audio*** representation is **double knock, double knock, knock, double knock, and knock.**

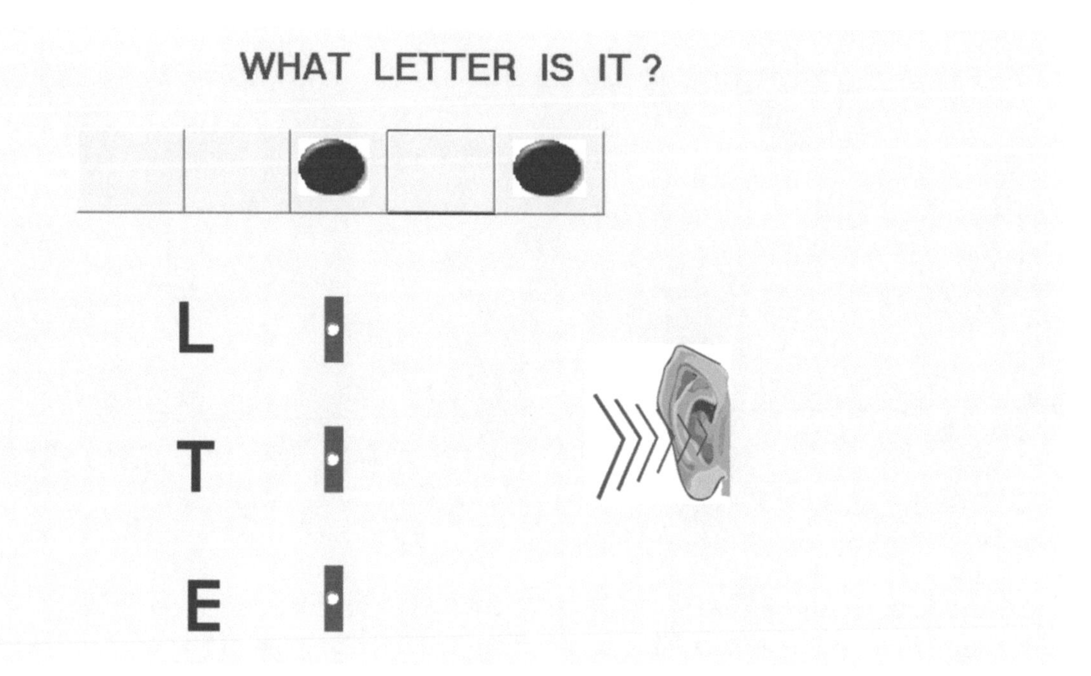

A knock represents a filled up position and a double knock represents an empty position. The clamping of the right hand represents a filled up position and the clamping of the left hand represents an empty position. The audio representation is corresponding to letter T. The ***kinesthetic*** representation for the letter T is clamping of left hand twice, right hand, left hand, and right hand.

EXERCISE 20.2 WHAT LETTERS ARE THESE?

In exercise 20.2, in ***visual*** display of alphabet we see two pictures of boxes. In the bottom we see letters which are probable answers. The first picture has the first, second, and fifth positions filled with symbols and the second picture has the third and fifth positions filled with symbols. The **audio** representation for the first picture is **knock, knock, double knock, double knock, and knock.** The **audio** representation for the second picture is **double knock, double knock, knock, double knock, and knock.**

From the probable answers we need to find the right one. The audio signals are:

1.) **knock, double knock, knock, knock, chick, double knock, knock, knock, and knock.**
2.) **knock, knock, double knock, double knock, knock, chick, double knock, double knock, knock, double knock, and knock.**

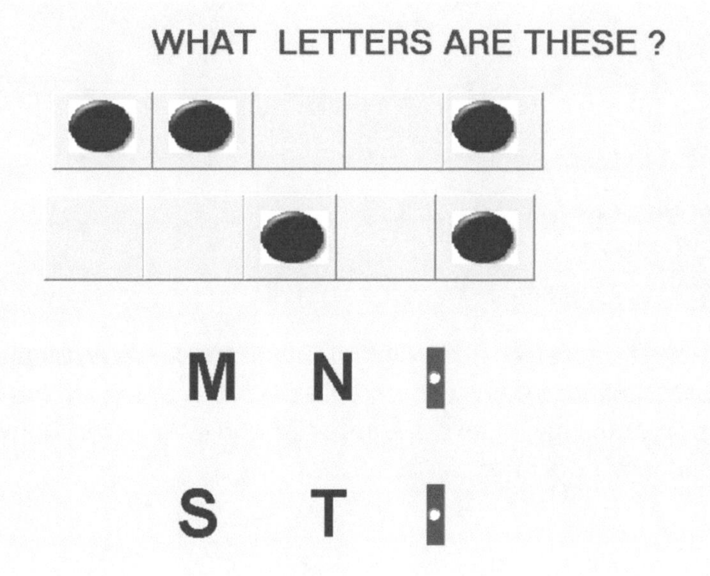

A knock represents a filled up position and a double knock represents an empty position. The clamping of the right hand represents a filled up position and the clamping of the left hand represents an empty position. The audio representation is corresponding to the letter S and letter T. The ***kinesthetic*** representation is clamping of the right hand twice, left hand twice, right hand, one hand shaking, left hand twice, right hand, left hand, and right hand.

INTRODUCTION TO ALPHABET

FIGURE 21

In figure 21, we have ***visual*** display of two pictures of boxes. The picture has the first, third, and fifth positions filled up with symbols. The decimal numbers 21 and letter U correspond to the picture. The letter U has the 21st position in the English alphabet.

The ***audio*** representation for the picture, number 21 and letter U, is **knock, double knock, knock, double knock, and knock.**

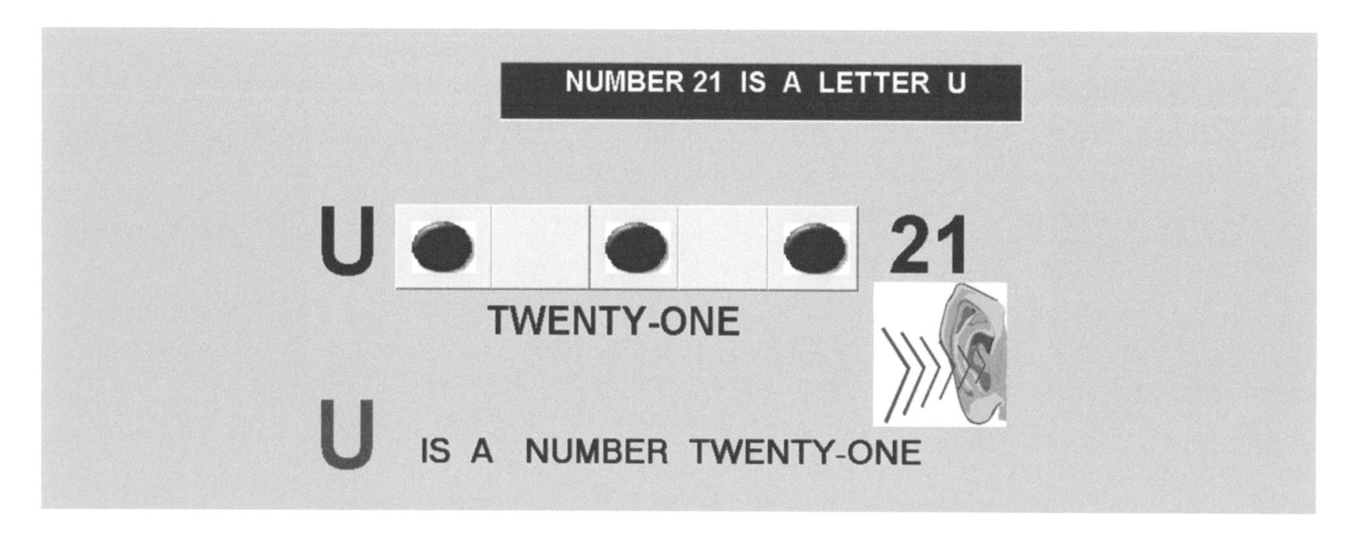

A knock represents a filled up position and a double knock represents an empty position. The clamping of the right hand represents a filled up position and the clamping of the left hand represents an empty position. The ***kinesthetic*** representation for the number 21 and letter U is clamping of the right hand, left hand, right hand, left hand, and right hand.

EXERCISE 21.1 WHAT LETTER IS IT?

In exercise 21.1, we have ***visual*** display of the letters A, U, and G. We have to select whether the audio representation is corresponding to the letter A, U, or G.

The ***audio*** representation is **knock, double knock, knock, double knock, and knock.**

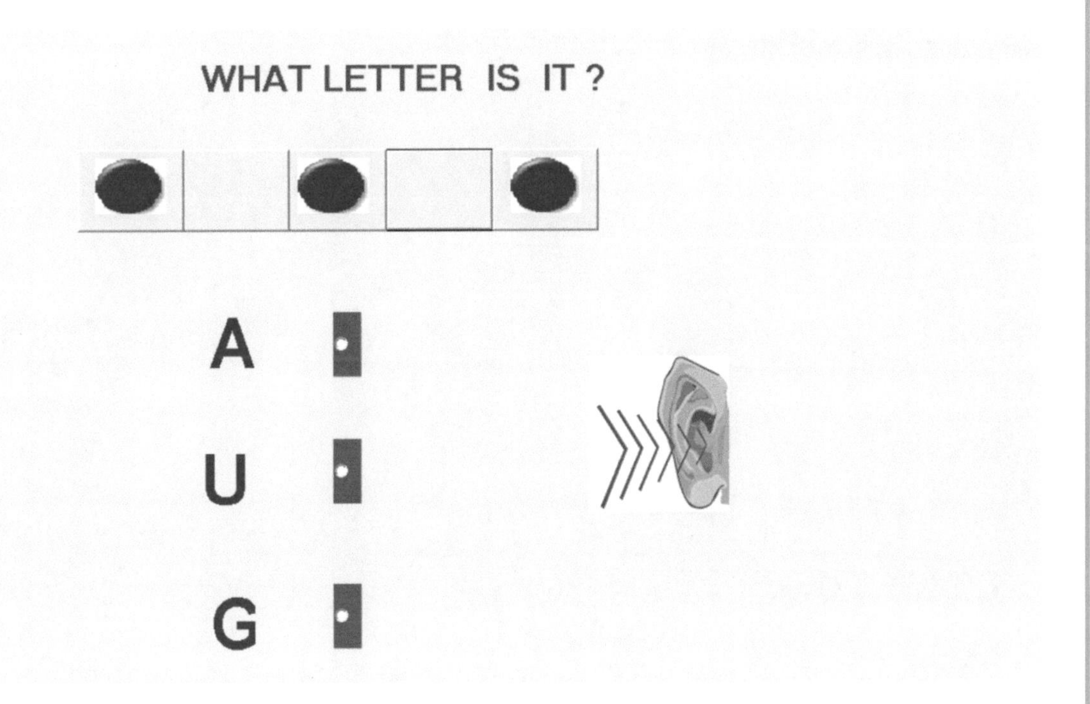

A knock represents a filled up position and a double knock represents an empty position. The clamping of the right hand represents a filled up position and the clamping of the left hand represents an empty position. The audio representation is corresponding to letter U. The ***kinesthetic*** representation for the letter U is clamping of the right hand, left hand, right hand, left hand, and right hand.

INTRODUCTION TO ALPHABET

FIGURE 22

In figure 22, we have ***visual*** display of a picture of boxes. The picture has the second, third, and fifth positions filled up with symbols. The decimal number 22 and letter V correspond to the picture. The letter V has the 22nd position in the English alphabet.

The ***audio*** representation for the picture, number 22 and letter V, is **double knock, knock, knock, double knock, and knock.**

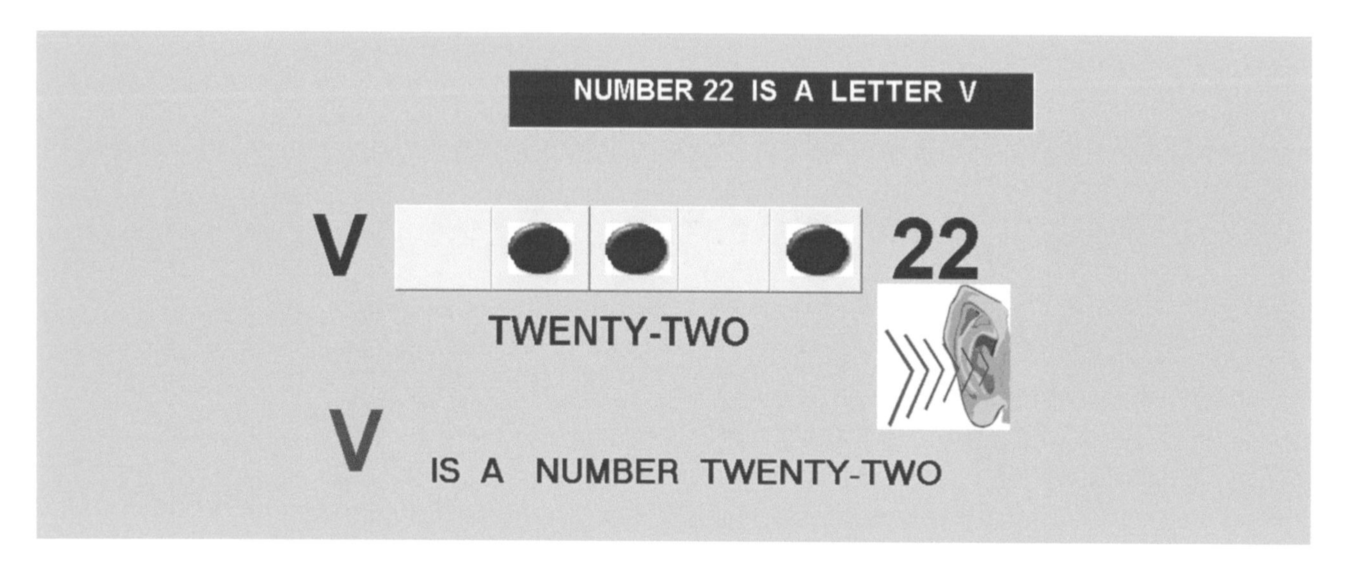

A knock represents a filled up position and a double knock represents an empty position. The clamping of the right hand represents a filled up position and the clamping of the left hand represents an empty position. The ***kinesthetic*** representation for the number 22 and letter V is clamping of left hand, right hand twice, left hand, and right hand.

EXERCISE 22.1 WHAT LETTER IS IT?

In exercise 22.1, we have ***visual*** display of the letters V, L, and G. We have to select whether the audio representation is corresponding to the letter V, L, or G.

The ***audio*** representation is **double knock, knock, knock, double knock, and knock.**

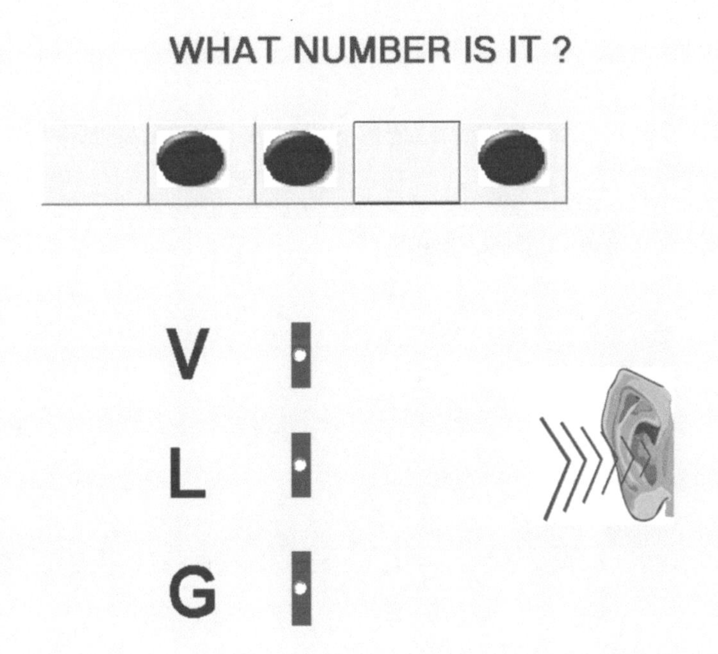

A knock represents a filled up position and a double knock represents an empty position. The clamping of the right hand represents a filled up position and the clamping of the left hand represents an empty position. The audio representation is corresponding to letter V. The ***kinesthetic*** representation for the letter V is clamping of left hand, right hand twice, left hand, and right hand.

WHAT LETTERS ARE THESE? EXERCISE 22.2

In exercise 22.2, in ***visual*** display of alphabet we see two pictures of boxes. In the bottom we see letters which are probable answers. The first picture has the first, third, and fifth positions filled with symbols and the second picture has the second, third, and fifth positions filled with symbols. The **audio** representation for the first picture is **knock, double knock, knock, double knock, and knock.** The **audio** representation for the second picture is **double knock, knock, knock, double knock, and knock.**

From the probable answers we need to find the right one. The audio signals are:

1.) **knock, double knock, knock, double knock, knock, chick, double knock, knock, knock, double knock, and knock.**
2.) **knock, double knock, double knock, double knock, knock, chick, double knock, knock, double knock, double knock, and knock.**

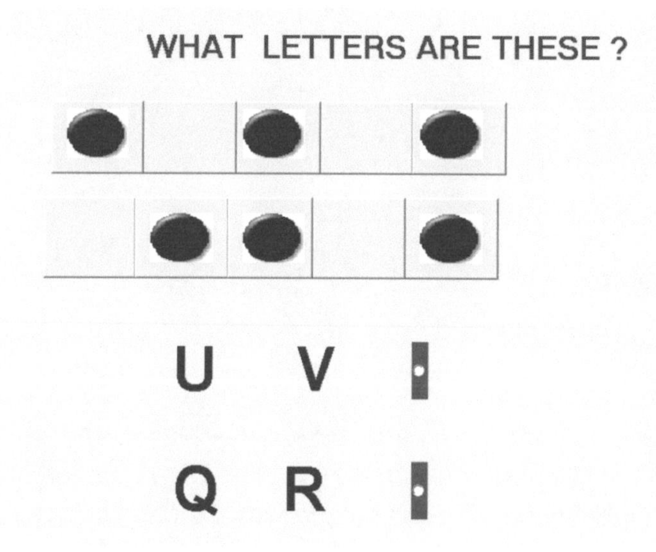

A knock represents a filled up position and a double knock represents an empty position. The clamping of the right hand represents a filled up position and the clamping of the left hand represents an empty position. The audio representation is corresponding to the letter U and letter V. The ***kinesthetic*** representation is clamping of the right hand, left hand, right hand, left hand, right hand, one hand shaking, left hand, right hand twice, left hand, and right hand.

FIGURE 23 INTRODUCTION TO ALPHABET

In figure 23, we have ***visual*** display of a picture of boxes. The picture has the first, second, third, and fifth positions filled up with symbols. The decimal numbers 23 and letter W correspond to the picture. The letter W has the 23rd position in the English alphabet.

The ***audio*** representation for the picture, number 23 and letter W, is **knock, knock, knock, double knock, and knock.**

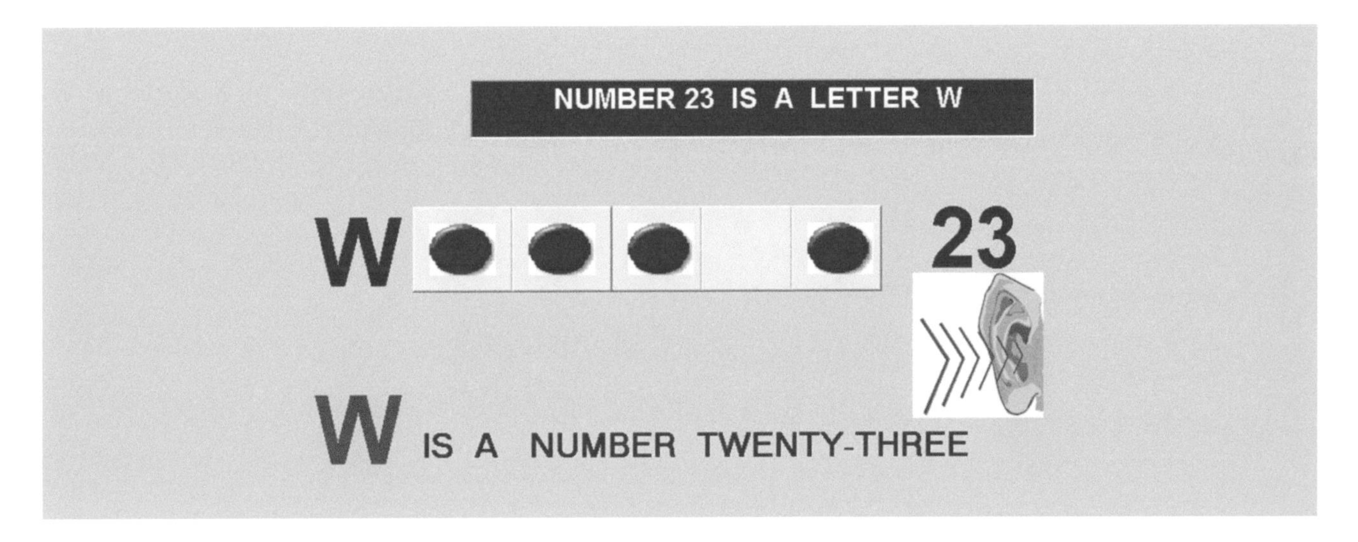

A knock represents a filled up position and a double knock represents an empty position. The clamping of the right hand represents a filled up position and the clamping of the left hand represents an empty position. The ***kinesthetic*** representation for the number 23 and letter W is clamping of the right hand thrice, left hand, and right hand.

WHAT LETTER IS IT?

EXERCISE 23.1

In exercise 23.1, we have ***visual*** display of the letters C, W, and F. We have to select whether the audio representation is corresponding to the letter C, W, or F.

The ***audio*** representation is **knock, knock, knock, double knock, and knock.**

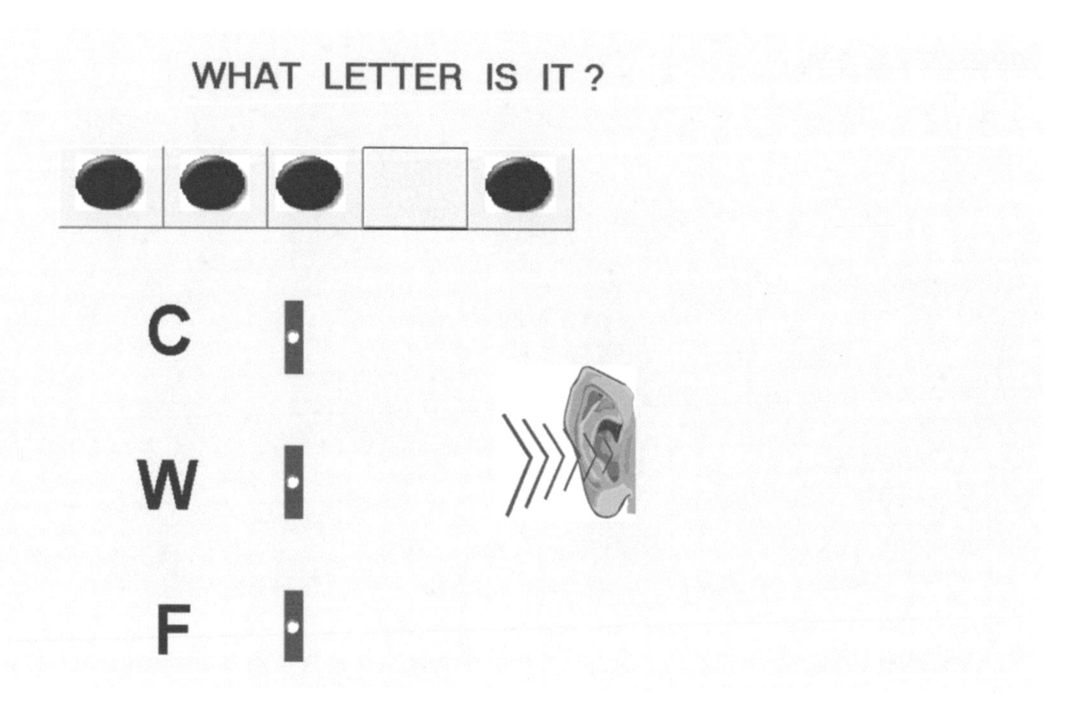

A knock represents a filled up position and a double knock represents an empty position. The clamping of the right hand represents a filled up position and the clamping of the left hand represents an empty position. The audio representation is corresponding to letter W. The ***kinesthetic*** representation for the letter W is clamping of the right hand thrice, left hand, and right hand.

FIGURE 24 INTRODUCTION TO ALPHABET

In figure 24, we have ***visual*** display of a picture of boxes. The picture has the fourth and fifth positions filled up with symbols. The decimal number 24 and letter X correspond to the picture. The letter X has the 24th position in the English alphabet.

The ***audio*** representation for the picture, number 24 and letter X, is **double knock, double knock, double knock, knock, and knock.**

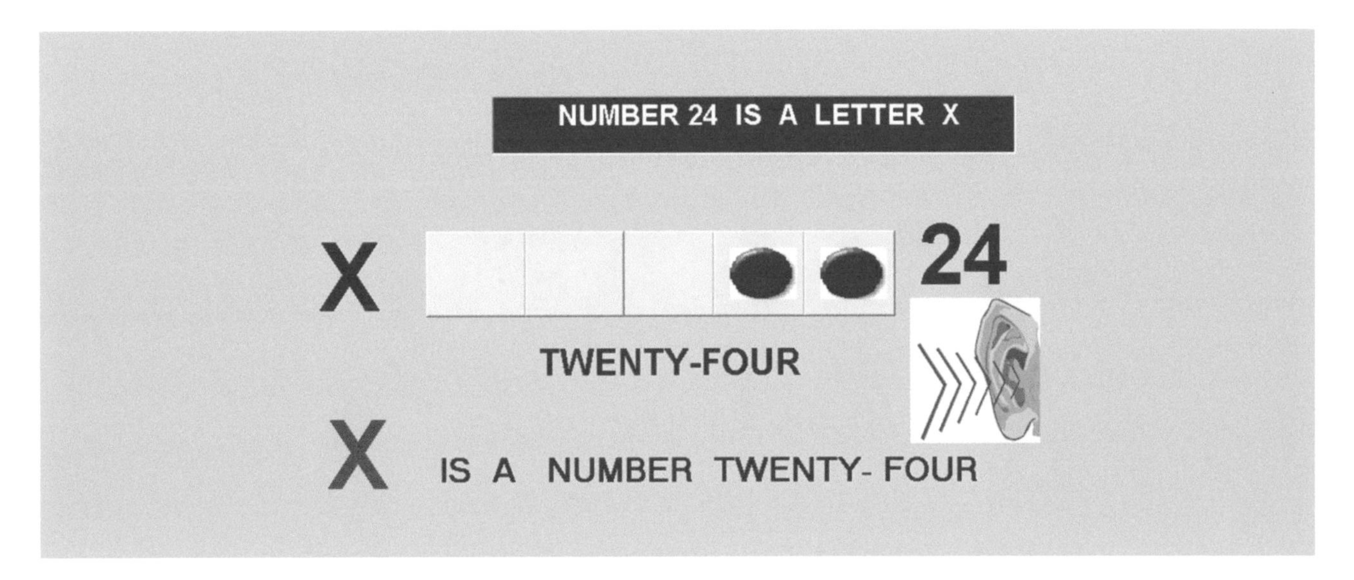

A knock represents a filled up position and a double knock represents an empty position. The clamping of the right hand represents a filled up position and the clamping of the left hand represents an empty position. The ***kinesthetic*** representation for the number 24 and letter X is clamping of left hand thrice and right hand twice.

WHAT LETTER IS IT?

EXERCISE 24.1

In exercise 24.1, we have ***visual*** display of the letters Q, X, and E. We have to select whether the audio representation is corresponding to the letter Q, X, or E.

The ***audio*** representation is **double knock, double knock, double knock, knock, and knock.**

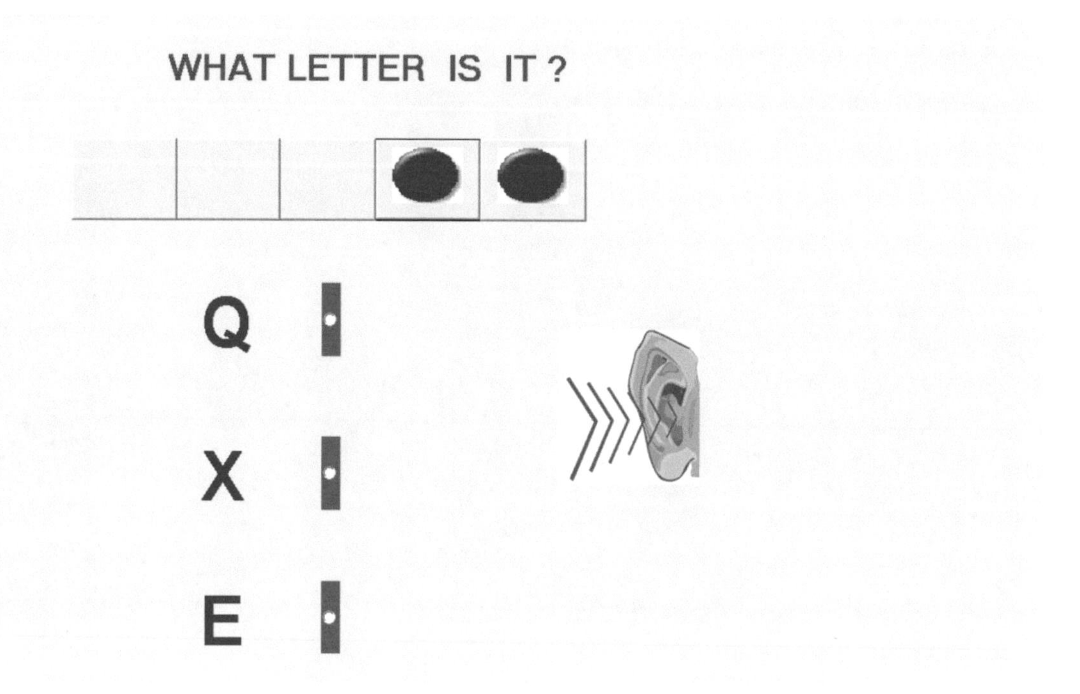

A knock represents a filled up position and a double knock represents an empty position. The clamping of the right hand represents a filled up position and the clamping of the left hand represents an empty position. The audio representation is corresponding to letter X. The ***kinesthetic*** representation for the letter X is clamping of left hand thrice and right hand twice.

EXERCISE 24.2 WHAT LETTERS ARE THESE?

In exercise 24.2, in ***visual*** display of alphabet we see two pictures of boxes. In the bottom we see letters which are probable answers. The first picture has the first, second, third, and fifth positions filled with symbols and the second picture has the fourth and fifth positions filled with symbols. The **audio** representation for the first picture is **knock, knock, knock, double knock, and knock.** The **audio** representation for the second picture is **double knock, double knock, double knock, knock, and knock.**

From the probable answers we need to find the right one. The audio signals are:

1.) **knock, knock, double knock, double knock, knock, chick, double knock, double knock, knock, double knock, and knock.**
2.) **knock, knock, knock, double knock, knock, chick, double knock, double knock, double knock, knock, and knock.**

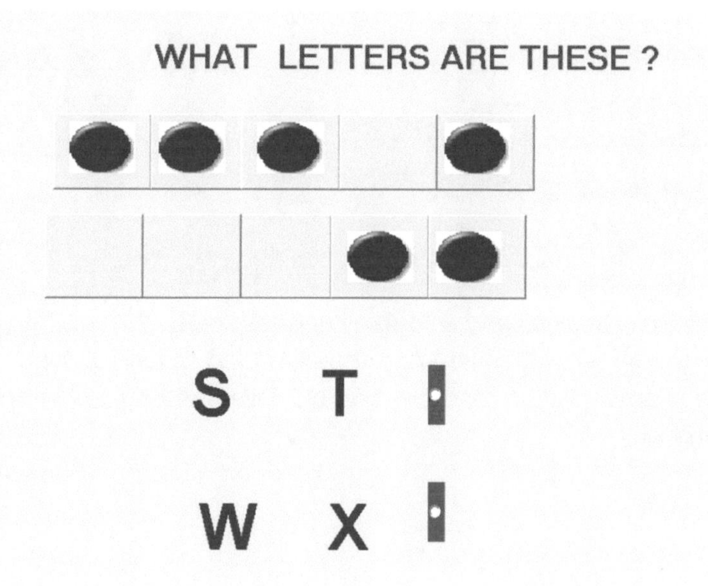

A knock represents a filled up position and a double knock represents an empty position. The clamping of the right hand represents a filled up position and the clamping of the left hand represents an empty position. The audio representation is corresponding to the letter W and letter X. The ***kinesthetic*** representation is clamping of the right hand thrice, left hand, right hand, one hand shaking, left hand thrice, and right hand twice.

INTRODUCTION TO ALPHABET

FIGURE 25

In figure 25, we have ***visual*** display of a picture of boxes. The picture has the first, fourth, and fifth positions filled up with symbols. The decimal numbers 25 and letter Y correspond to the picture. The letter Y has the 25th position in the English alphabet.

The ***audio*** representation for the picture, number 25 and letter Y, is **knock, double knock, double knock, knock, and knock.**

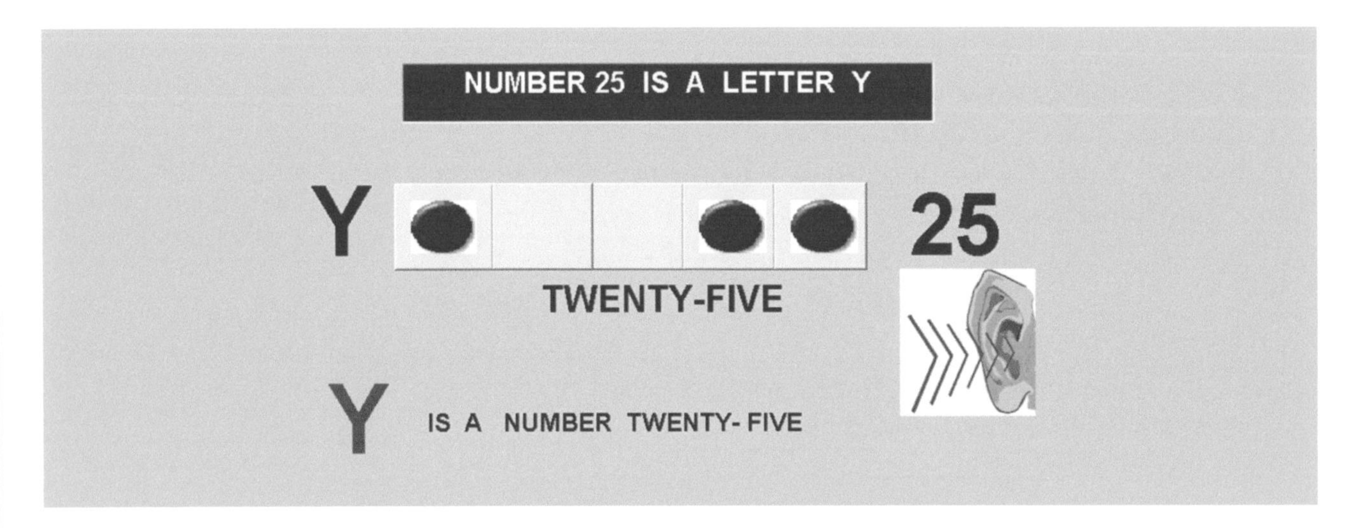

A knock represents a filled up position and a double knock represents an empty position. The clamping of the right hand represents a filled up position and the clamping of the left hand represents an empty position. The ***kinesthetic*** representation for the number 25 and letter Y is clamping of the right hand, left hand twice, and right hand.

EXERCISE 25.1 WHAT LETTER IS IT?

In exercise 25.1, we have ***visual*** display of the letters Y, J, and K. We have to select whether the audio representation is corresponding to the letter Y, J, or K.

The ***audio*** representation is **knock, double knock, double knock, knock, and knock.**

A knock represents a filled up position and a double knock represents an empty position. The clamping of the right hand represents a filled up position and the clamping of the left hand represents an empty position. The audio representation is corresponding to letter Y. The ***kinesthetic*** representation for the letter Y is clamping of the right hand, left hand twice, and right hand.

INTRODUCTION TO ALPHABET

FIGURE 26

In figure 26, we have ***visual*** display of two pictures of boxes. The picture has the second, fourth, and fifth positions filled up with symbols. The decimal number 26 and letter Z correspond to the picture. The letter Z has the 26th position in the English alphabet.

The ***audio*** representation for the picture, number 26 and letter Z, is **double knock, knock, double knock, knock, and knock.**

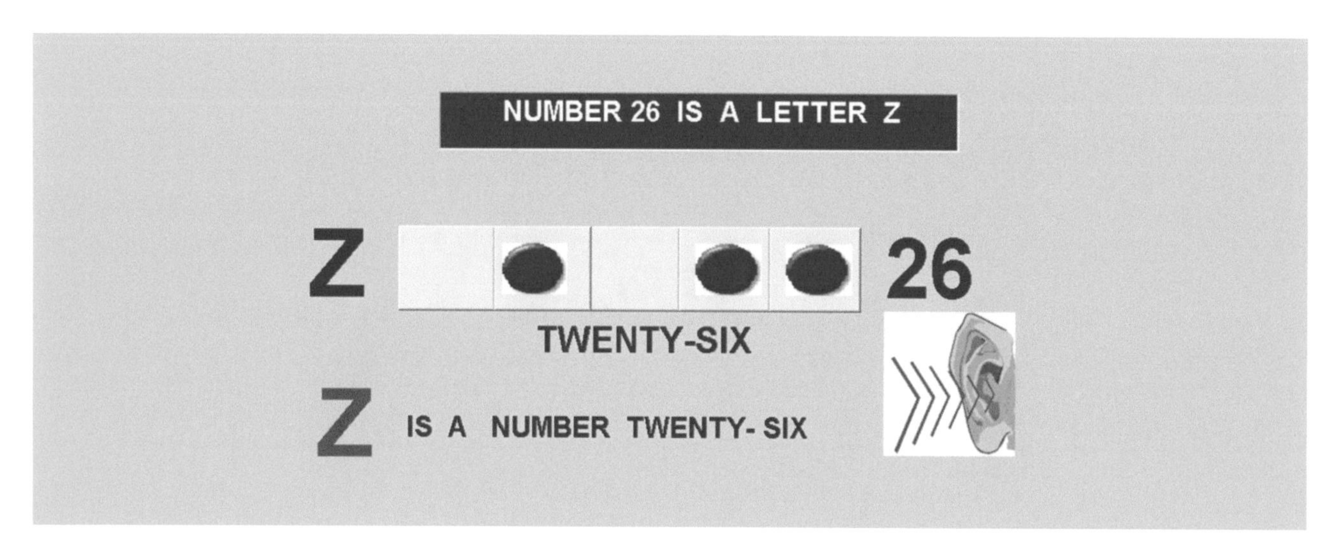

A knock represents a filled up position and a double knock represents an empty position. The clamping of the right hand represents a filled up position and the clamping of the left hand represents an empty position. The ***kinesthetic*** representation for the number 26 and letter Z is clamping of left hand, right hand, left hand, and right hand twice.

EXERCISE 26.1 WHAT LETTER IS IT?

In exercise 26.1, we have ***visual*** display of the letters H, Z, and F. We have to select whether the audio representation is corresponding to the letter H, Z, or F.

The ***audio*** representation is **double knock, knock, double knock, knock, and knock.**

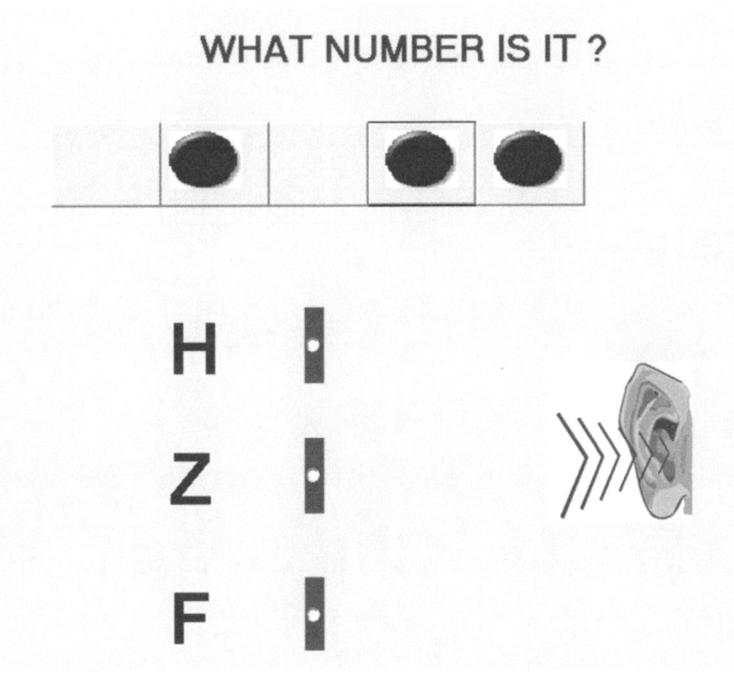

A knock represents a filled up position and a double knock represents an empty position. The clamping of the right hand represents a filled up position and the clamping of the left hand represents an empty position. The audio representation is corresponding to letter Z. The ***kinesthetic*** representation for the letter Z is clamping of left hand, right hand, left hand, and right hand twice.

WHAT LETTERS ARE THESE?

EXERCISE 26.2

In exercise 26.2, in ***visual*** display of alphabet we see two pictures of boxes. In the bottom we see letters which are probable answers. The first picture has the first, fourth, and fifth positions filled with symbols and the second picture has the second, fourth, and fifth positions filled with symbols. The **audio** representation for the first picture is **knock, double knock, double knock, knock, and knock.** The **audio** representation for the second picture is **double knock, knock, double knock, knock, and knock.**

From the probable answers we need to find the right one. The audio signals are:

1.) **knock, double knock, double knock, knock, knock, chick, double knock, knock, double knock, knock, and knock.**
2.) **knock, double knock, knock, double knock, knock, chick, double knock, knock, knock, double knock, and knock.**

A knock represents a filled up position and a double knock represents an empty position. The clamping of the right hand represents a filled up position and the clamping of the left hand represents an empty position. The audio representation is corresponding to the letter Y and letter Z. The ***kinesthetic*** representation is clamping of the right hand, left hand twice, right hand twice, one hand shaking, left hand, right hand, left hand, and right hand twice.

FIGURE 1 INTRODUCTION TO ALPHABET

In figure 1, we have ***visual*** display of two pictures of boxes. The left picture has the first position filled up with a symbol. The decimal numbers 1 and letter A correspond to the first picture.

The ***audio*** representation for the left picture, number 1 and letter A, is **knock and double knock.**

The second picture on the right has the second position filled up with a symbol. The decimal number 2 and letter B correspond to the second picture.

The ***audio*** representation for the right picture, number 2 and letter B, is **double knock and knock.**

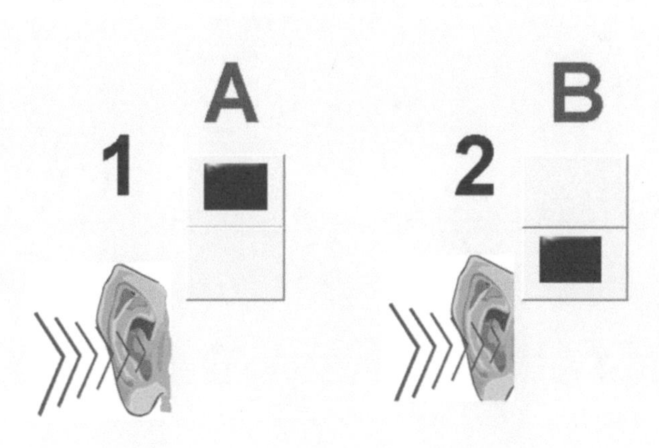

A knock represents a filled up position and a double knock represents an empty position. The clamping of the right hand represents a filled up position and the clamping of the left hand represents an empty position. The ***kinesthetic*** representation for the number 1 and letter A is clamping of the right hand and left hand. The ***kinesthetic*** representation for the number 2 and letter B is clamping of left hand and right hand.

INTRODUCTION TO ALPHABET

FIGURE 1

In figure 1, we have ***visual*** display of decimal numbers 1 and 2 on the left side, while on the right side we have alphabetical letters A and B which correspond to those numbers.

The ***audio*** representation for the number 1 and letter A is **knock.**

The ***audio*** representation for the number 2 and letter B is **double knock and knock.**

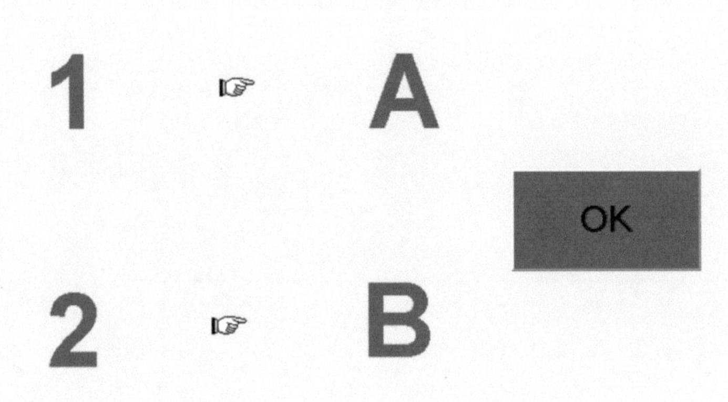

The knock represents a filled up position and a double knock represents an empty position. The clamping of the right hand represents a filled up position and the clamping of the left hand represents an empty position. The ***kinesthetic*** representation for the number 1 and letter A is clamping of the right hand. The ***kinesthetic*** representation for the number 2 and letter B is clamping of left hand and right hand.

EXAMPLE 1.1 WHAT LETTER IS IT?

In example 1.1, in visual display we see picture where the first position is empty while the second position is filled up with a symbol. We are given two choices of letters, A and B, which are supposed to correspond to the visual display.

The ***audio*** representation of the picture is **double knock and knock.**

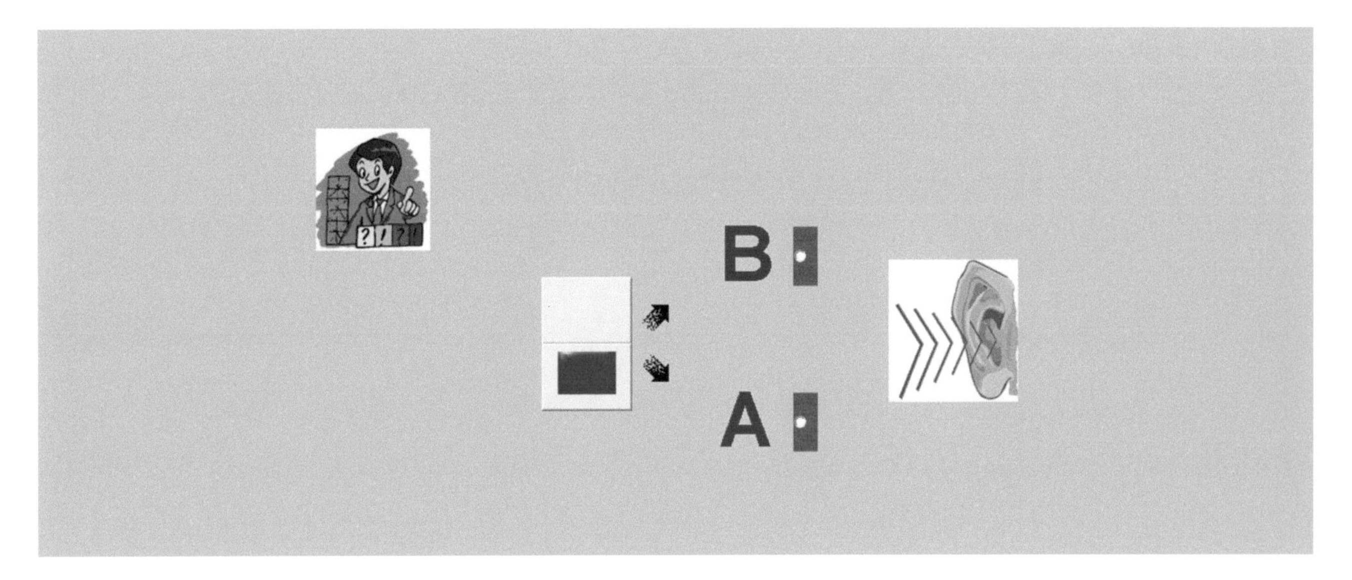

The knock represents a filled up position and a double knock represents an empty position. In our example double knock and knock are corresponding to the letter B. The clamping of the right hand represents a filled up position and the clamping of the left hand represents an empty position. In our example the ***kinesthetic*** representation for the answer is clamping of the left hand and then right hand. For letter A, the kinesthetic representation is clamping of right hand and then left hand.

WHAT LETTER IS IT? EXAMPLE 1

In example 1, in visual display we see picture where the first position is filled up with a symbol and the second position is empty. We are given two choices of letters, A and B, which are supposed to correspond to the visual display.

The ***audio*** representation of the picture is **knock and double knock.**

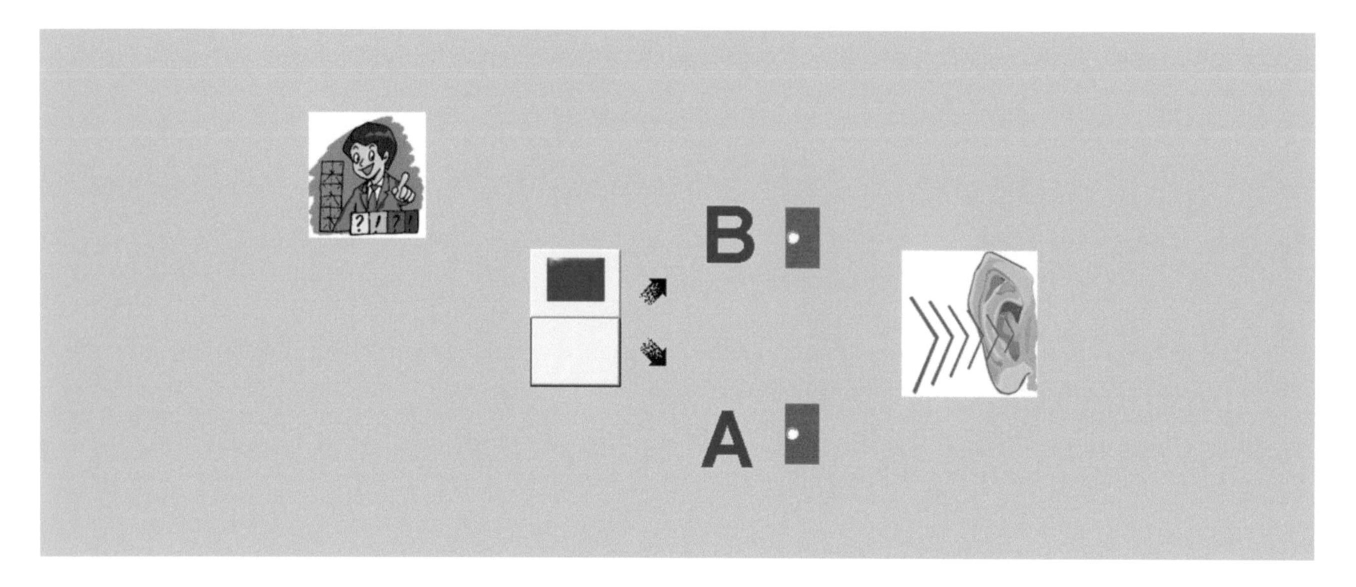

The knock represents a filled up position and a double knock represents an empty position. In our example knock and double knock are corresponding to the letter A. The clamping of the right hand represents a filled up position and the clamping of the left hand represents an empty position. In our example the ***kinesthetic*** representation for the answer is clamping of the right hand and then left hand. For letter B, the kinesthetic representation is clamping of left hand and then right hand.

FIGURE 2 INTRODUCTION TO ALPHABET

In figure 2, we have ***visual*** display of two pictures of boxes. The left picture has both positions filled up with symbols. The decimal numbers 3 and letter C correspond to the first picture.

The ***audio*** representation for the left picture, number 3 and letter C, is **knock and knock.**

The second picture on the right has the third position filled up with a symbol. The decimal number 4 and letter D correspond to the second picture.

The ***audio*** representation for the right picture, number 4 and letter D, is **double knock, double knock, and knock.**

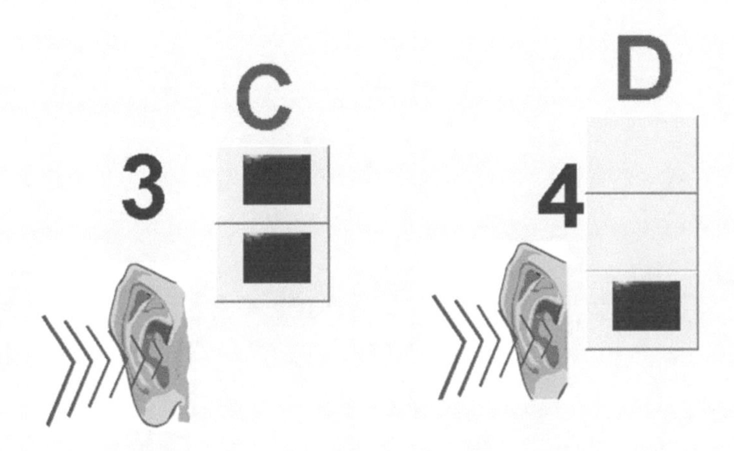

A knock represents a filled up position and a double knock represents an empty position. The clamping of the right hand represents a filled up position and the clamping of the left hand represents an empty position. The ***kinesthetic*** representation for the number 3 and letter C is clamping of the right hand twice. The ***kinesthetic*** representation for the number 4 and letter D is clamping of left hand twice and right hand.

INTRODUCTION TO ALPHABET

FIGURE 2

In figure 2, we have ***visual*** display of decimal numbers 3 and 4 on the left side, while on the right side we have alphabetical letters C and D which correspond to those numbers.

The ***audio*** representation for the number 3 and letter C is **knock and knock.**

The ***audio*** representation for the number 4 and letter D is **double knock, double knock, and knock.**

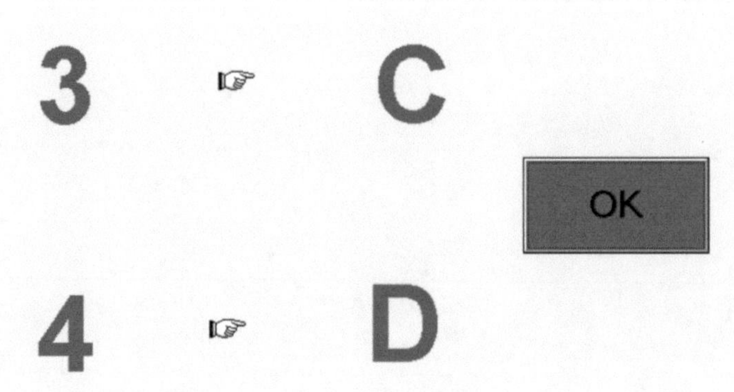

The knock represents a filled up position and a double knock represents an empty position. The clamping of the right hand represents a filled up position and the clamping of the left hand represents an empty position. The ***kinesthetic*** representation for the number 3 and letter C is clamping of the right hand twice. The ***kinesthetic*** representation for the number 4 and letter D is clamping of left hand twice and right hand.

EXAMPLE 2.1 WHAT LETTER IS IT?

In example 2.1, in visual display we see picture where both positions are filled up with a symbol. We are given two choices of letters, A and C, which are supposed to correspond to the visual display.

The ***audio*** representation of the picture is **knock and knock.**

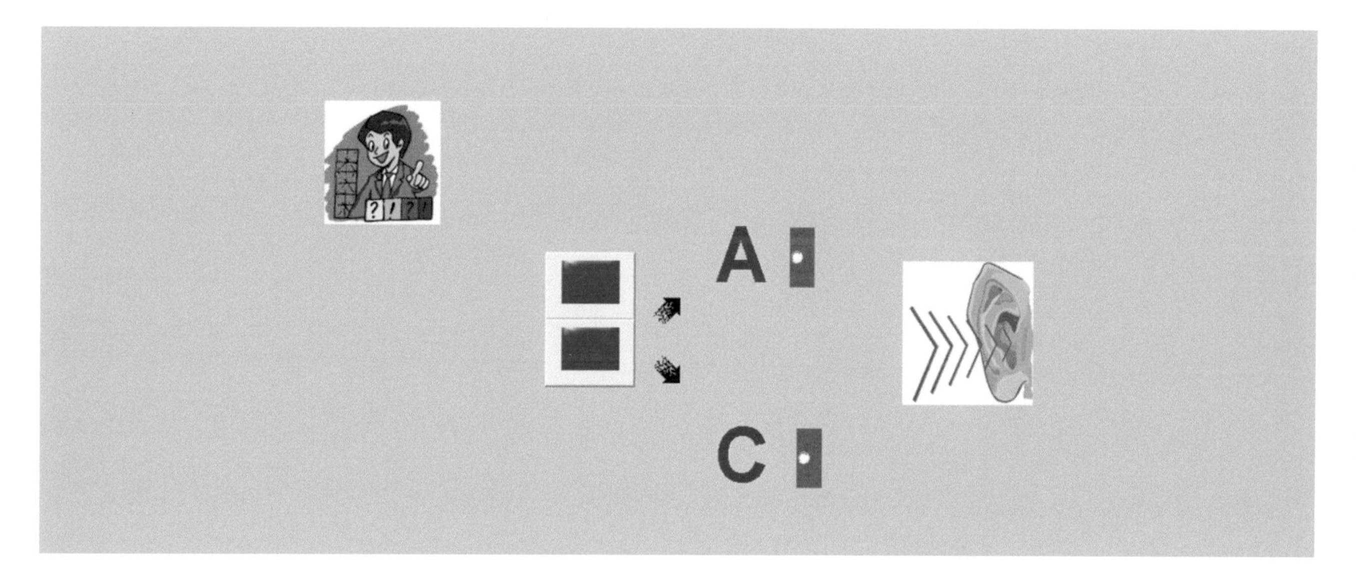

The knock represents a filled up position and a double knock represents an empty position. In our example knock and knock are corresponding to the letter C. The clamping of the right hand represents a filled up position and the clamping of the left hand represents an empty position. In our example the ***kinesthetic*** representation for the answer is clamping of the right hand and twice. For letter A, the kinesthetic representation is clamping of right hand and then left hand.

WHAT LETTER IS IT? EXAMPLE 2.2

In example 2.2, in visual display we see picture where the first two positions are empty while the last position is filled up with a symbol. We are given two choices of letters, B and D, which are supposed to correspond to the visual display.

The ***audio*** representation of the picture is **double knock, double knock, and knock.**

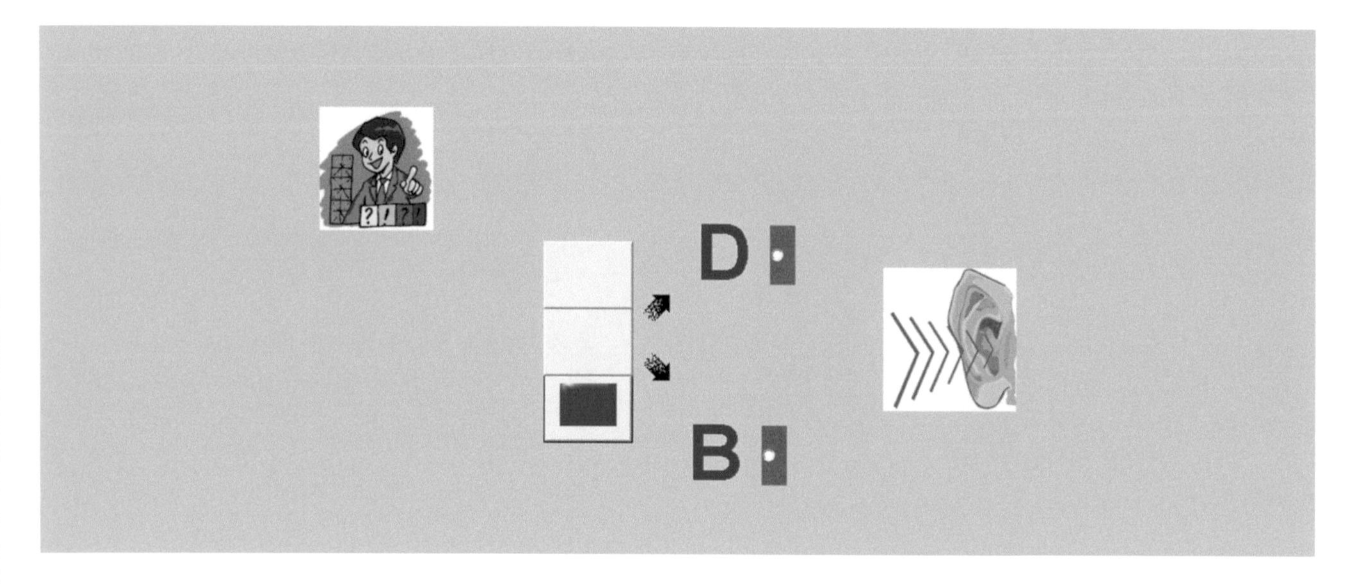

The knock represents a filled up position and a double knock represents an empty position. In our example double knock, double knock, and knock are corresponding to the letter D. The clamping of the right hand represents a filled up position and the clamping of the left hand represents an empty position. In our example the ***kinesthetic*** representation for the answer is clamping of the left hand twice and the right hand. For letter B, the kinesthetic representation is clamping of left hand and then right hand.

FIGURE 3 INTRODUCTION TO ALPHABET

In figure 3, we have ***visual*** display of two pictures of boxes. The left picture has the first and third positions filled up with symbols. The decimal numbers 5 and letter E correspond to the first picture.

The ***audio*** representation for the left picture, number 5 and letter E, is **knock, double knock, and knock.**

The second picture on the right has the second and third position filled up with a symbol. The decimal number 6 and letter F correspond to the second picture.

The ***audio*** representation for the right picture, number 6 and letter F, is **double knock, knock, and knock.**

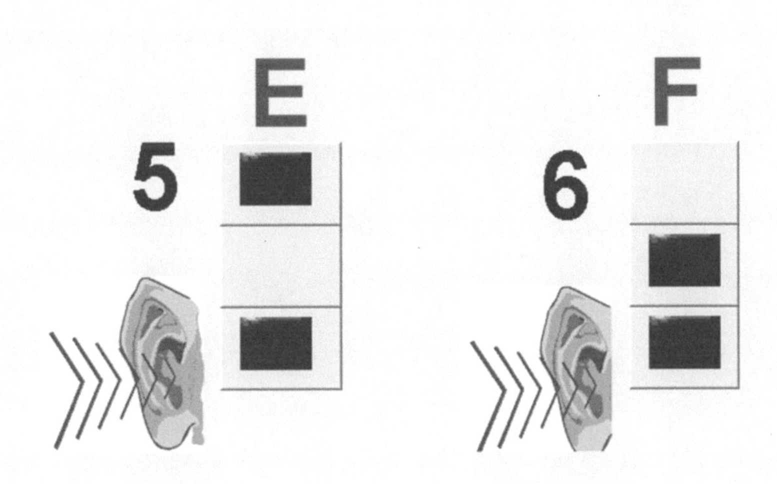

A knock represents a filled up position and a double knock represents an empty position. The clamping of the right hand represents a filled up position and the clamping of the left hand represents an empty position. The ***kinesthetic*** representation for the number 5 and letter E is clamping of the right hand, left hand, and right hand. The ***kinesthetic*** representation for the number 6 and letter F is clamping of left hand and right hand twice.

INTRODUCTION TO ALPHABET

FIGURE 3

In figure 3, we have ***visual*** display of decimal numbers 5 and 6 on the left side, while on the right side we have alphabetical letters E and F which correspond to those numbers.

The ***audio*** representation for the number 5 and letter E is **knock, double knock, and knock.**

The ***audio*** representation for the number 6 and letter F is **double knock, knock, and knock.**

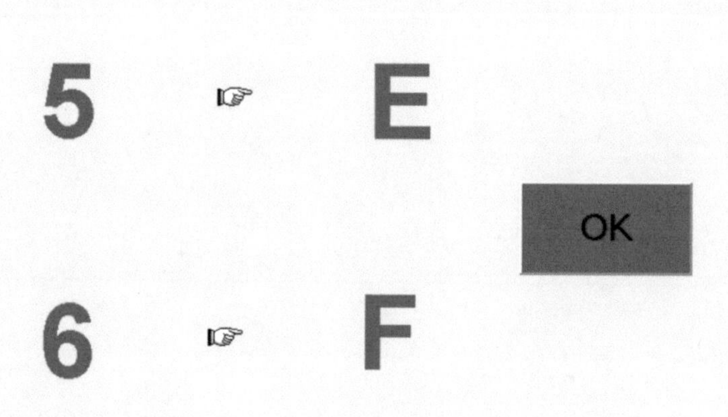

The knock represents a filled up position and a double knock represents an empty position. The clamping of the right hand represents a filled up position and the clamping of the left hand represents an empty position. The ***kinesthetic*** representation for the number 5 and letter E is clamping of the right hand, left hand, and right hand. The ***kinesthetic*** representation for the number 6 and letter F is clamping of left hand and right hand twice.

EXAMPLE 3.1 WHAT LETTER IS IT?

In example 3.1, in visual display we see a picture where the first position is filled up with a symbol, the second position is empty, and the last position is filled up with a symbol. We are given two choices of letters, B and E, which are supposed to correspond to the visual display.

The ***audio*** representation of the picture is **knock, double knock, and knock.**

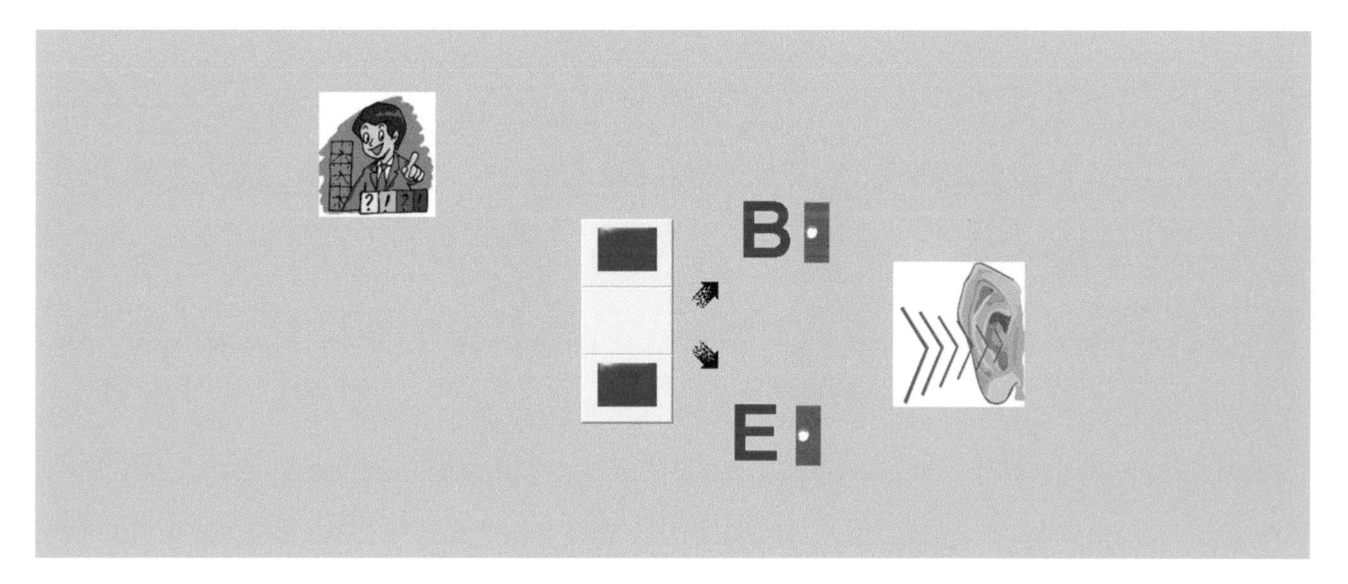

The knock represents a filled up position and a double knock represents an empty position. In our example knock, double knock, and knock are corresponding to the letter E. The clamping of the right hand represents a filled up position and the clamping of the left hand represents an empty position. In our example the ***kinesthetic*** representation for the answer is clamping of the right hand, left hand, and the right hand. For letter B, the kinesthetic representation is clamping of left hand and then right hand.

WHAT LETTER IS IT?

EXAMPLE 3.3

In example 3.3, in visual display we see a picture where the first position is empty while the last two positions are filled up with a symbol. We are given two choices of letters, C and F, which are supposed to correspond to the visual display.

The ***audio*** representation of the picture is **double knock, knock, and knock.**

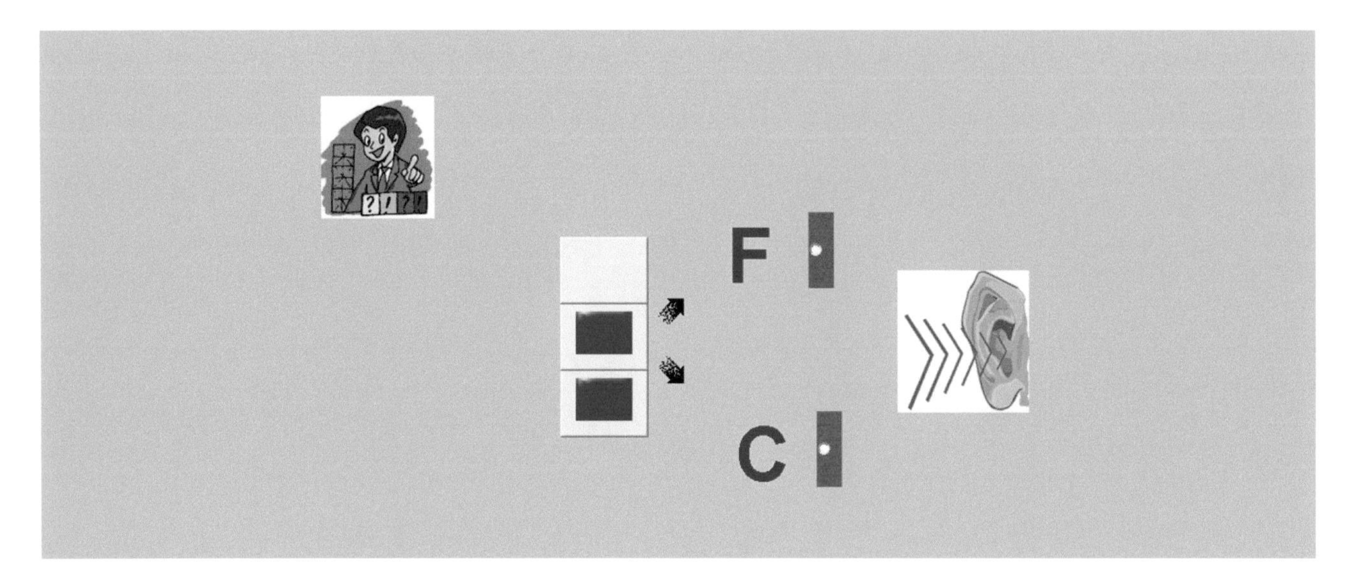

The knock represents a filled up position and a double knock represents an empty position. In our example double knock, knock, and knock are corresponding to the letter F. The clamping of the right hand represents a filled up position and the clamping of the left hand represents an empty position. In our example the ***kinesthetic*** representation for the answer is clamping of the left hand and right hand twice. For letter C, the kinesthetic representation is clamping of left hand and right hand twice.

FIGURE 4 INTRODUCTION TO ALPHABET

In figure 4, we have ***visual*** display of decimal numbers 7 and 8 on the left side, while on the right side we have alphabetical letters G and H which correspond to those numbers.

The ***audio*** representation for the number 7 and letter G is **knock, knock, and knock.**

The ***audio*** representation for the number 8 and letter H is **double knock, double knock, double knock, and knock.**

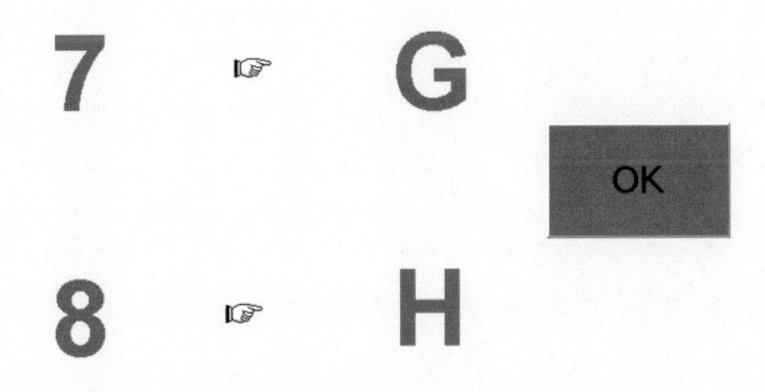

The knock represents a filled up position and a double knock represents an empty position. The clamping of the right hand represents a filled up position and the clamping of the left hand represents an empty position. The ***kinesthetic*** representation for the number 7 and letter G is clamping of the right hand thrice. The ***kinesthetic*** representation for the number 8 and letter H is clamping of left hand thrice and right hand.

INTRODUCTION TO ALPHABET

FIGURE 4

In figure 4, we have ***visual*** display of two pictures of boxes. The left picture has all three positions filled up with symbols. The decimal numbers 7 and letter G correspond to the first picture.

The ***audio*** representation for the left picture, number 7 and letter G, is **knock, knock, and knock.**

The second picture on the right has the fourth position filled up with a symbol. The decimal number 8 and letter H correspond to the second picture.

The ***audio*** representation for the right picture, number 8 and letter H, is **double knock, double knock, double knock, and knock.**

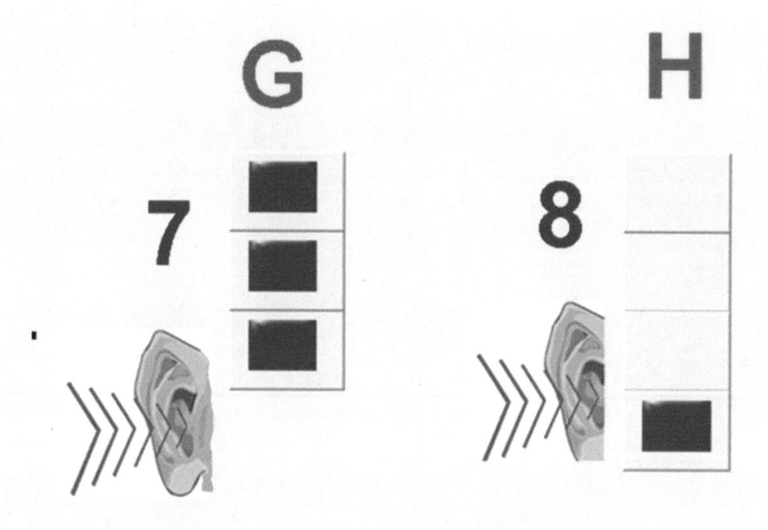

A knock represents a filled up position and a double knock represents an empty position. The clamping of the right hand represents a filled up position and the clamping of the left hand represents an empty position. The ***kinesthetic*** representation for the number 7 and letter G is clamping of the right hand thrice. The ***kinesthetic*** representation for the number 8 and letter H is clamping of left hand thrice and right hand.

EXAMPLE 4.1 WHAT LETTER IS IT?

In example 4.1, in visual display we see picture where all three positions are filled up with a symbol. We are given two choices of letters, F and G, which are supposed to correspond to the visual display.

The ***audio*** representation of the picture is **knock, knock, and knock.**

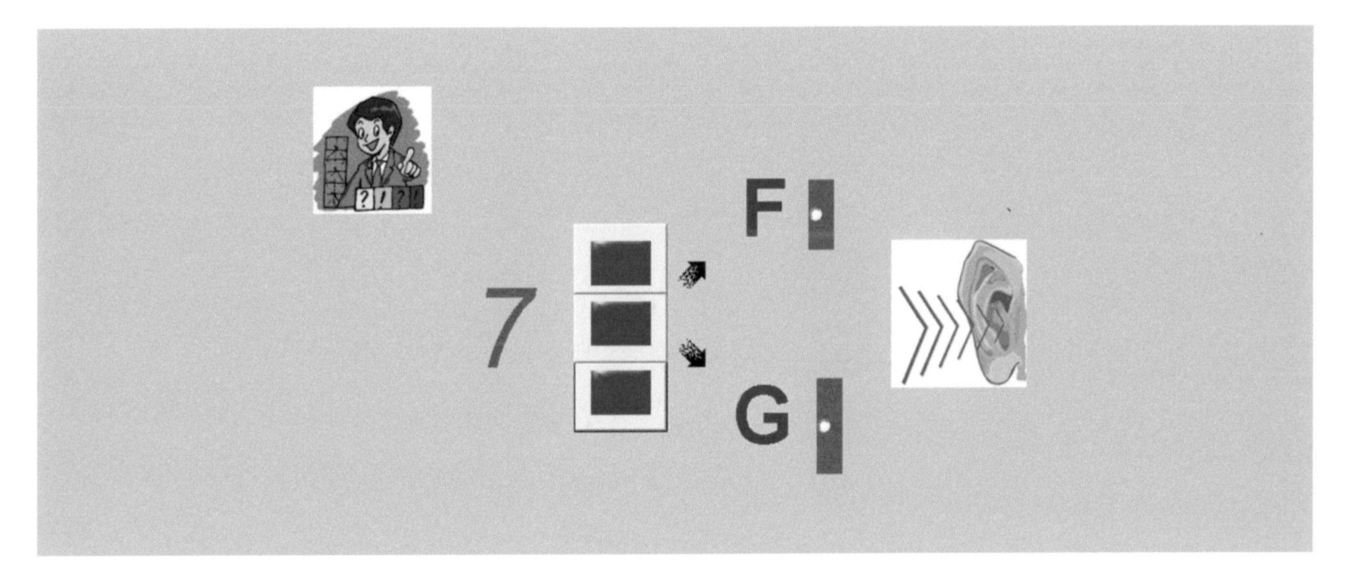

The knock represents a filled up position and a double knock represents an empty position. In our example knock and double knock are corresponding to the letter G. The clamping of the right hand represents a filled up position and the clamping of the left hand represents an empty position. In our example the ***kinesthetic*** representation for the answer is clamping of the right hand thrice. For letter F, the kinesthetic representation is clamping of left hand and right hand twice.

WHAT LETTER IS IT? EXAMPLE 4.2

In example 4.2, in visual display we see picture where the first three positions are empty while the last position is filled up with a symbol. We are given two choices of letters, B and H, which are supposed to correspond to the visual display.

The ***audio*** representation of the picture is **double knock, double knock, double knock, and knock.**

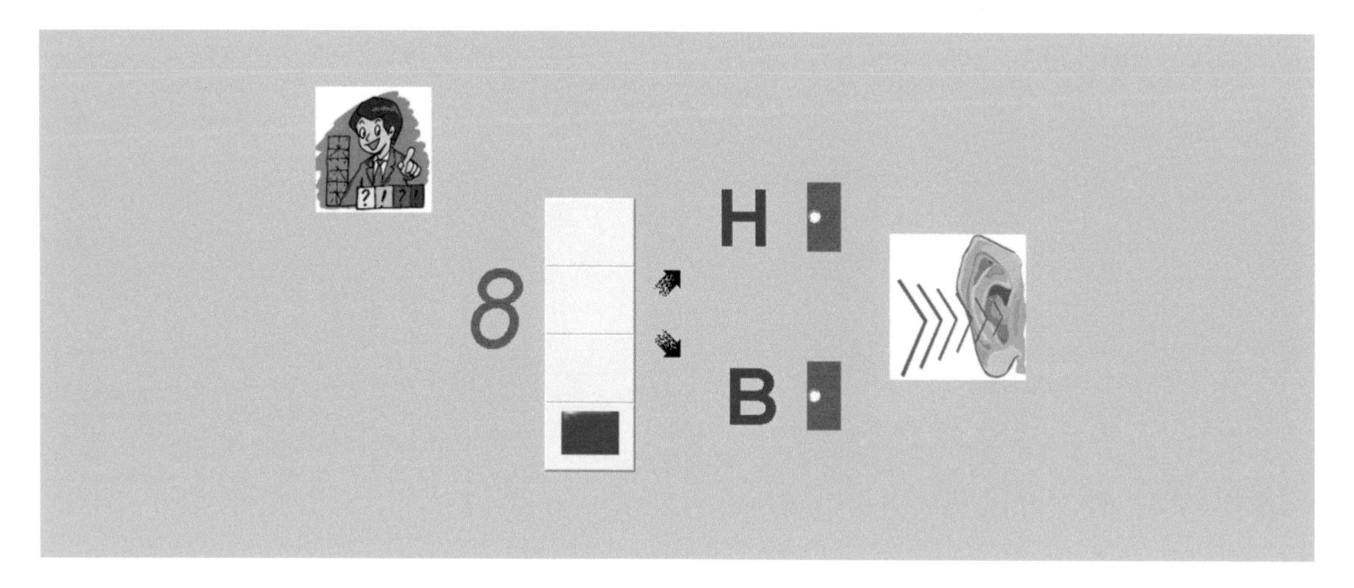

The knock represents a filled up position and a double knock represents an empty position. In our example knock and double knock are corresponding to the letter H. The clamping of the right hand represents a filled up position and the clamping of the left hand represents an empty position. In our example the ***kinesthetic*** representation for the answer is clamping of the left hand thrice and the right hand. For letter H, the kinesthetic representation is clamping of left hand and right hand.

FIGURE 5 INTRODUCTION TO ALPHABET

In figure 5, we have ***visual*** display of two pictures of boxes. The left picture has the first and fourth positions filled up with symbols. The decimal numbers 9 and letter I correspond to the first picture.

The ***audio*** representation for the left picture, number 9 and letter I, is **knock, double knock, double knock, and knock.**

The second picture on the right has the second and fourth positions filled up with symbols. The decimal number 10 and letter J correspond to the second picture.

The ***audio*** representation for the right picture, number 10 and letter J, is **double knock, knock, double knock, and knock.**

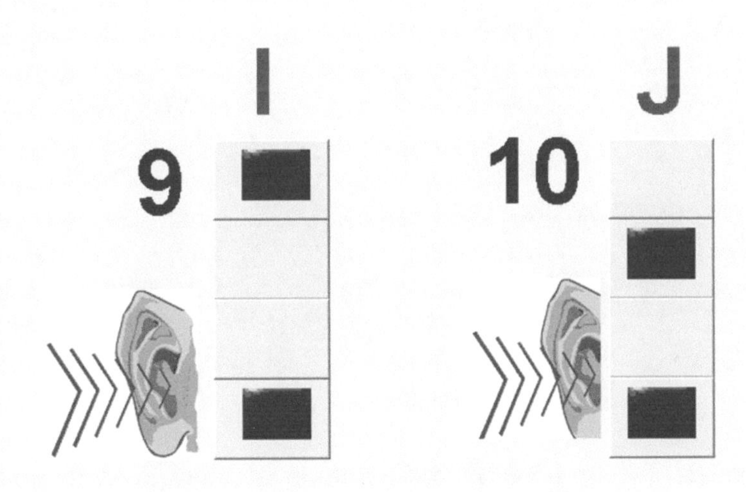

A knock represents a filled up position and a double knock represents an empty position. The clamping of the right hand represents a filled up position and the clamping of the left hand represents an empty position. The ***kinesthetic*** representation for the number 9 and letter I is clamping of the right hand, left hand twice, and right hand. The ***kinesthetic*** representation for the number 10 and letter J is clamping of left hand, right hand, left hand, and right hand.

INTRODUCTION TO ALPHABET

FIGURE 5

In figure 5, we have ***visual*** display of decimal numbers 9 and 10 on the left side, while on the right side we have alphabetical letters I and J which correspond to those numbers.

The ***audio*** representation for the number 9 and letter I is **knock, double knock, double knock, and knock.**

The ***audio*** representation for the number 10 and letter J is **double knock, knock, double knock, and knock.**

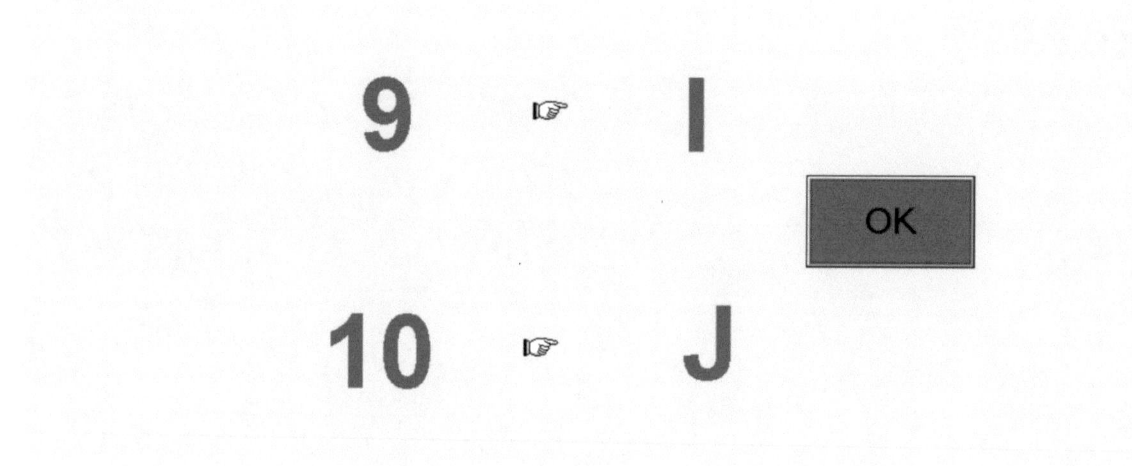

The knock represents a filled up position and a double knock represents an empty position. The clamping of the right hand represents a filled up position and the clamping of the left hand represents an empty position. The ***kinesthetic*** representation for the number 9 and letter I is clamping of the right hand, left hand twice, and right hand. The ***kinesthetic*** representation for the number 10 and letter J is clamping of left hand, right hand, left hand, and right hand.

EXAMPLE 5.1 WHAT LETTER IS IT?

In example 2.1, in visual display we see picture where the first position is filled up with a symbol, the next two positions are empty, and the last position is filled up with a symbol. We are given two choices of letters, A and I, which are supposed to correspond to the visual display.

The ***audio*** representation of the picture is **knock, double knock, double knock, and knock.**

The knock represents a filled up position and a double knock represents an empty position. In our example knock and double knock are corresponding to the letter I. The clamping of the right hand represents a filled up position and the clamping of the left hand represents an empty position. In our example the ***kinesthetic*** representation for the answer is clamping of the right hand, left hand twice, and the right hand. For letter A, the kinesthetic representation is clamping right hand and left hand.

WHAT LETTER IS IT? EXAMPLE 5.2

In example 5.2, in visual display we see picture where the first position is empty, the second position is filled up with a symbol, the third position is empty, and the last position is filled up with a symbol. We are given two choices of letters, D and J, which are supposed to correspond to the visual display.

The ***audio*** representation of the picture is **double knock, knock, double knock, and knock.**

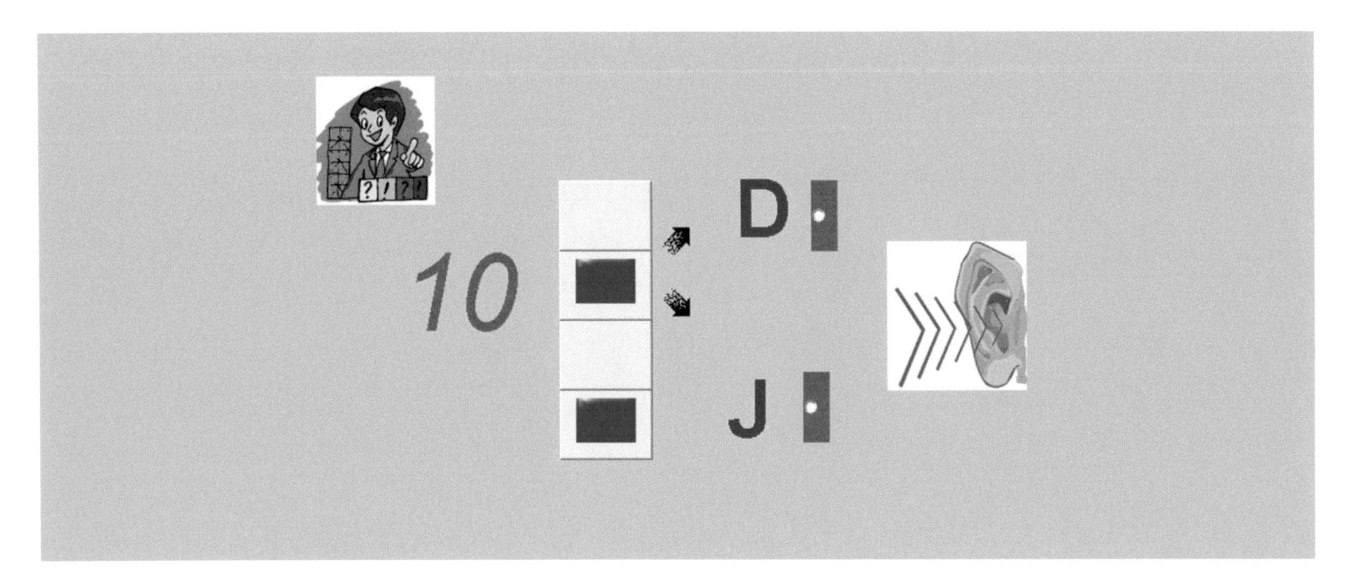

The knock represents a filled up position and a double knock represents an empty position. In our example knock and double knock are corresponding to the letter J. The clamping of the right hand represents a filled up position and the clamping of the left hand represents an empty position. In our example the ***kinesthetic*** representation for the answer is clamping of the left hand, right hand, left hand, and right hand. For letter D, the kinesthetic representation is clamping left hand twice and right hand.

FIGURE 6 INTRODUCTION TO ALPHABET

In figure 6, we have ***visual*** display of decimal numbers 11 and 12 on the left side, while on the right side we have alphabetical letters K and L which correspond to those numbers.

The ***audio*** representation for the number 11 and letter K is **knock, knock, double knock, and knock.**

The ***audio*** representation for the number 12 and letter L is **double knock, double knock, knock, and knock.**

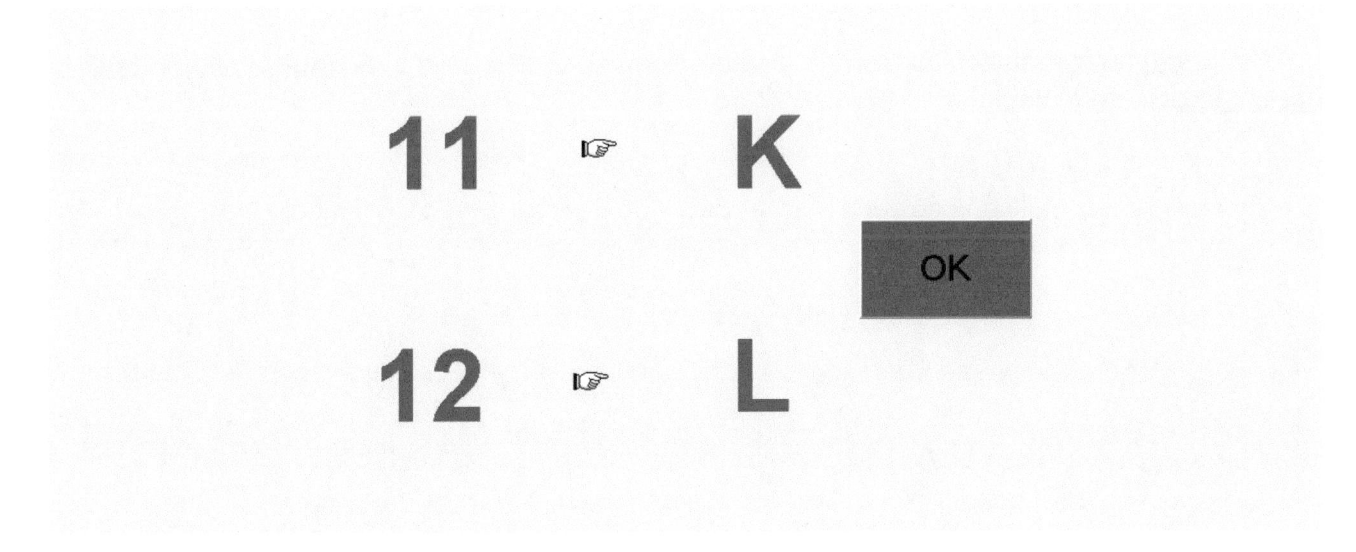

The knock represents a filled up position and a double knock represents an empty position. The clamping of the right hand represents a filled up position and the clamping of the left hand represents an empty position. The ***kinesthetic*** representation for the number 11 and letter K is clamping of the right hand twice, left hand, and right hand. The ***kinesthetic*** representation for the number 12 and letter L is clamping of left hand twice and right hand twice.

INTRODUCTION TO ALPHABET

FIGURE 6

In figure 6, we have ***visual*** display of two pictures of boxes. The left picture has the first, second, and fourth positions filled up with symbols. The decimal numbers 11 and letter K correspond to the first picture.

The ***audio*** representation for the left picture, number 11 and letter K, is **knock, knock, double knock, and knock.**

The second picture on the right has the third and fourth positions filled up with symbols. The decimal number 12 and letter L correspond to the second picture.

The ***audio*** representation for the right picture, number 12 and letter L, is **double knock, knock, double knock, and knock.**

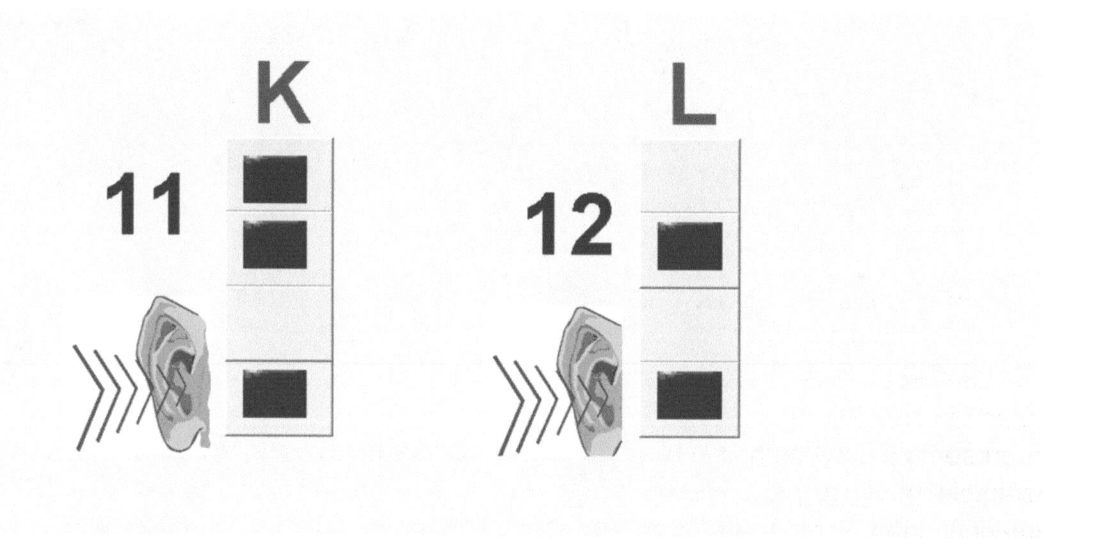

A knock represents a filled up position and a double knock represents an empty position. The clamping of the right hand represents a filled up position and the clamping of the left hand represents an empty position. The ***kinesthetic*** representation for the number 11 and letter K is clamping of the right hand twice, left hand, and right hand. The ***kinesthetic*** representation for the number 12 and letter L is clamping of left hand twice and right hand twice.

EXAMPLE 6.1 WHAT LETTER IS IT?

In example 6.1, in visual display we see picture where the first two positions are filled up with a symbol, the next position is empty, and the last position is filled up with a symbol. We are given two choices of letters, A and K, which are supposed to correspond to the visual display.

The ***audio*** representation of the picture is **knock, knock, double knock, and knock.**

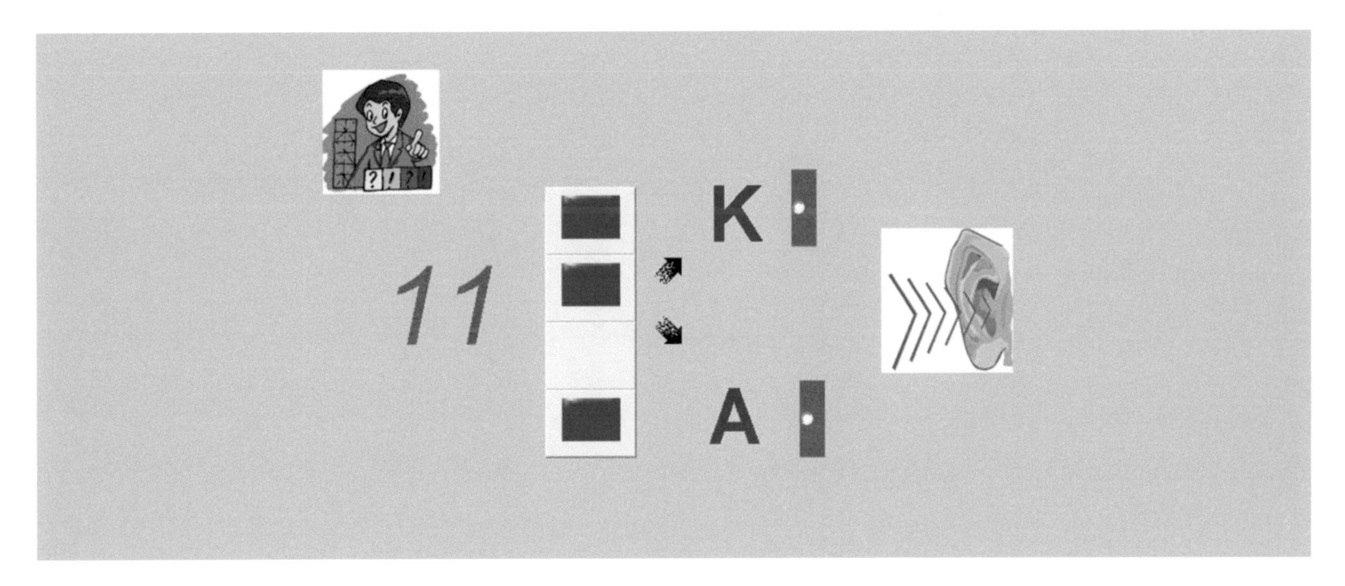

The knock represents a filled up position and a double knock represents an empty position. In our example knock and double knock are corresponding to the letter K. The clamping of the right hand represents a filled up position and the clamping of the left hand represents an empty position. In our example the ***kinesthetic*** representation for the answer is clamping of the right hand twice, left hand, and the right hand. For letter A, the kinesthetic representation is clamping right hand and left hand.

WHAT LETTER IS IT? EXAMPLE 6.2

In example 6.3, in visual display we see picture where the first two positions are empty and the next two positions are filled up with a symbol. We are given two choices of letters, F and L, which are supposed to correspond to the visual display.

The ***audio*** representation of the picture is **double knock, double knock, knock, and knock.**

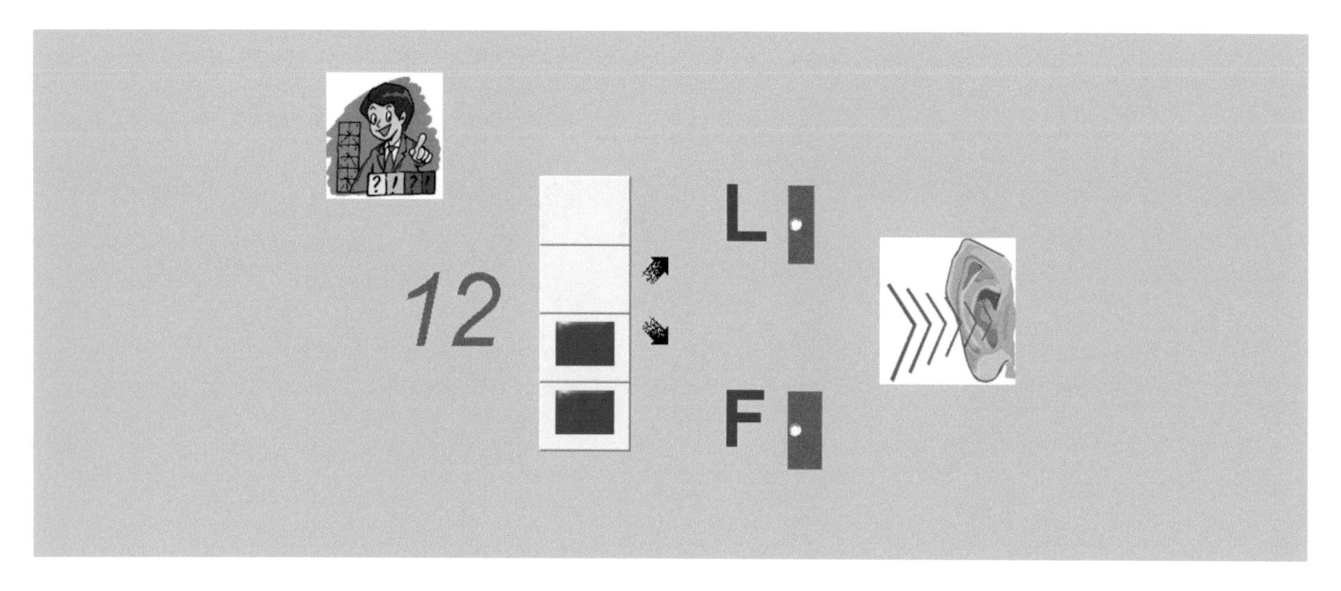

The knock represents a filled up position and a double knock represents an empty position. In our example knock and double knock are corresponding to the letter L. The clamping of the right hand represents a filled up position and the clamping of the left hand represents an empty position. In our example the ***kinesthetic*** representation for the answer is clamping of the left hand twice right hand twice. For letter F, the kinesthetic representation is clamping left hand and right hand twice.

FIGURE 7 INTRODUCTION TO ALPHABET

In figure 7, we have ***visual*** display of two pictures of boxes. The left picture has the first, third, and fourth positions filled up with symbols. The decimal numbers 13 and letter M correspond to the first picture.

The ***audio*** representation for the left picture, number 13 and letter M, is **knock, double knock, knock, and knock.**

The second picture on the right has the second, third, and fourth positions filled up with symbols. The decimal number 14 and letter N correspond to the second picture.

The ***audio*** representation for the right picture, number 14 and letter N, is **double knock, knock, knock, and knock.**

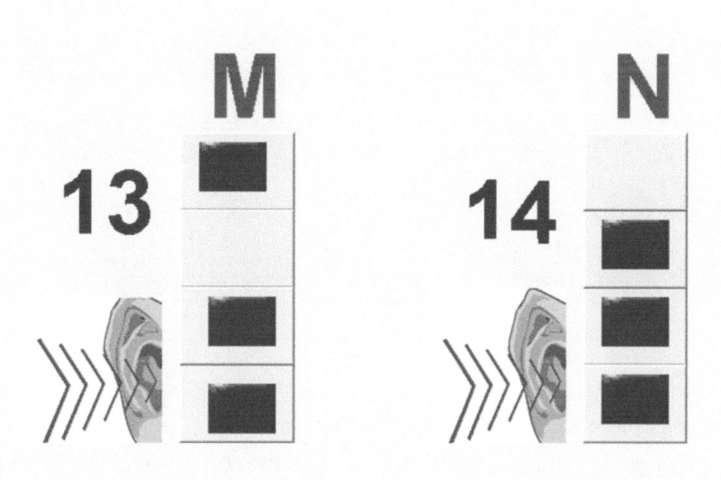

A knock represents a filled up position and a double knock represents an empty position. The clamping of the right hand represents a filled up position and the clamping of the left hand represents an empty position. The ***kinesthetic*** representation for the number 13 and letter M is clamping of the right hand, left hand, and right hand twice. The ***kinesthetic*** representation for the number 14 and letter N is clamping of left hand and right hand thrice.

INTRODUCTION TO ALPHABET

FIGURE 7

In figure 7, we have ***visual*** display of decimal numbers 13 and 14 on the left side, while on the right side we have alphabetical letters M and N which correspond to those numbers.

The ***audio*** representation for the number 13 and letter M is **knock, double knock, knock, and knock.**

The ***audio*** representation for the number 14 and letter N is **double knock, knock, knock, and knock.**

The knock represents a filled up position and a double knock represents an empty position. The clamping of the right hand represents a filled up position and the clamping of the left hand represents an empty position. The ***kinesthetic*** representation for the number 13 and letter M is clamping of the right hand, left hand, and right hand twice. The ***kinesthetic*** representation for the number 14 and letter N is clamping of left hand and right hand thrice.

EXAMPLE 1 WHAT LETTER IS IT?

In example 1, in visual display we see picture where the first position is filled up with a symbol, the second position is empty, and the last two positions are filled up with a symbol. We are given two choices of letters, B and M, which are supposed to correspond to the visual display.

The ***audio*** representation of the picture is **knock, double knock, knock, and knock.**

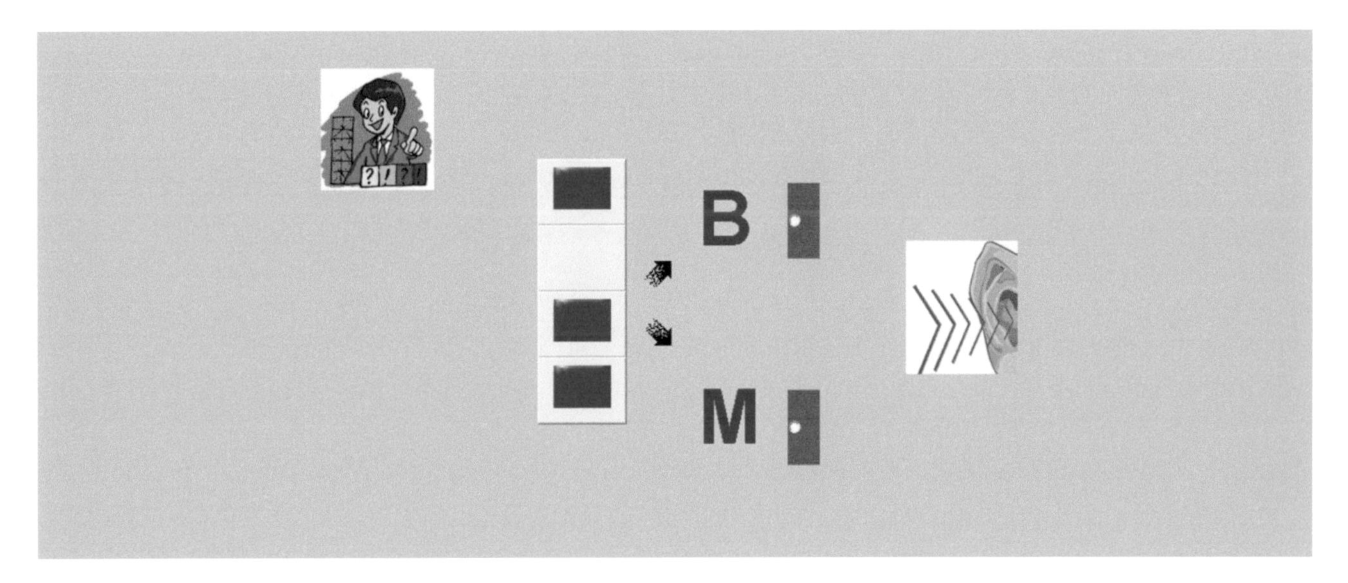

The knock represents a filled up position and a double knock represents an empty position. In our example knock and double knock are corresponding to the letter M. The clamping of the right hand represents a filled up position and the clamping of the left hand represents an empty position. In our example the ***kinesthetic*** representation for the answer is clamping of the right hand, left hand, and right hand twice. For letter B, the kinesthetic representation is clamping of left hand and then right hand.

WHAT LETTER IS IT? EXAMPLE 7.1

In example 7.1, in visual display we see picture where the first position is empty and the last three positions are filled up with a symbol. We are given two choices of letters, E and N, which are supposed to correspond to the visual display.

The ***audio*** representation of the picture is **double knock, knock, knock, and knock.**

The knock represents a filled up position and a double knock represents an empty position. In our example knock and double knock are corresponding to the letter N. The clamping of the right hand represents a filled up position and the clamping of the left hand represents an empty position. In our example the ***kinesthetic*** representation for the answer is clamping of the left hand and right hand thrice. For letter E, the kinesthetic representation is clamping of right hand, left hand, and right hand.

FIGURE 8 INTRODUCTION TO ALPHABET

In figure 8, we have ***visual*** display of two pictures of boxes. The left picture has the all four positions filled up with symbols. The decimal numbers 15 and letter O correspond to the first picture.

The ***audio*** representation for the left picture, number 15 and letter M, is **knock, knock, knock, and knock.**

The second picture on the right has the fifth position filled up with a symbol. The decimal number 16 and letter P correspond to the second picture.

The ***audio*** representation for the right picture, number 16 and letter P, is **double knock, double knock, double knock, double knock, and knock.**

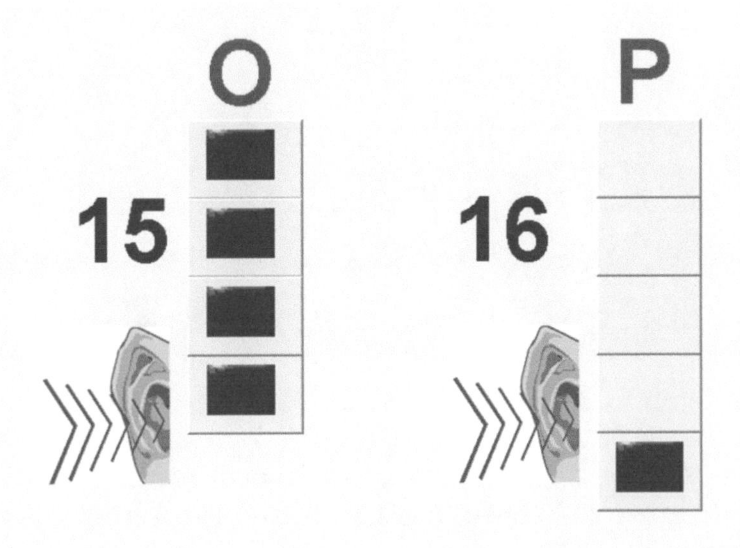

A knock represents a filled up position and a double knock represents an empty position. The clamping of the right hand represents a filled up position and the clamping of the left hand represents an empty position. The ***kinesthetic*** representation for the number 15 and letter O is clamping of the right hand four times. The ***kinesthetic*** representation for the number 16 and letter P is clamping of left hand four times and right hand.

INTRODUCTION TO ALPHABET FIGURE 8

In figure 8, we have ***visual*** display of decimal numbers 15 and 16 on the left side, while on the right side we have alphabetical letters O and P which correspond to those numbers.

The ***audio*** representation for the number 15 and letter O is **knock, knock, knock, and knock.**

The ***audio*** representation for the number 16 and letter P is **double knock, double knock, double knock, double knock, and knock.**

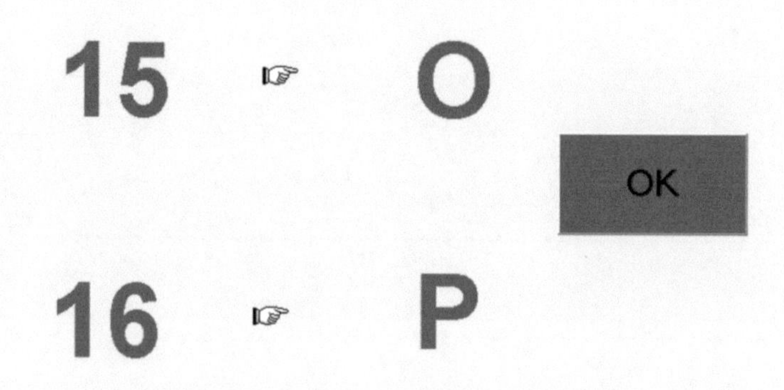

The knock represents a filled up position and a double knock represents an empty position. The clamping of the right hand represents a filled up position and the clamping of the left hand represents an empty position. The ***kinesthetic*** representation for the number 15 and letter O is clamping of the right hand four times. The ***kinesthetic*** representation for the number 16 and letter P is clamping of left hand four times and right hand.

EXAMPLE 7.1 WHAT LETTER IS IT?

In example 2.1, in visual display we see picture where all the positions are filled up with a symbol. We are given two choices of letters, C and O, which are supposed to correspond to the visual display.

The ***audio*** representation of the picture is **knock, knock, knock, and knock.**

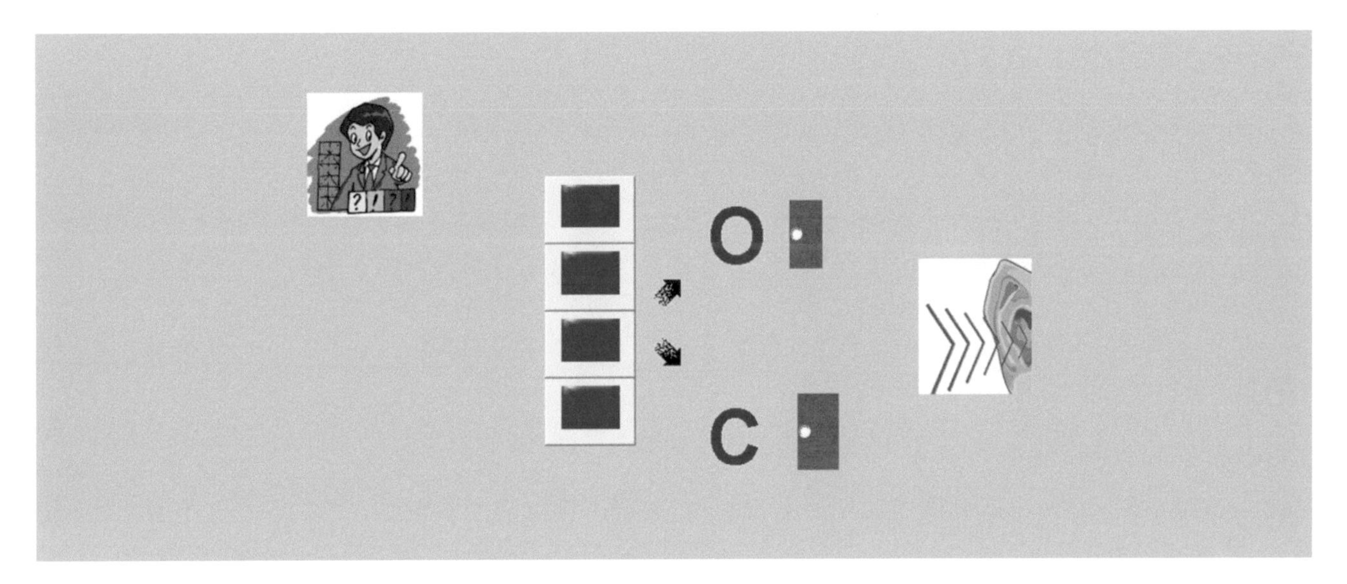

The knock represents a filled up position and a double knock represents an empty position. In our example knock and double knock are corresponding to the letter O. The clamping of the right hand represents a filled up position and the clamping of the left hand represents an empty position. In our example the ***kinesthetic*** representation for the answer is clamping of the right hand four times. For letter C, the kinesthetic representation is clamping of right hand twice.

WHAT LETTER IS IT? EXAMPLE 7.2

In example 7.2, in visual display we see picture where the first four positions are empty and the last position is filled up with a symbol. We are given two choices of letters, D and P, which are supposed to correspond to the visual display.

The ***audio*** representation of the picture is **double knock, double knock, double knock, double knock, and knock.**

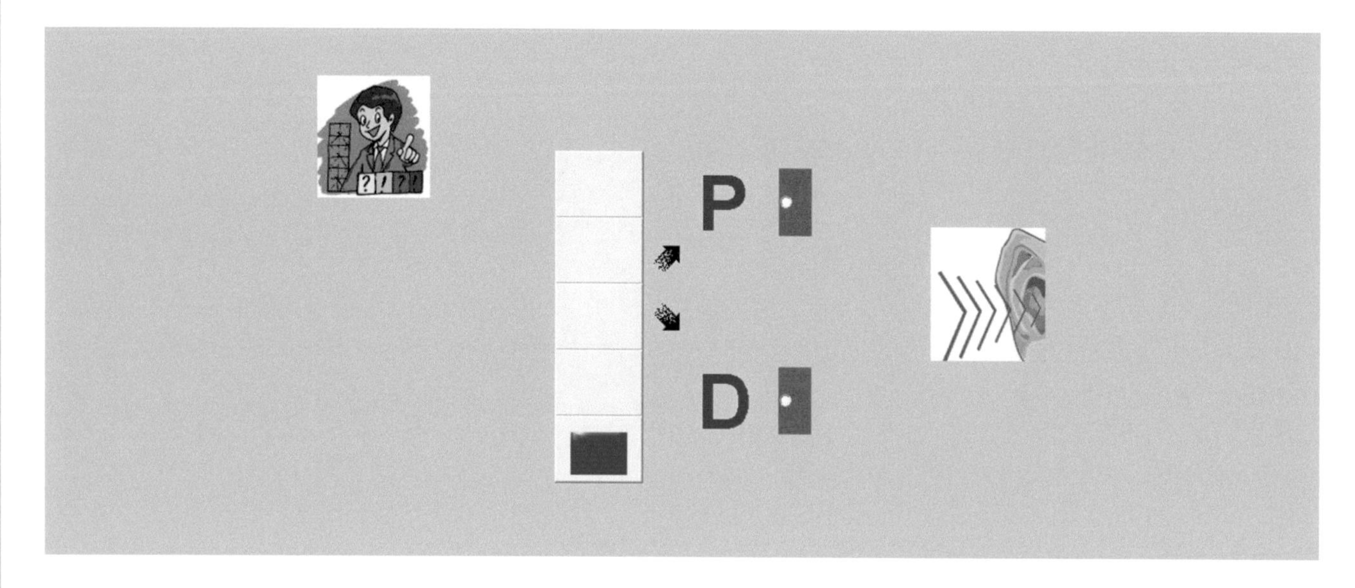

The knock represents a filled up position and a double knock represents an empty position. In our example knock and double knock are corresponding to the letter P. The clamping of the right hand represents a filled up position and the clamping of the left hand represents an empty position. In our example the ***kinesthetic*** representation for the answer is clamping of the left hand four times and right hand. For letter D, the kinesthetic representation is clamping of left hand twice and right hand.

FIGURE 9 INTRODUCTION TO ALPHABET

In figure 9, we have ***visual*** display of two pictures of boxes. The left picture has the first and fifth positions filled up with symbols. The decimal numbers 17 and letter Q correspond to the first picture.

The ***audio*** representation for the left picture, number 17 and letter Q, is **knock, double knock, double knock, double knock, and knock.**

The second picture on the right has the second and fifth positions filled up with symbols. The decimal number 18 and letter R correspond to the second picture.

The ***audio*** representation for the right picture, number 18 and letter R, is **double knock, knock, double knock, double knock, and knock.**

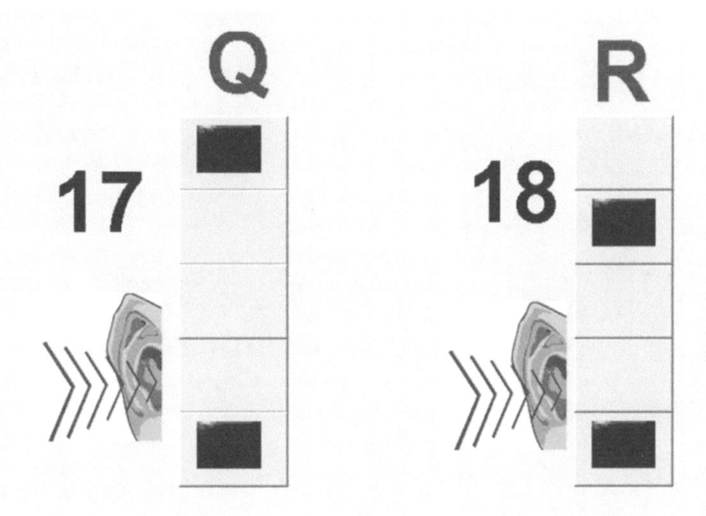

A knock represents a filled up position and a double knock represents an empty position. The clamping of the right hand represents a filled up position and the clamping of the left hand represents an empty position. The ***kinesthetic*** representation for the number 17 and letter Q is clamping of the right hand, left hand thrice, and right hand. The ***kinesthetic*** representation for the number 18and letter R is clamping of left hand, right hand, left hand twice, and right hand.

INTRODUCTION TO ALPHABET

FIGURE 9

In figure 9, we have ***visual*** display of decimal numbers 17 and 18 on the left side, while on the right side we have alphabetical letters Q and R which correspond to those numbers.

The ***audio*** representation for the number 17 and letter Q is **knock, double knock, double knock, double knock, and knock.**

The ***audio*** representation for the number 18 and letter R is **double knock, knock, double knock, double knock, and knock.**

The knock represents a filled up position and a double knock represents an empty position. The clamping of the right hand represents a filled up position and the clamping of the left hand represents an empty position. The ***kinesthetic*** representation for the number 17 and letter Q is clamping of the right hand, left hand thrice, and right hand. The ***kinesthetic*** representation for the number 18 and letter R is clamping of left hand, right hand, left hand twice, and right hand.

EXAMPLE 9.1 WHAT LETTER IS IT?

In example 9.1, in visual display we see picture where the first position is filled up with a symbol, the next three positions are empty, and the last position is filled up with a symbol. We are given two choices of letters, O and Q, which are supposed to correspond to the visual display.

The ***audio*** representation of the picture is **knock, double knock, double knock, double knock, and knock.**

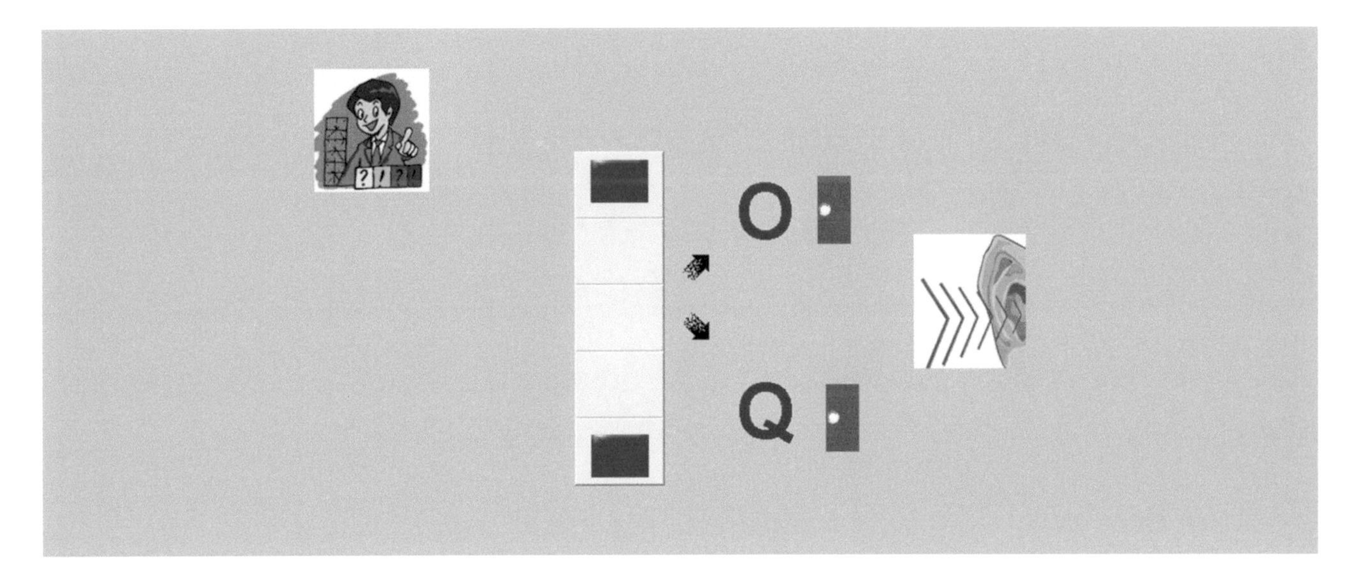

The knock represents a filled up position and a double knock represents an empty position. In our example knock and double knock are corresponding to the letter Q. The clamping of the right hand represents a filled up position and the clamping of the left hand represents an empty position. In our example the ***kinesthetic*** representation for the answer is clamping of the right hand, left hand thrice, and right hand. For letter O, the kinesthetic representation is clamping of right hand four times.

WHAT LETTER IS IT? EXAMPLE 9.2

In example 9.2, in visual display we see picture where the first position is empty, the second position is filled up with a symbol, the next two positions are empty, and the last position is filled up with a symbol. We are given two choices of letters, M and R, which are supposed to correspond to the visual display.

The ***audio*** representation of the picture is **double knock, knock, double knock, double knock, and knock.**

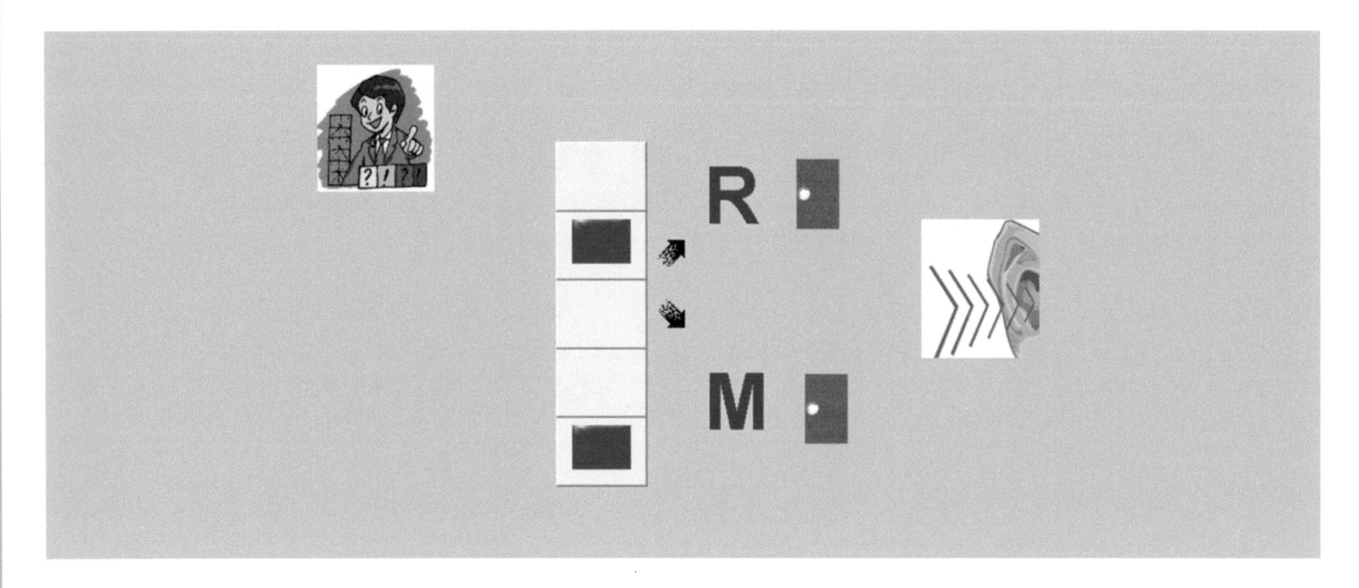

The knock represents a filled up position and a double knock represents an empty position. In our example knock and double knock are corresponding to the letter R. The clamping of the right hand represents a filled up position and the clamping of the left hand represents an empty position. In our example the ***kinesthetic*** representation for the answer is clamping of the left hand, right hand, left hand twice, and right hand. For letter M, the kinesthetic representation is clamping of right hand, left hand, and right hand twice.

FIGURE 10 INTRODUCTION TO ALPHABET

In figure 10, we have ***visual*** display of two pictures of boxes. The left picture has the first, second, and fifth positions filled up with symbols. The decimal numbers 19 and letter S correspond to the first picture.

The ***audio*** representation for the left picture, number 19 and letter S, is **knock, knock, double knock, double knock, and knock.**

The second picture on the right has the third and fifth positions filled up with symbols. The decimal number 20 and letter T correspond to the second picture.

The ***audio*** representation for the right picture, number 20 and letter T, is **double knock, double knock, knock, double knock, and knock.**

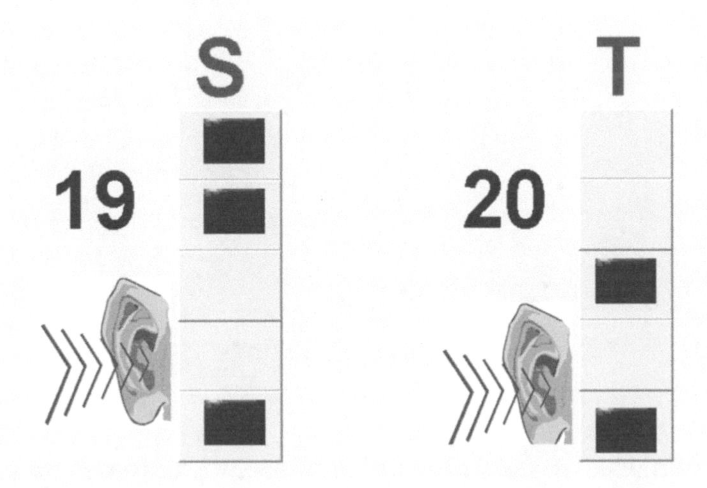

A knock represents a filled up position and a double knock represents an empty position. The clamping of the right hand represents a filled up position and the clamping of the left hand represents an empty position. The ***kinesthetic*** representation for the number 19 and letter S is clamping of the right hand twice, left hand twice, and right hand. The ***kinesthetic*** representation for the number 20 and letter T is clamping of left hand twice, right hand, left hand, and right hand.

INTRODUCTION TO ALPHABET

FIGURE 10

In figure 10, we have ***visual*** display of decimal numbers 19 and 20 on the left side, while on the right side we have alphabetical letters S and T which correspond to those numbers.

The ***audio*** representation for the number 19 and letter S is **knock, knock, double knock, double knock, and knock.**

The ***audio*** representation for the number 20 and letter T is **double knock, double, knock, knock, double knock, and knock.**

The knock represents a filled up position and a double knock represents an empty position. The clamping of the right hand represents a filled up position and the clamping of the left hand represents an empty position. The ***kinesthetic*** representation for the number 19 and letter S is clamping of the right hand twice, left hand twice, and right hand. The ***kinesthetic*** representation for the number 20 and letter T is clamping of left hand twice, right hand, left hand, and right hand.

EXAMPLE 1 WHAT LETTER IS IT?

In example 1, in visual display we see picture where the first two positions are filled with a symbol, the third and fourth positions are empty, and the last position is filled up with a symbol. We are given two choices of letters, Q and S, which are supposed to correspond to the visual display.

The ***audio*** representation of the picture is **knock, knock, double knock, double knock, and knock.**

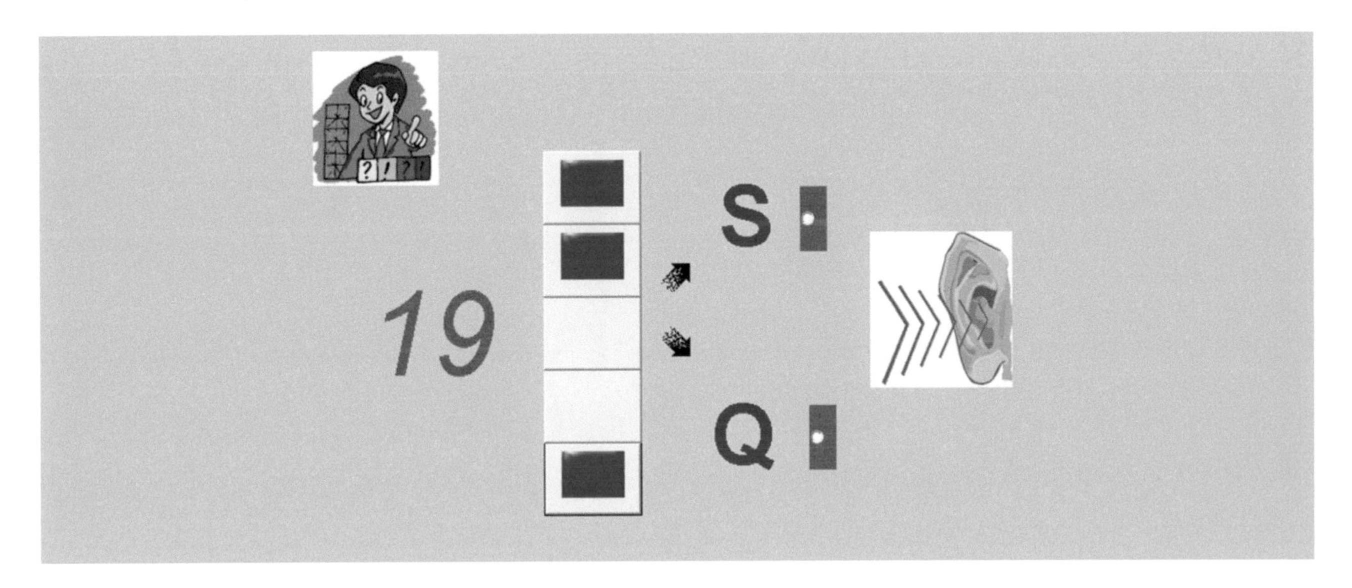

The knock represents a filled up position and a double knock represents an empty position. In our example knock and double knock are corresponding to the letter S. The clamping of the right hand represents a filled up position and the clamping of the left hand represents an empty position. In our example the ***kinesthetic*** representation for the answer is clamping of the right hand twice, left hand twice, and right hand. For letter Q, the kinesthetic representation is clamping of right hand, left hand thrice, and right hand.

WHAT LETTER IS IT? EXAMPLE 10.1

In example 1.1, in visual display we see picture where the first two positions are empty, the position is filled up with a symbol, the fourth position is empty, and the last position is filled up with a symbol. We are given two choices of letters, Land T, which are supposed to correspond to the visual display.

The ***audio*** representation of the picture is **double knock, double knock, knock, double knock, and knock.**

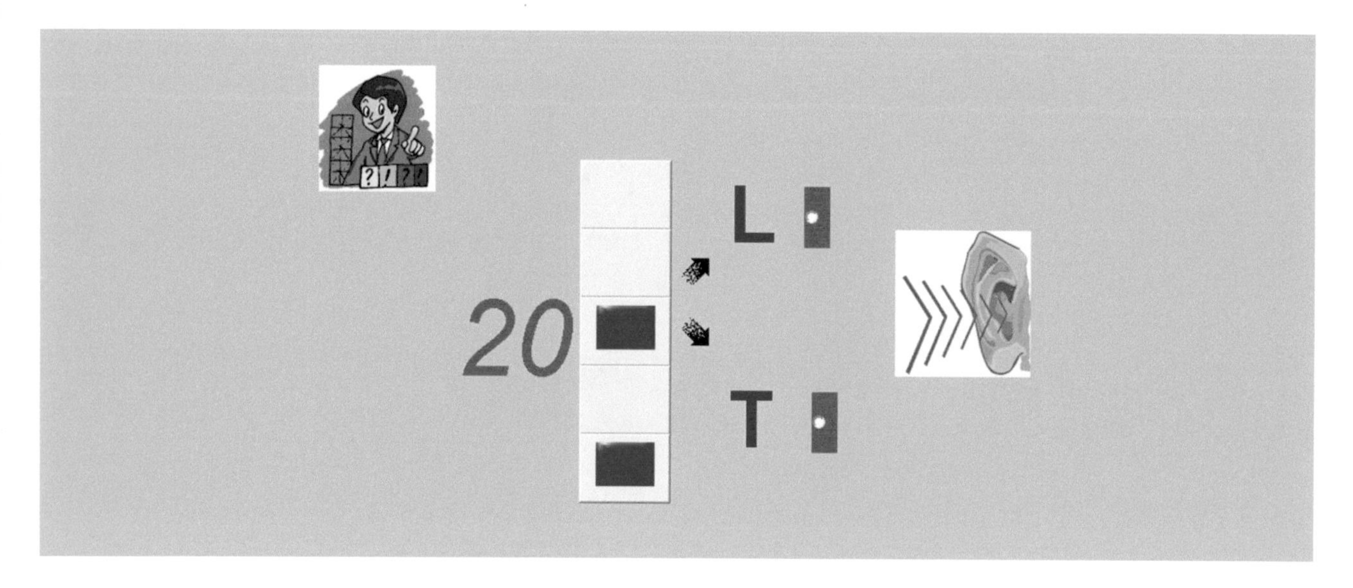

The knock represents a filled up position and a double knock represents an empty position. In our example knock and double knock are corresponding to the letter T. The clamping of the right hand represents a filled up position and the clamping of the left hand represents an empty position. In our example the ***kinesthetic*** representation for the answer is clamping of the left hand twice, right hand, left hand, and right hand. For letter L, the kinesthetic representation is clamping of left hand twice and right hand twice.

FIGURE 11 INTRODUCTION TO ALPHABET

In figure 11, we have ***visual*** display of two pictures of boxes. The left picture has the first, third, and fifth positions filled up with symbols. The decimal numbers 21 and letter U correspond to the first picture.

The ***audio*** representation for the left picture, number 21 and letter U, is **knock, double knock, knock, double knock, and knock.**

The second picture on the right has the second, third, and fifth positions filled up with symbols. The decimal number 22 and letter V correspond to the second picture.

The ***audio*** representation for the right picture, number 22 and letter V, is **double knock, knock, knock, double knock, and knock.**

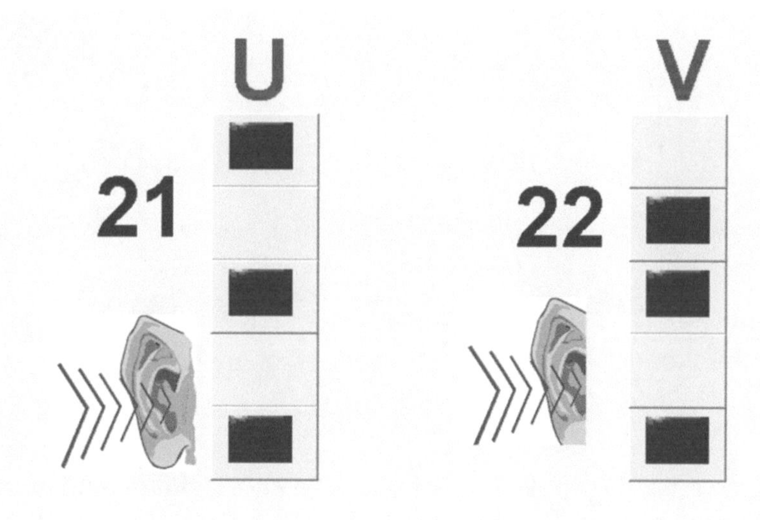

A knock represents a filled up position and a double knock represents an empty position. The clamping of the right hand represents a filled up position and the clamping of the left hand represents an empty position. The ***kinesthetic*** representation for the number 21 and letter U is clamping of the right hand, left hand, right hand, left hand, and right hand. The ***kinesthetic*** representation for the number 22 and letter V is clamping of left hand, right hand twice, left hand, and right hand.

INTRODUCTION TO ALPHABET

FIGURE 11

In figure 11, we have ***visual*** display of decimal numbers 21 and 22 on the left side, while on the right side we have alphabetical letters U and V which correspond to those numbers.

The ***audio*** representation for the number 21 and letter U is **knock, double knock, knock, double knock, and knock.**

The ***audio*** representation for the number 22 and letter V is **double knock, knock, knock, double knock, and knock.**

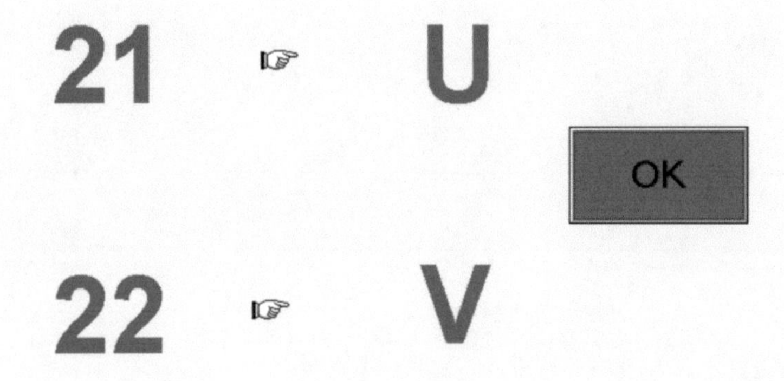

The knock represents a filled up position and a double knock represents an empty position. The clamping of the right hand represents a filled up position and the clamping of the left hand represents an empty position. The ***kinesthetic*** representation for the number 21 and letter U is clamping of the right hand, left hand, right hand, left hand, and right hand. The ***kinesthetic*** representation for the number 22 and letter V is clamping of left hand, right hand twice, left hand, and right hand.

EXAMPLE 11.1 WHAT LETTER IS IT?

In example 2.1, in visual display we see picture where the first position is filled up with a symbol, the second position is empty, the third position is filled up with a symbol, the fourth position is empty, and the last position is filled up with a symbol. We are given two choices of letters, S and U, which are supposed to correspond to the visual display.

The ***audio*** representation of the picture is **double knock, double knock, knock, double knock, and knock.**

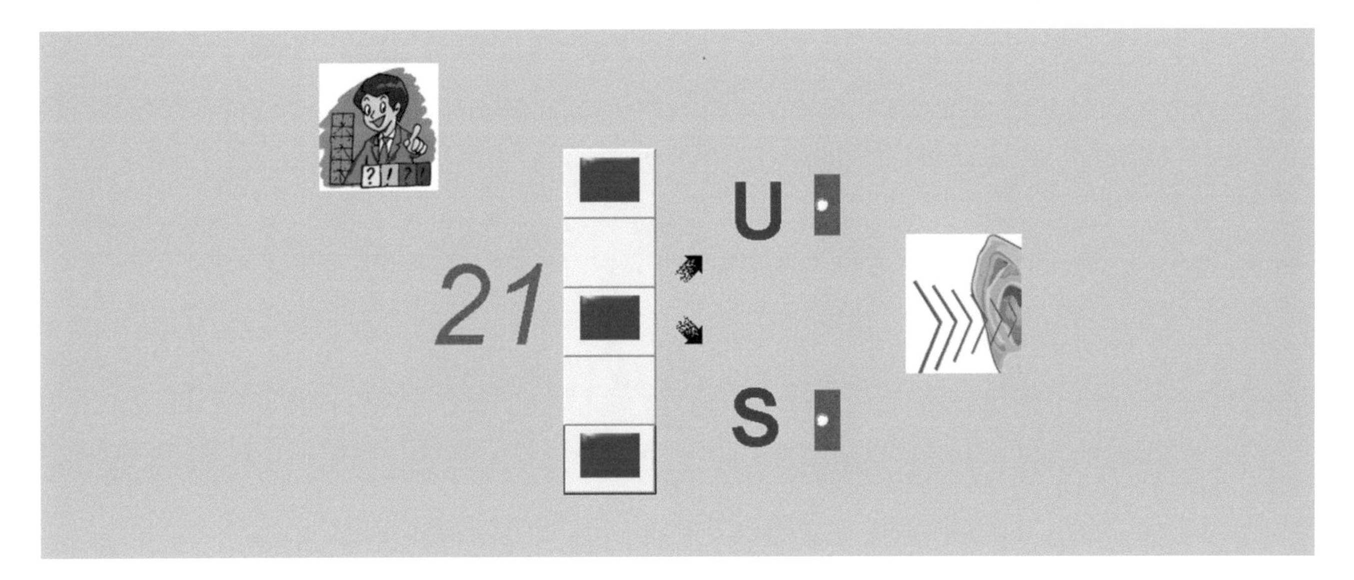

The knock represents a filled up position and a double knock represents an empty position. In our example knock and double knock are corresponding to the letter U. The clamping of the right hand represents a filled up position and the clamping of the left hand represents an empty position. In our example the ***kinesthetic*** representation for the answer is clamping of the right hand, left hand, right hand, left hand, and right hand. For letter S, the kinesthetic representation is clamping of right hand twice, left hand twice, and right hand.

WHAT LETTER IS IT? EXAMPLE 11.2

In example 2.2, in visual display we see picture where the first position is empty, the next two positions are filled up with a symbol, the fourth position is empty, and the last position is filled up with a symbol. We are given two choices of letters, T and V, which are supposed to correspond to the visual display.

The ***audio*** representation of the picture is **double knock, knock, knock, double knock, and knock.**

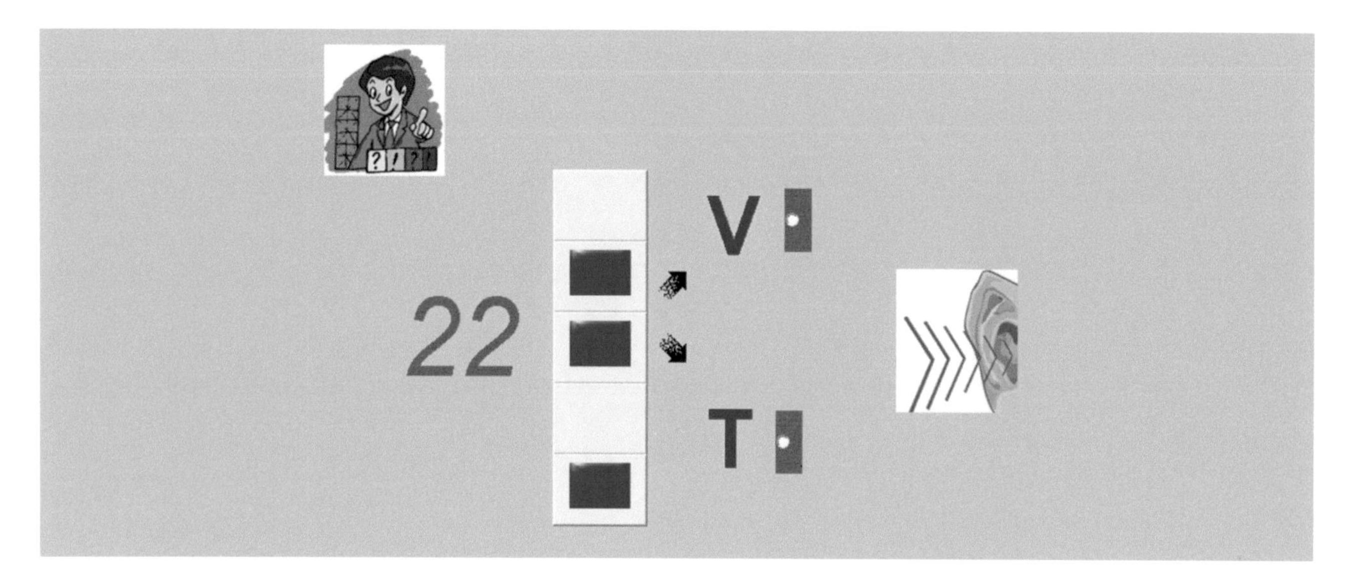

The knock represents a filled up position and a double knock represents an empty position. In our example knock and double knock are corresponding to the letter V. The clamping of the right hand represents a filled up position and the clamping of the left hand represents an empty position. In our example the ***kinesthetic*** representation for the answer is clamping of the left hand, right hand twice, left hand, and right hand. For letter T, the kinesthetic representation is clamping of left hand twice, right hand, left hand, and right hand.

FIGURE 12 INTRODUCTION TO ALPHABET

In figure 12, we have ***visual*** display of two pictures of boxes. The left picture has the first, second, third, and fifth positions filled up with symbols. The decimal numbers 23 and letter W correspond to the first picture.

The ***audio*** representation for the left picture, number 23 and letter W, is **knock, knock, knock, double knock, and knock.**

The second picture on the right has the fourth and fifth positions filled up with symbols. The decimal number 24 and letter X correspond to the second picture.

The ***audio*** representation for the right picture, number 24 and letter X, is **double knock, double knock, double knock, knock, and knock.**

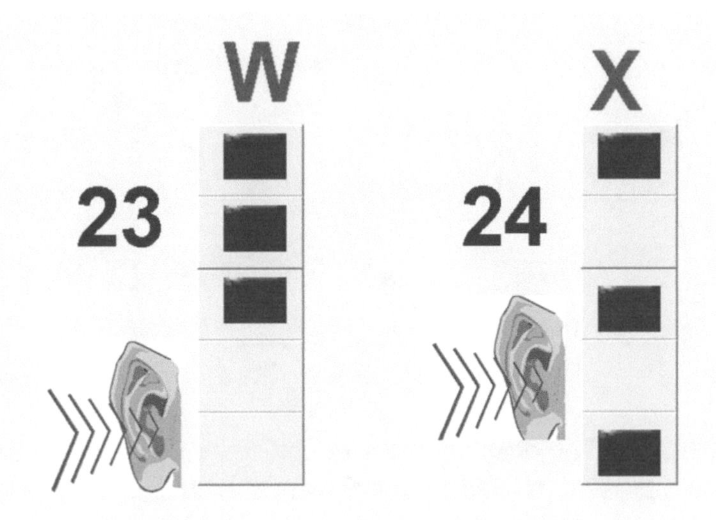

A knock represents a filled up position and a double knock represents an empty position. The clamping of the right hand represents a filled up position and the clamping of the left hand represents an empty position. The ***kinesthetic*** representation for the number 23 and letter W is clamping of the right hand thrice, left hand, and right hand. The ***kinesthetic*** representation for the number 24 and letter X is clamping of left hand thrice and right hand twice.

INTRODUCTION TO ALPHABET

FIGURE 12

In figure 12, we have ***visual*** display of decimal numbers 23 and 24 on the left side, while on the right side we have alphabetical letters W and X which correspond to those numbers.

The ***audio*** representation for the number 23 and letter W is **knock, knock, knock, double knock, and knock.**

The ***audio*** representation for the number 24 and letter X is **double knock, double knock, double knock, knock, and knock.**

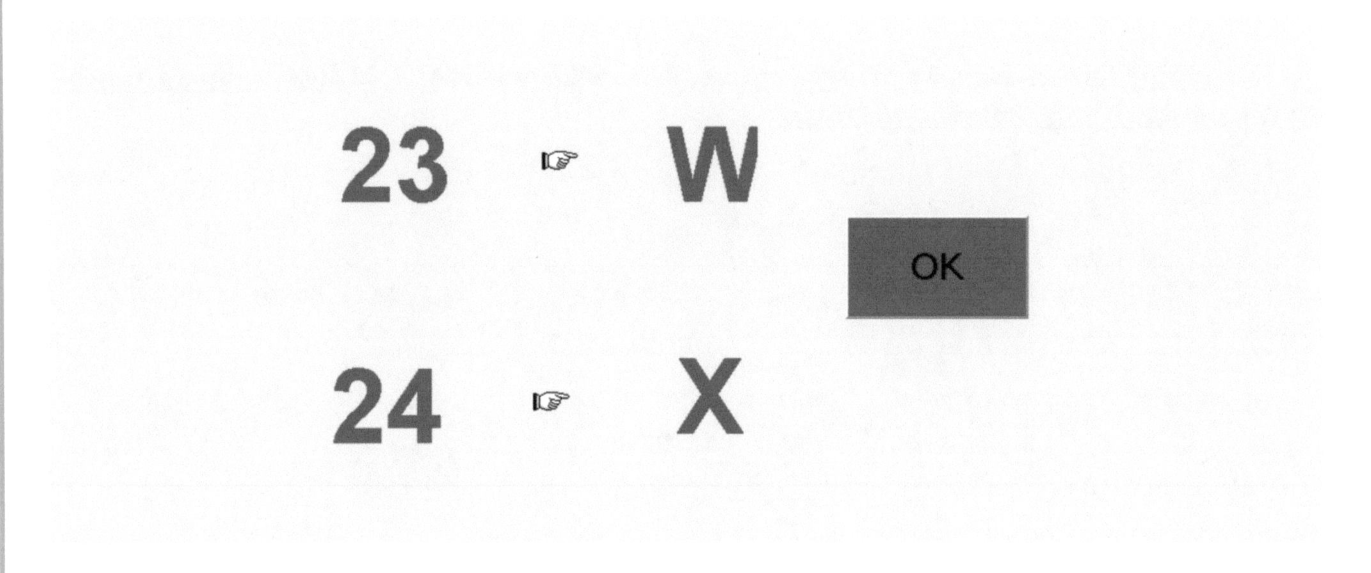

The knock represents a filled up position and a double knock represents an empty position. The clamping of the right hand represents a filled up position and the clamping of the left hand represents an empty position. The ***kinesthetic*** representation for the number 23 and letter W is clamping of the right hand thrice, left hand, and right hand. The ***kinesthetic*** representation for the number 24 and letter X is clamping of left hand thrice and right hand.

EXAMPLE 12.1 WHAT LETTER IS IT?

In example 3.1, in visual display we see picture where the first position three positions are filled up with a symbol, the fourth position is empty, and the last position is filled up with a symbol. We are given two choices of letters, K and W, which are supposed to correspond to the visual display.

The ***audio*** representation of the picture is **knock, knock, knock, double knock, and knock.**

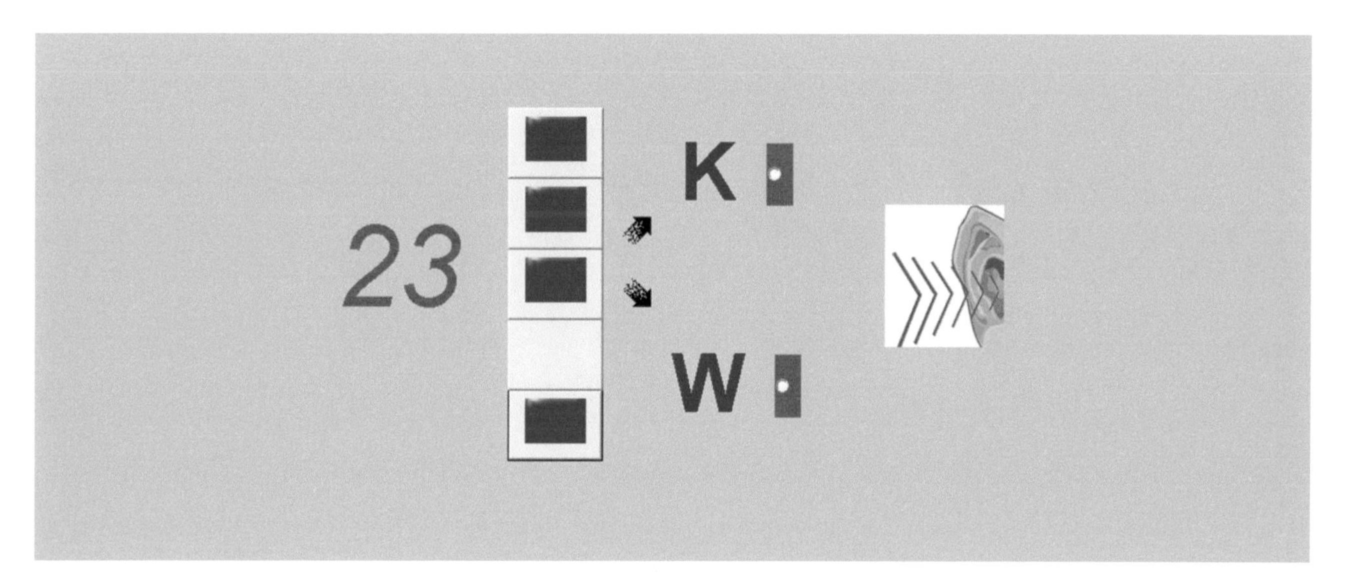

The knock represents a filled up position and a double knock represents an empty position. In our example knock and double knock are corresponding to the letter K. The clamping of the right hand represents a filled up position and the clamping of the left hand represents an empty position. In our example the ***kinesthetic*** representation for the answer is clamping of the right hand thrice, left hand, and right hand. For letter K, the kinesthetic representation is clamping of right hand twice, left hand, and right hand

WHAT LETTER IS IT? EXAMPLE 12.3

In example 3.3, in visual display we see picture where the first position three positions empty and the last two positions are filled up with a symbol. We are given two choices of letters, T and X, which are supposed to correspond to the visual display.

The ***audio*** representation of the picture is **double knock, double knock, double knock, knock, and knock.**

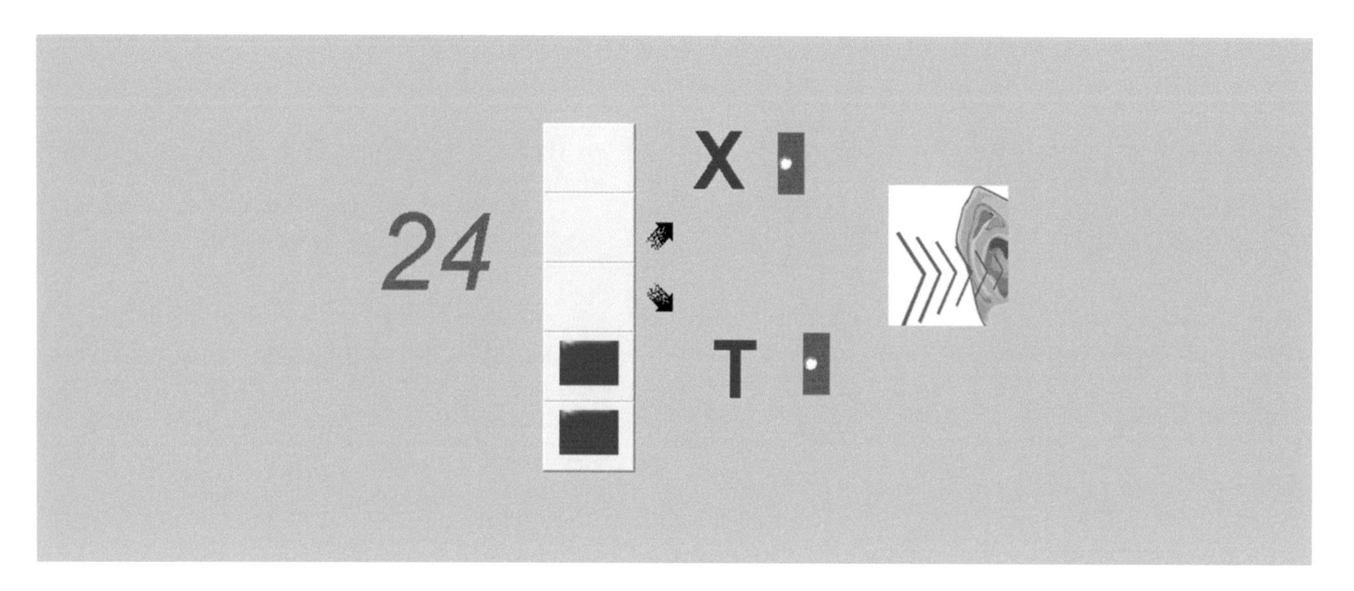

The knock represents a filled up position and a double knock represents an empty position. In our example knock and double knock are corresponding to the letter T. The clamping of the right hand represents a filled up position and the clamping of the left hand represents an empty position. In our example the ***kinesthetic*** representation for the answer is clamping of the left hand thrice and right hand twice. For letter T, the kinesthetic representation is clamping of left hand twice, right hand, left hand, and right hand

FIGURE 13 INTRODUCTION TO ALPHABET

In figure 13, we have ***visual*** display of two pictures of boxes. The left picture has the first, fourth, and fifth positions filled up with symbols. The decimal numbers 25 and letter Y correspond to the first picture.

The ***audio*** representation for the left picture, number 25 and letter Y, is **knock, double knock, double knock, knock, and knock.**

The second picture on the right has the second, fourth, and fifth positions filled up with symbols. The decimal number 26 and letter Z correspond to the second picture.

The ***audio*** representation for the right picture, number 26 and letter Z, is **double knock, knock, double knock, knock, and knock.**

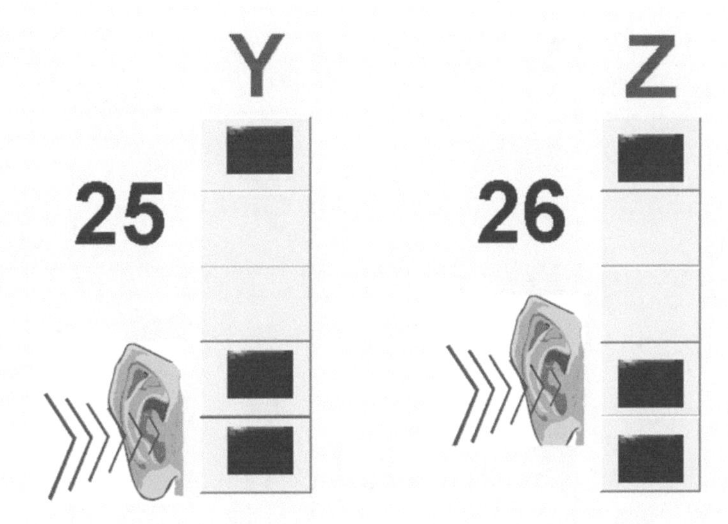

A knock represents a filled up position and a double knock represents an empty position. The clamping of the right hand represents a filled up position and the clamping of the left hand represents an empty position. The ***kinesthetic*** representation for the number 25 and letter Y is clamping of the right hand, left hand twice, and right hand. The ***kinesthetic*** representation for the number 26 and letter Z is clamping of left hand, right hand, left hand, and right hand twice.

INTRODUCTION TO ALPHABET

FIGURE 13

In figure 13, we have ***visual*** display of decimal numbers 25 and 26 on the left side, while on the right side we have alphabetical letters Y and Z which correspond to those numbers.

The ***audio*** representation for the number 25 and letter Y is **knock, double knock, double knock, knock, and knock.**

The ***audio*** representation for the number 26 and letter Z is **double knock, knock, double knock, knock, and knock.**

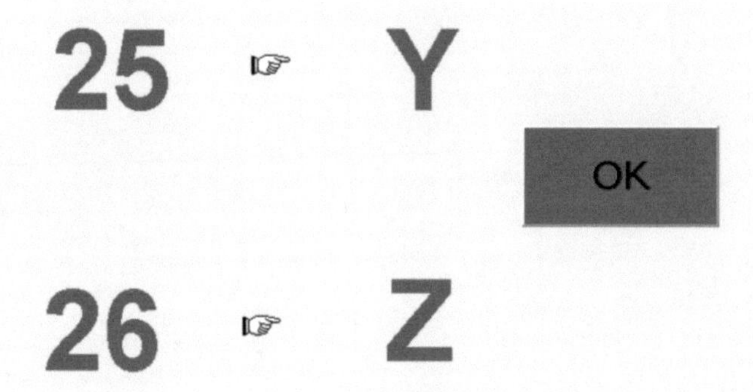

The knock represents a filled up position and a double knock represents an empty position. The clamping of the right hand represents a filled up position and the clamping of the left hand represents an empty position. The ***kinesthetic*** representation for the number 25 and letter Y is clamping of the right hand, left hand twice, and right hand twice. The ***kinesthetic*** representation for the number 26 and letter Z is clamping of left hand, right hand, left hand, and right hand twice.

EXAMPLE 13.1 WHAT LETTER IS IT?

In example 4.1, in visual display we see picture where the first position is filled up with a symbol, the next two positions are empty, and the last two positions are filled up with a symbol. We are given two choices of letters, K and Y, which are supposed to correspond to the visual display.

The ***audio*** representation of the picture is **knock, double knock, double knock, knock, and knock.**

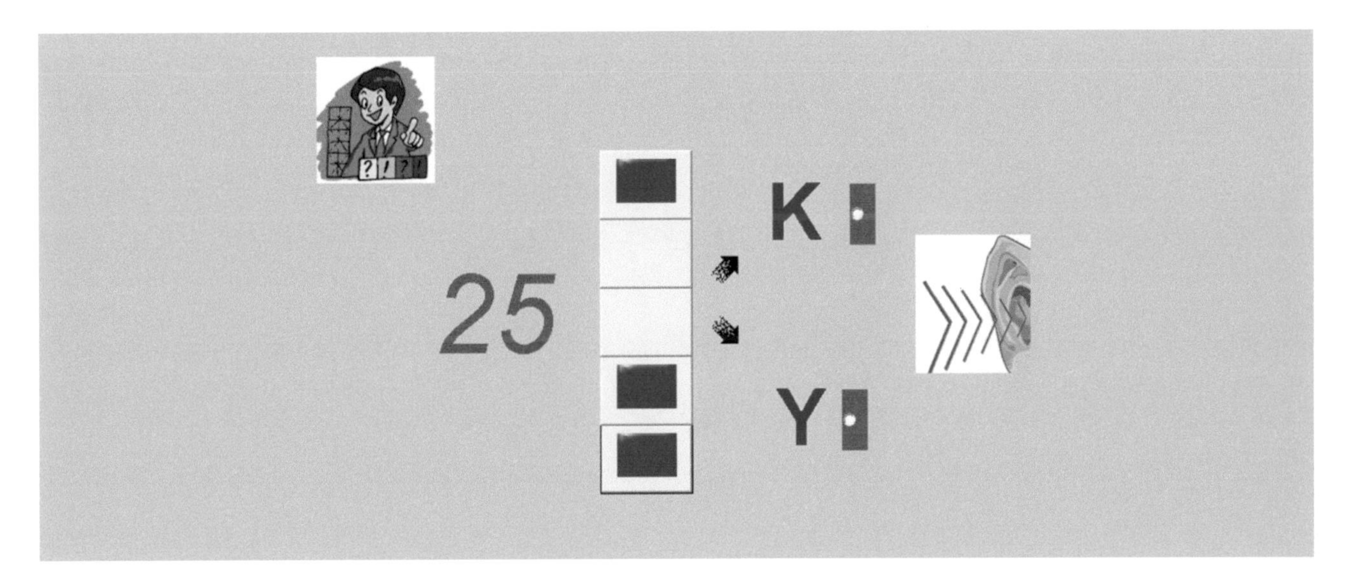

The knock represents a filled up position and a double knock represents an empty position. In our example knock and double knock are corresponding to the letter K. The clamping of the right hand represents a filled up position and the clamping of the left hand represents an empty position. In our example the ***kinesthetic*** representation for the answer is clamping of the right hand, left hand twice, and right hand twice. For letter K, the kinesthetic representation is clamping of left hand twice and right hand twice.

WHAT LETTER IS IT?

EXAMPLE 13.2

In example 4.4, in visual display we see picture where the first position is empty, the next position is filled up with a symbol, the next position is empty, and the last two positions are filled up with a symbol. We are given two choices of letters, N and Z, which are supposed to correspond to the visual display.

The ***audio*** representation of the picture is **double knock, knock, double knock, knock, and knock.**

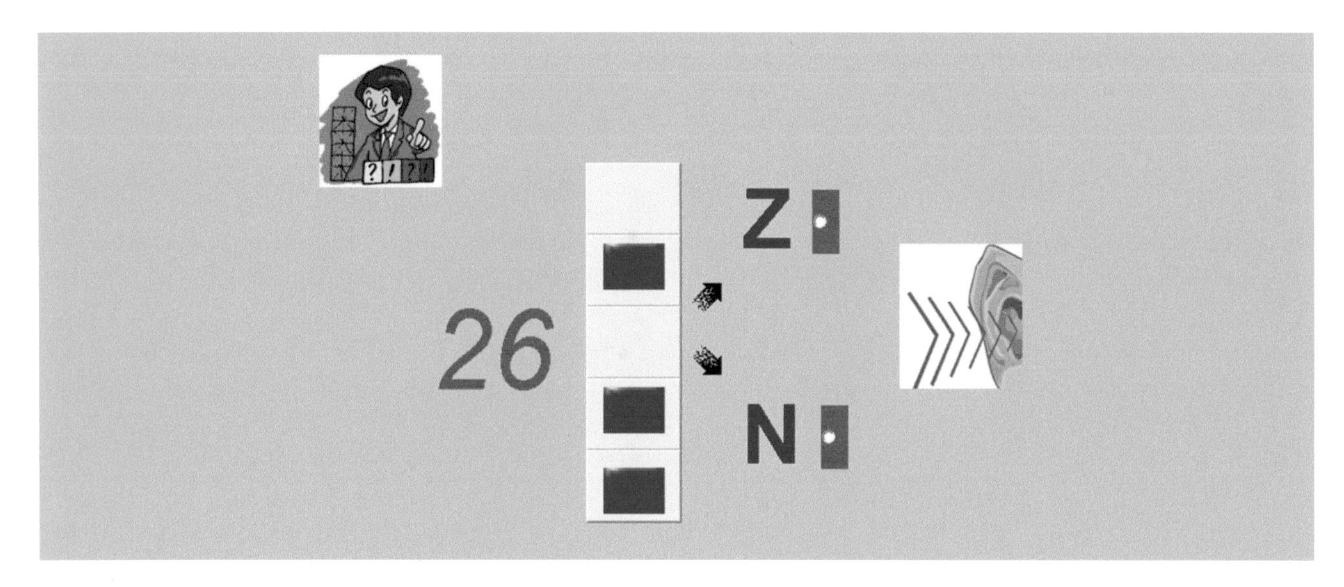

The knock represents a filled up position and a double knock represents an empty position. In our example knock and double knock are corresponding to the letter N. The clamping of the right hand represents a filled up position and the clamping of the left hand represents an empty position. In our example the ***kinesthetic*** representation for the answer is clamping of the left hand, right hand, left hand, and right hand twice. For letter N, the kinesthetic representation is clamping of left hand and right hand thrice.

WORDS – JUG, EXHIBIT 20

In exhibit 20, in ***visual*** display of the alphabet we see three pictures of boxes. The first picture has the second and fourth positions filled with a symbol; thus corresponding to the letter J and number 10. The second picture has the first, third, and fifth positions filled with a symbol; thus corresponding to the letter U and number 21. The third picture has the first, second, and third positions filled with a symbol; thus corresponding to the letter G and number 7. Together these three pictures make up the word JUG. The ***audio*** representation for the word consumes of a **double knock, knock, double knock, knock, chick, knock, double knock, knock, double knock, knock, chick, knock, knock, and knock.**

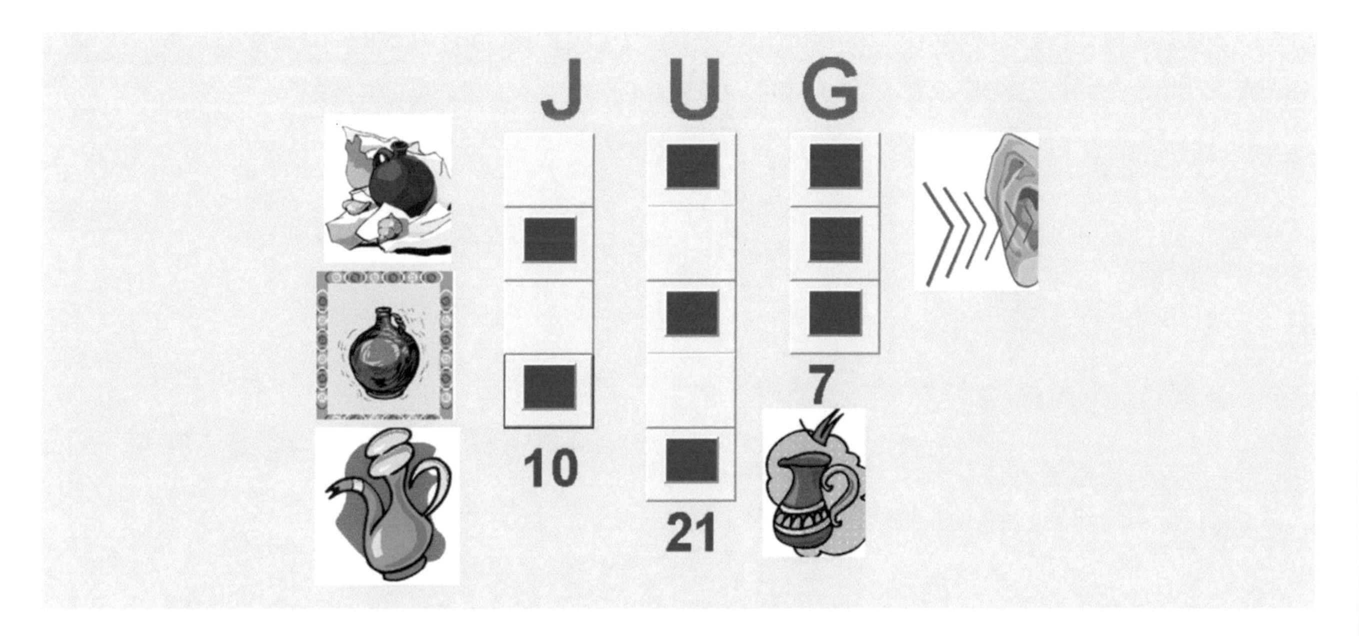

The ***kinesthetic*** representation of the filled up position is announced by clamping of the right hand. An empty position is represented by the clamping of the left hand. The chick represents sign of space between letters. The kinesthetic representation of the chick is one of the hands shaking. The corresponding numbers for the letters J-U-G are 10-21-7. The kinesthetic representation for the word JUG is the clamping of the left hand, right hand, left hand, right hand, one hand shaking, right hand, left hand, right hand, left hand, right hand, one hand shaking, and right hand three times.

ALPHABET

WORDS – SCALE, EXHIBIT 20

In exhibit 20, in ***visual*** display of the alphabet we see five pictures of boxes. The first picture has the first, second, and fifth positions filled with a symbol; thus corresponding to the letter S and number 19. The second picture has the first, and second positions filled with a symbol; thus corresponding to the letter C and number 3. The third picture has the first position filled with a symbol; thus corresponding to the letter A and number 1. The fourth picture has the first, third, and fourth positions filled with a symbol; thus corresponding to the letter L and number 12. The fifth picture has the first, and third positions filled with a symbol; thus corresponding to the letter E and number 5. Together these five pictures make up the word Scale. The ***audio*** representation for the word consumes of a **knock, knock, double knock, double knock, knock, chick, knock, knock, chick, knock, double knock, chick, knock, double knock, knock, knock, double knock, chick and knock, double knock, knock.**

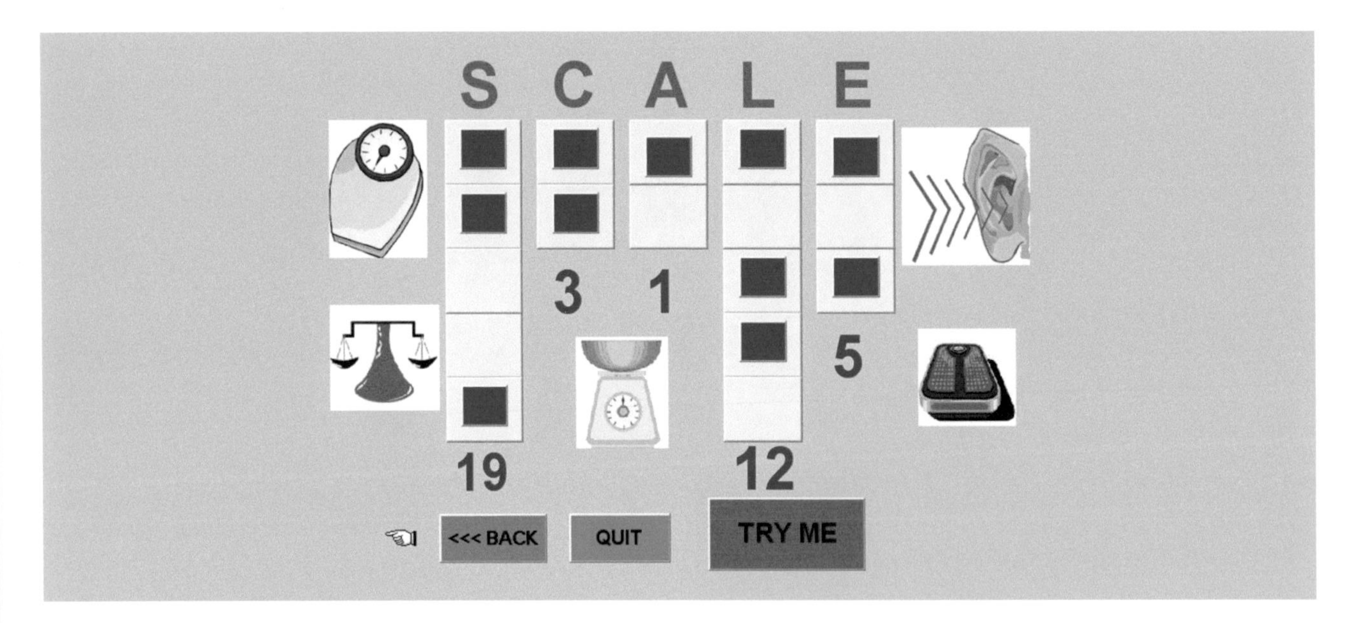

The ***kinesthetic*** representation of the filled up position is announced by clamping of the right hand. An empty position is represented by the clamping of the left hand. The chick represents space between letters. The kinesthetic representation of the chick is one of the hands shaking. The corresponding numbers for the letters S-C-A-L-E are 19-3-1-12-5. The kinesthetic representation for the word JUG is the clamping of the left hand, right hand, left hand, right hand, one hand shaking, right hand, left hand, right hand, left hand, right hand, one hand shaking, and right hand twice.

WORDS – SNAKE, EXHIBIT 20

In exhibit 20, in ***visual*** display of the alphabet we see three pictures of boxes. The first picture has the first, second and fifth positions filled with a symbol; thus corresponding to the letter S and number 19. The second picture has the second, third, and fourth positions filled with a symbol; thus corresponding to the letter N and number 14. The third picture has the first, second, and fourth positions filled with a symbol; thus corresponding to the letter A and number 1. The fourth picture has the first, second and fourth position filled with a symbol; thus corresponding to letter K and number 14. The fifth picture has the first and third position filled with a symbol; thus corresponding letter E and number 5. Together these three pictures make up the word SNAKE. The ***audio*** representation for the word consumes of a **knock, knock, double knock,double knock, knock chick, double knock, knock, knock, knock, chick, knock, double knock, chick, knock, knock, double knock, knock, chick, knock, double knock, knock.**

The ***kinesthetic*** representation of the filled up position is announced by clamping of the right hand. An empty position is represented by the clamping of the left hand. The chick represents sign of space between letters. The kinesthetic representation of the chick is one of the hands shaking. The corresponding numbers for the letters S-N-A-K-E are 19-14-1-11-5. The kinesthetic representation for the word SNAKE is the clamping of the left hand, left hand, right hand, right hand, left hand, one hand shaking, right hand, left hand, left hand, left hand, one hand shaking, left hand, right hand, one hand shaking, left hand, left hand, right hand, left hand, one hand shaking, left hand, right hand, left hand, one hand shaking.

WORDS – SOCK, EXHIBIT 20

In exhibit 20, in ***visual*** display of the alphabet we see three pictures of boxes. The first picture has the first, second and fifth positions filled with a symbol; thus corresponding to the letter S and number 19. The second picture has the first,second, third, and fourth positions filled with a symbol; thus corresponding to the letter O and number 15. The third picture has the first and second positions filled with a symbol; thus corresponding to the letter C and number 3 The fourth picture has the first, second, and fourth positions filled with a symbol; thus corresponding to the letter K and number 15. Together these three pictures make up the word JUG. The ***audio*** representation for the word consumes of a **knock, knock, double knock, double knock, knock, chick, knock, knock, knock, knock, chick, knock, knock, chick, knock, knock, double knock, knock, chick.**

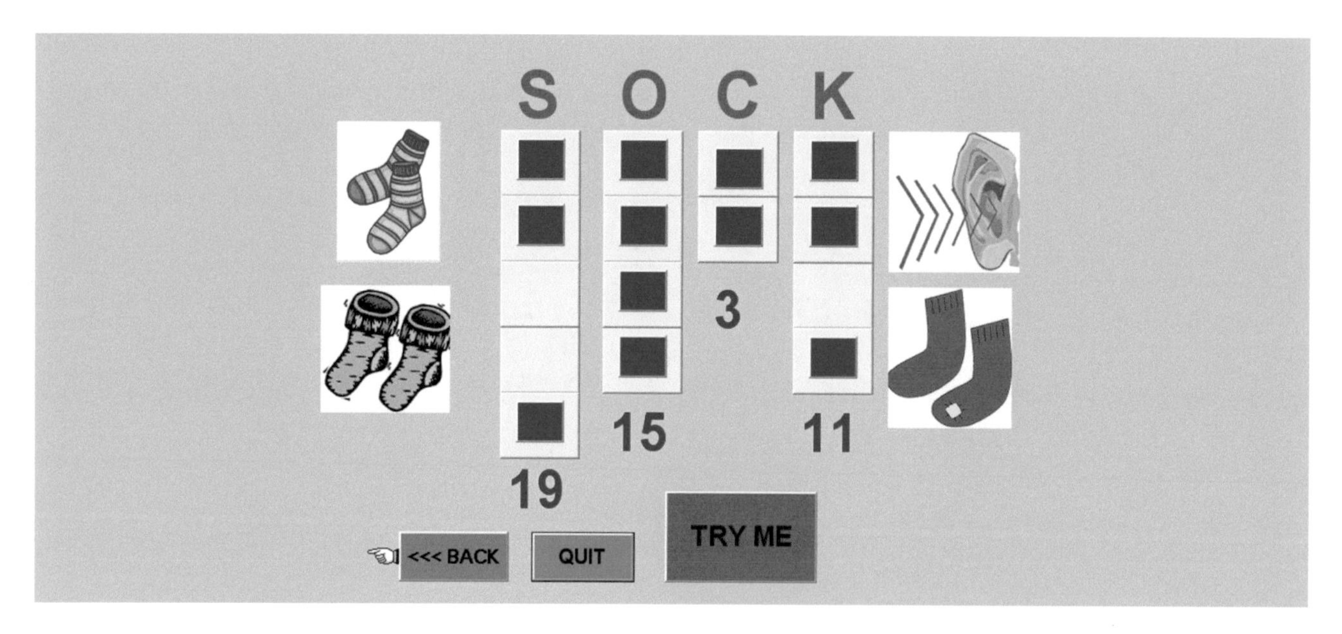

The ***kinesthetic*** representation of the filled up position is announced by clamping of the right hand. An empty position is represented by the clamping of the left hand. The chick represents sign of space between letters. The kinesthetic representation of the chick is one of the hands shaking. The corresponding numbers for the letters S-O-C-K are 19-15-3-11. The kinesthetic representation for the word SOCK is the clamping of the left hand, left hand, right hand, right hand, left hand, one hand shaking, left hand, left hand, left hand, left hand, one hand shaking, left hand, left hand, one hand shaking, left hand, left hand, right hand, left hand, one hand shaking.

WORDS – SUN, EXHIBIT 20

In exhibit 20, in ***visual*** display of the alphabet we see three pictures of boxes. The first picture has the second and fourth positions filled with a symbol; thus corresponding to the letter S and number 19. The second picture has the first, third, and fifth positions filled with a symbol; thus corresponding to the letter U and number 21. The third picture has the first, second, and third positions filled with a symbol; thus corresponding to the letter N and number 14. Together these three pictures make up the word JUG. The ***audio*** representation for the word consumes of a **knock, knock, double knock, double knock, knock, chick, knock, double knock, knock, double knock, chick, double knock, knock, knock,knock, chick.**

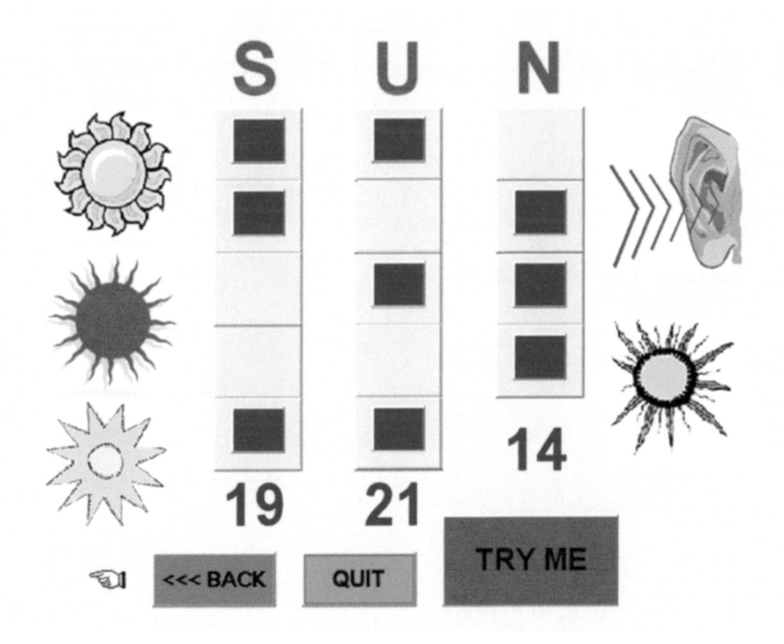

The ***kinesthetic*** representation of the filled up position is announced by clamping of the right hand. An empty position is represented by the clamping of the left hand. The chick represents sign of space between letters. The kinesthetic representation of the chick is one of the hands shaking. The corresponding numbers for the letters S-U-N are 19-21-14. The kinesthetic representation for the word SUN is the clamping of the left hand, left hand, right hand, right hand, left hand, one hand shaking, left hand, right hand, left hand, right hand, left hand, one hand shaking, right hand, left hand, left hand, left hand, one hand shaking.

ALPHABET

WORDS – TABLE, EXHIBIT 20

In exhibit 20, in ***visual*** display of the alphabet we see three pictures of boxes. The first picture has the first, second and fifth positions filled with a symbol; thus corresponding to the letter T and number 20. The second picture has the first position filled with a symbol; thus corresponding to the letter A and number 1. The third picture has the second position filled with a symbol; thus corresponding to the letter B and number 2. The fourth picture has the third and fourth positions filled with a symbol; thus corresponding to the letter L with the number 12. The fifth position has first and third positions filled with a symbol; thus corresponding to the letter E and number 15. Together these 5 pictures make up the word TABLE. The ***audio*** representation for the word consumes of a **knock, knock, double knock, double knock, knock, chick, knock, double knock, chick, double knock, knock, chick, double knock, double knock, knock, knock, chick, double knock, double knock, knock, knock, chick, knock, double knock, knock, chick.**

The ***kinesthetic*** representation of the filled up position is announced by clamping of the right hand. An empty position is represented by the clamping of the left hand. The chick represents sign of space between letters. The kinesthetic representation of the chick is one of the hands shaking. The corresponding numbers for the letters T-A-B-L-E are 20-1-2-12-18. The kinesthetic representation for the word TABLE is the clamping of the left hand, left hand, right hand, right hand, left hand, one hand shaking, left hand, right hand, one hand shaking, right hand, left hand, one hand shaking, right hand, right hand, left hand, left hand, one hand shaking, left hand, right hand, left hand, one hand shaking,

WORDS – TEETH, EXHIBIT 20

In exhibit 20, in ***visual*** display of the alphabet we see three pictures of boxes. The first picture has the third and fifth positions filled with a symbol; thus corresponding to the letter T and number 20. The second picture has the first and third positions filled with a symbol; thus corresponding to the letter E and number 5. The third picture has the first and third positions filled with a symbol; thus corresponding to the letter E and number 5. The fourth picture has the third and fifth positions filled with a symbol; thus corresponding to the letter T and number 20. The fifth position had the second and fifth positions filled with a symbol; thus corresponding to the letter H and number 18. Together these five pictures make up the word Teeth. The ***audio*** representation for the word consumes of a **double knock, double knock, knock, double knock, chick, knock, double knock, knock, chick, knock, double knock, knock, chick, double knock, double knock, knock, double knock, knock, chick, double knock, knock, double knock, double knock, knock, chick.**

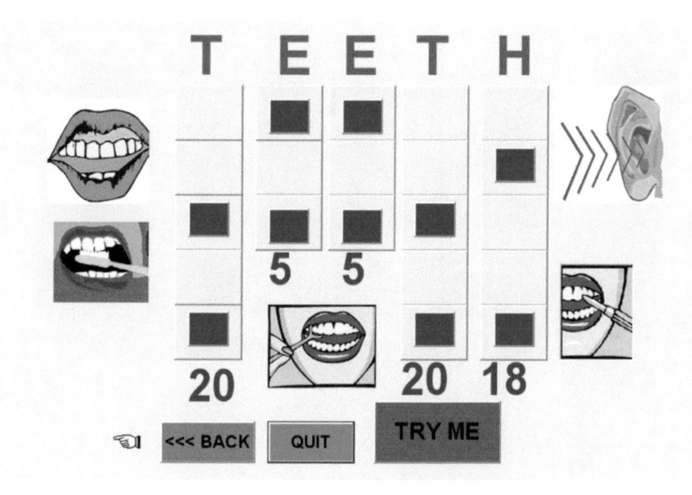

The ***kinesthetic*** representation of the filled up position is announced by clamping of the right hand. An empty position is represented by the clamping of the left hand. The chick represents sign of space between letters. The kinesthetic representation of the chick is one of the hands shaking. The corresponding numbers for the letters T-E-E-T-H are 20-5-5-20-18. The kinesthetic representation for the word TEETH is the clamping of the right hand, right hand, left hand, right hand, left hand, one hand shaking, left hand, right hand, left hand, one hand shaking, left hand, right hand, left hand, one hand shaking, right hand, right hand, left hand, right hand, left hand, one hand shaking, right hand, left hand, right hand, right hand, left hand, one hand shaking.

ALPHABET

WORDS – TIE EXHIBIT 20

In exhibit 20, in ***visual*** display of the alphabet we see three pictures of boxes. The first picture has the third and fifth positions filled with a symbol; thus corresponding to the letter T and number 20. The second picture has the first, and fourth positions filled with a symbol; thus corresponding to the letter I and number 9. The third picture has the first and third positions filled with a symbol; thus corresponding to the letter E and number 5. Together these three pictures make up the word JUG. The ***audio*** representation for the word consumes of a **double knock, double knock, knock, double knock, knock, chick, knock, double knock, double knock, knock, chick, knock, double knock, knock.**

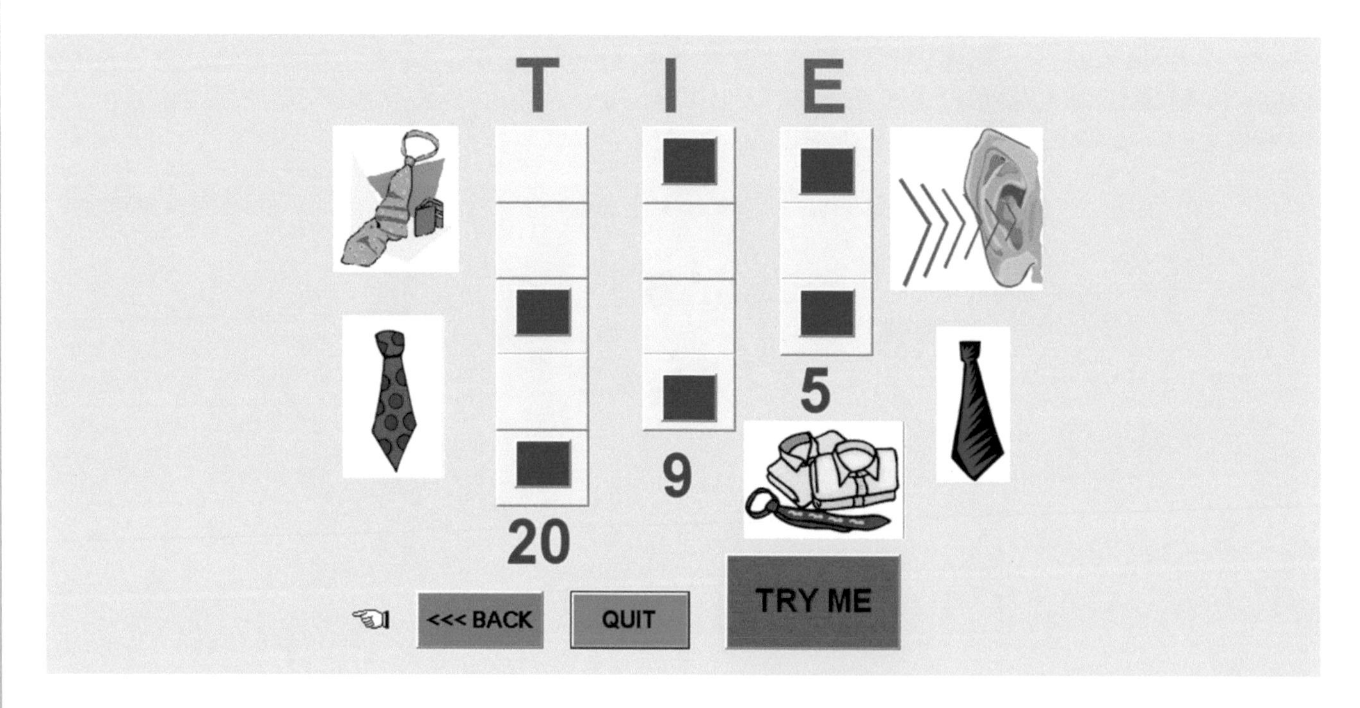

The ***kinesthetic*** representation of the filled up position is announced by clamping of the right hand. An empty position is represented by the clamping of the left hand. The chick represents sign of space between letters. The kinesthetic representation of the chick is one of the hands shaking. The corresponding numbers for the letters T-I-E are 20-9-5. The kinesthetic representation for the word TIE is clamping of the right hand, right hand, left hand, right hand, left hand, one hand shaking, left hand, right hand, right hand, left hand, one hand shaking, left hand, right hand, left hand.

WORDS – TIGER, EXHIBIT 20

In exhibit 20, in ***visual*** display of the alphabet we see three pictures of boxes. The first picture has the third and fifth positions filled with a symbol; thus corresponding to the letter T and number 20. The second picture has the first and fourth positions filled with a symbol; thus corresponding to the letter I and number 9. The third picture has the first, second, and third positions filled with a symbol; thus corresponding to the letter G and number 7. The fourth picture has the first and third positions filled with a symbol; thus corresponding to the letter E and number 5. The fifth picture has the second, and fifth positions filled with a symbol; thus corresponding to the letter R and number 18. Together these five pictures make up the word TIGER. The ***audio*** representation for the word consumes of a **double knock, double knock, knock, double knock, chick, knock, double knock, double knock, knock, chick, knock, knock, knock, chick, knock, double knock, knock, chick, double knock, knock, double knock, double knock, knock.**

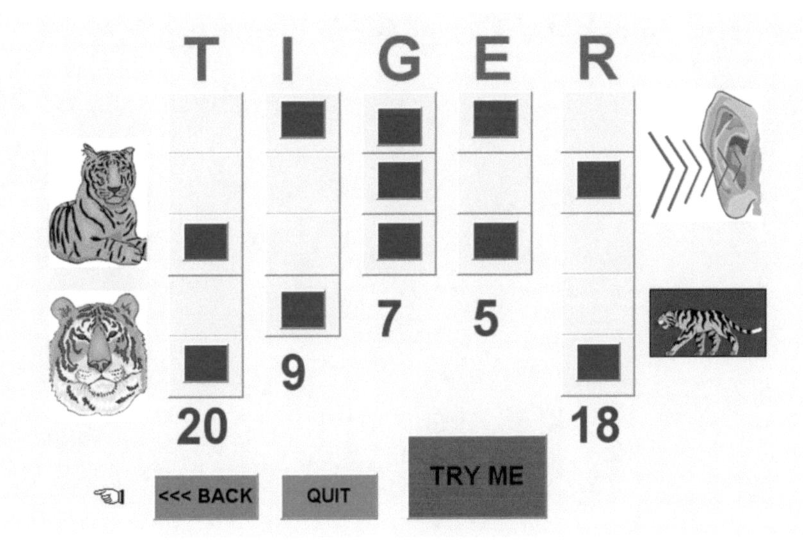

The ***kinesthetic*** representation of the filled up position is announced by clamping of the right hand. An empty position is represented by the clamping of the left hand. The chick represents sign of space between letters. The kinesthetic representation of the chick is one of the hands shaking. The corresponding numbers for the letters T-I-G-E-R are 20-9-7-5-18. The kinesthetic representation for the word TIGER is the clamping of the right hand, right hand, left hand, right hand, left hand, one hand shaking, left hand, right hand, right hand, left hand, one hand shaking, left hand, left hand left hand, one hand shaking, left hand, right hand, left hand, one hand shaking, right hand, left hand, right hand, right hand, left hand.

ALPHABET

WORDS – TURTLE, EXHIBIT 20

In exhibit 20, in ***visual*** display of the alphabet we see three pictures of boxes. The first picture has the third and fifth positions filled with a symbol; thus corresponding to the letter T and number 20. The second picture has the first, third, and fifth positions filled with a symbol; thus corresponding to the letter U and number 21. The third picture has the second and fifth positions filled with a symbol; thus corresponding to the letter R and number 18. The fourth picture has the third and fifth positions filled with a symbol; thus corresponding to the letter T and number 20. The fifth picture has the third and fourth positions filled with a symbol; thus corresponding to the letter L and number 12. The sixth picture has the first and third positions filled with a symbol; thus corresponding to the letter E and number 5. Together these six pictures make up the word TURTLE. The ***audio*** representation for the word consumes of a **double knock, double knock, knock, double knock, knock, chick, knock, double knock, knock, double knock, knock, chick, double knock, knock, double knock, double knock, knock, chick, double knock, double knock, knock, double knock, knock, chick, double knock, double knock, knock, knock, chick knock, double knock, knock.**

The ***kinesthetic*** representation of the filled up position is announced by clamping of the right hand. An empty position is represented by the clamping of the left hand. The chick represents sign of space between letters. The kinesthetic representation of the chick is one of the hands shaking. The corresponding numbers for the letters T-U-R-T-L-E are 20-21-18-20-12-5. The kinesthetic representation for the word TURTLE is the clamping of the right hand, right hand, left hand, right hand, left hand, one hand shaking, left hand, right hand, left hand, right hand, left hand, one hand shaking, right hand, left hand, right hand, right hand, left hand, one hand shaking, right hand, left hand, right hand, right hand, left hand, one hand shaking, right hand, right hand, left hand, left hand, one hand shaking, left hand, right hand, left hand.

WORDS – UMBRELLA, EXHIBIT 20

In exhibit 20, in ***visual*** display of the alphabet we see three pictures of boxes. The first picture has the first, third and fifth positions filled with a symbol; thus corresponding to the letter U and number 21. The second picture has the first, third, and fourth positions filled with a symbol; thus corresponding to the letter M and number 13. The third picture has the second positions filled with a symbol; thus corresponding to the letter B and number 2. The fourth picture had the first and third positions filled with a symbol; thus corresponding to the letter R and number 18. The fifth picture has the first and third positions filled with a symbol; thus corresponding to the letter E and number 5. The sixth picture has the second and fourth positions filled with a symbol; thus corresponding to the letter L and number 12.The seventh picture has the second and third positions filled with a symbol; thus corresponding to the letter L and number 12. The eighth picture has the first position filled with a symbol; thus corresponding to the letter A and number 1. Together these eight pictures make up the word U-M-B-R-E-L-L-A. The ***audio*** representation for the word consumes of a **knock, double knock, knock, double knock, knock, chick, knock, double knock, knock, knock, chick, double knock, knock, chick, double knock, knock, double knock, double knock, knock, chick, knock, double knock, knock, chick, double knock, knock, double knock, knock, chick, double knock, knock, double knock, knock, chick, knock, double knock.**

The ***kinesthetic*** representation of the filled up position is announced by clamping of the right hand. An empty position is represented by the clamping of the left hand. The chick represents sign of space between letters. The kinesthetic representation of the chick is one of the hands shaking. The corresponding numbers for the letters U-M-B-R-E-L-L-A are 21-13-2-18-5-12-12-1. The kinesthetic representation for the word UMBRELLA is the clamping of the left hand, right hand, left hand, right hand, left hand, one hand shaking, left hand, right hand, left hand, left hand, one hand shaking, right hand, left hand, one hand shaking, right hand, left hand, right hand, right hand, left hand, one hand shaking left hand, right hand, left hand, one hand shaking, right hand, left hand, right hand, left hand, one hand shaking, right hand, left hand, right hand, left hand, one hand shaking, left hand, right hand.

WORDS – JUG, EXHIBIT 20

In exhibit 20, in ***visual*** display of the alphabet we see three pictures of boxes. The first picture has the first, third and fifth positions filled with a symbol; thus corresponding to the letter U and number 21. The second picture has the first and fifth positions filled with a symbol; thus corresponding to the letter S and number 19. The third picture has the first, positions filled with a symbol; thus corresponding to the letter A and number 1. Together these three pictures make up the word USA. The ***audio*** representation for the word consumes of a **knock, double knock, knock, double knock, knock, chick, knock, double knock, double knock, double knock, knock, chick, knock, double knock.**

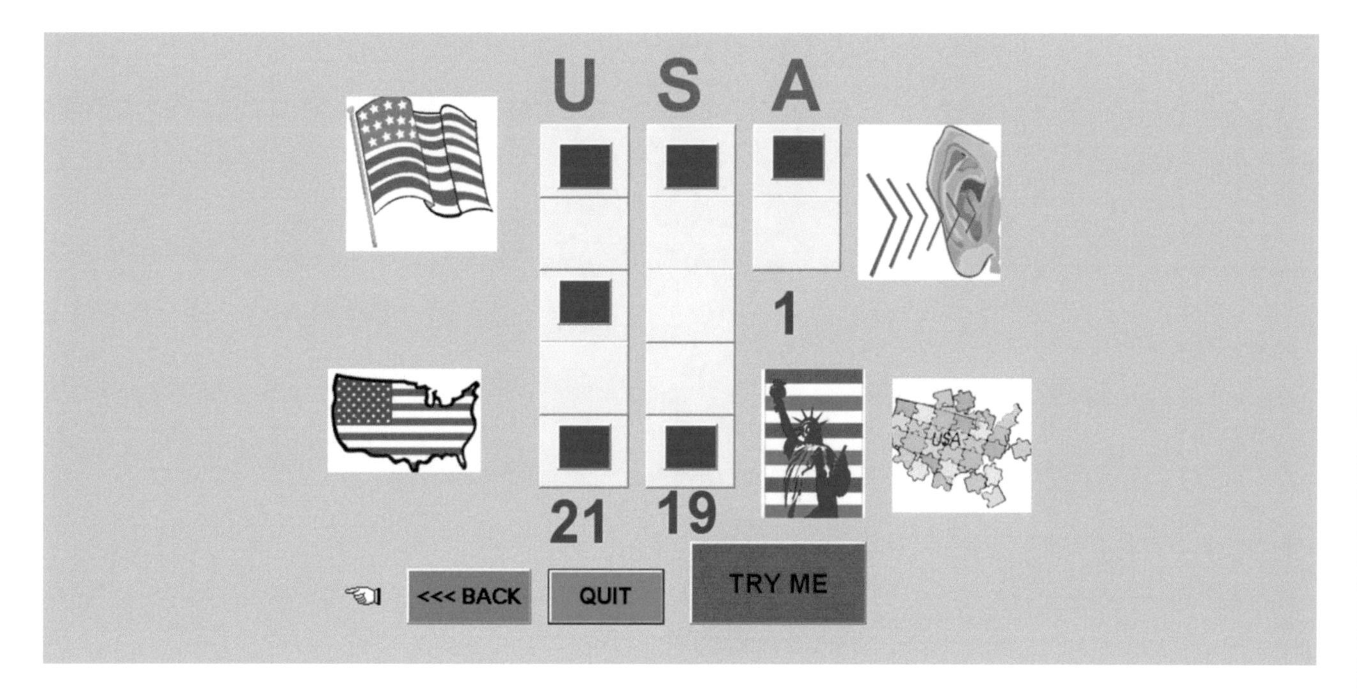

The ***kinesthetic*** representation of the filled up position is announced by clamping of the right hand. An empty position is represented by the clamping of the left hand. The chick represents sign of space between letters. The kinesthetic representation of the chick is one of the hands shaking. The corresponding numbers for the letters U-S-A are 21-19-1. The kinesthetic representation for the word USA is the clamping of the left hand, right hand, left hand, right hand, left hand, one hand shaking, left hand, right hand, right hand, right hand, left hand, one hand shaking, left hand, right hand.

Psychoconduction

Chester Litvin, Ph.D.

Dedicated to my nephew, David Gimelfarb,
Lost in Costa Rico in 2009.

Psychoconduction is a new neuro- psychological discovery that targets the impairment of the brain and provides remedies. Psychoconduction has four parts:

1) By using psychoconduction we calibrate, balance, and align different parts of the brain against each other. We use patterns of video, audio, kinesthetic, and olfactory expressions, which have the same meaning. The same meanings of patterns we express in different forms. The audio patterns we express in sounds. The video patterns we express in visual symbols. The kinesthetic patterns we express in the body or an object's movements. The olfactory patterns we express using different smells. We translate the same patterns to different modes of expression. In treatment we use the variety of the stimulus. For example, for treatment of dyslexia we focus on audio and visual patterns, but to treat the coordination or movement we emphasize kinesthetic patterns.

2) By using psychoconduction, we are able to modify the simple brain cells to provide replacement of damaged complex cells, and find the right direction for needed complex cells. We employ the simple cell to function in the same capacity as the complex cells. We help the non-functional complex brain cells to provide the needed function. Our goal is not to replace the damaged complex cells, but to assist the partially damage complex cells to function better by using simple cells. Sometimes we can not process the information by complex cells because the information was sent to the wrong address. The brain cannot process the requested information in the wrong area. The simple cells act as guides and help the stimuli to find the right place. The simple cells send the stimuli to the correct address, where the request will be correctly processed. We developed Litvin's Code to have references and create the different modes for patterns of the simple symbols.

3) By using the theory of psychoconduction we can do therapy to balance our emotional experiences. We process our audio, visual, kinesthetic, and olfactory experiences and comparing them with our life experiences in similar areas.

4) By using Litvin's Code psychoconduction we increase the brain capacity in processing of information. With Litvin's Code we create a system of patterns to balance the brain and to restructure the brain cells to be able to correctly process complex information. Psychoconduction has enormous capacities to create patterns by utilizing many existing systems. Litvin's Code is one of the systems to improve the brain's ability to process information. Litvin's Code represents digits, letters, addition, subtraction, multiplication, division, words and sentences by using binary numbers. With Litvin's Code we create many different patterns, which are easily absorbed by the brain, in the visual, audio, kinesthetic and olfactory modes.

Contents

Theory

Psychoconduction is a comprehensive system for the correction and enhancement of brain capacity. With this process we can enhance the existing structure, restructure, tune, calibrate and recalibrate the different areas of the brain without any pharmacological intervention. All the brain cells generally can be distinguished as complex or simple, even a brain that has different purposed cells to process diverse information. In the brain all cells are purposed cells. The brain contains the book of addresses, where to direct the received information. In a brain the simple cells have a great capability to help or even to replace the complex cells to process information. The simple cells reestablish lost address to the place in the brain, where the purposed information is processed. To stimulate the brain and to build the stable references we translate the sequences of simple symbols from one mode of expression to another. The mode of expression could be visual, audio, kinesthetic, tactile, olfactory, etc. We do not count the amount of simple and complex cells but look at the effect of treatment. When complex cells can not process information correctly, it means that complex cells are damaged or the request did not reach them.

Psychoconduction is a comprehensive system for correction and enhancement of brain capacity. With this process we can enhance the existing structure, restructure, tune, calibrate and recalibrate the different areas of the brain without any pharmacological intervention. Psychoconduction allows simple cells to function as complex cells within the different areas of the brain. Originally, the brain's structure uses simple brain cells to process simple stimuli and complex brain cells to process more complex stimuli. Psychoconduction also establishes addresses to reach the complex cells to process appropriate stimuli.

Psychoconduction focuses on simple stimuli to utilize simple brain cells in different capacities. We have many disorders because of the deficiency of the brain to correctly process visual, audio, kinesthetic and olfactory information. As a result, we have confusion, misunderstanding, anxiety, despair, anger, overreaction and learning disabilities. The main reason for the difficulties described above are the inability of the impaired or underdeveloped complex brain cells to correctly process complex information. Normally the visual, audio, kinesthetic or olfactory stimuli trigger the release of chemical messenger molecules such as dopamine, SSRI, and non-epinephrine to get a correct reaction. When a stimulus is impossible to process because of incorrect or ambivalent addresses, the impulse travels down to incorrect places. It ends up in a different area from where it was intended. This is why even the undamaged complex brain cells cannot be provided with the right information.

The areas of the brain, which are involved in vision, are the occipital, parietal and temporal cortexes. The area where the brain is processing the auditory information includes the primary auditory cortex

(PAC) in the superior temporal gyrus of the temporal lobe. The kinesthetic stimulus is processed in the cerebellum, which is responsible for the appropriate execution of motor commands. The olfactory stimulus is processed in the olfactory mucosa (OM), which processes different odors. We are using only simple cells in those areas, which are responsible for the processing of the simple stimulus. In psychoconduction we are stimulating areas of the brain, which usually process complex symbol, with a simple visual, audio, kinesthetic or olfactory symbol.

We are enhancing the brain's capacity by stimulating the brain while gradually increase the complexity of the simple symbols. We use the simple cells in the brain to process complex information. Psychoconduction stimulates with the same patterns the different parts of the brain. For example, we translate kinesthetic symbols to audio and visual. During the translation we attune the various areas of the brain. We can process the kinesthetic symbol in an area which is processing the visual and audio symbol. When we are processing the same pattern of the simple symbol, we are achieving congruence of the chemicals released by the different areas of the brain. This is similar to biofeedback of the brain while the brain corrects itself and creates the equilibrium between different areas of the brain in response to the same stimulus.

Psychoconduction helps to correct inappropriate responses to the stimulus. When we use only one area of the brain to process information, the information can be distorted. The inadequate response is an overreaction or underestimation of the information. To correct those problem we use the patterns created by Litvin's Code when we are stimulating different parts of the brain with simple symbols. We enhance the learning capacity of the brain. By using Litvin's Code we create the connection between different parts of the brain. The impulses from the same patterns of simple symbols trigger the release of equivalent chemicals in different parts of the brain. We create the equilibrium between different parts of the brain.

Psychoconduction relies on the simple symbols in assigned positions. We use the patterns created with Litvin's Code.

When the brain is not processing particular information, the reason is that the complex cells, which were purposed to process this kind of information, are partially or completely damaged, or the request did not come to the right place. To get to the right place and to be able to process information, we are introducing the simple symbol (visual - dot, audio - knock, kinesthetic - simple movement). We are putting the simple cells in the sequences, which are coded with the complex information, which the simple cells are able to process. The subjects, who have difficulty processing complex information through the coded complex cells, are able to process the same complex information through simple cells. We are introducing a simple symbol in the complex patterns and translating them between different modes of expression.

With psychoconduction we stimulate the areas of the brain, which are responsible for processing audio, visual, kinesthetic and olfactory information. Psychoconduction is the method to increase capacity for communication between different areas of the brain and enabling them to process the complex patterns. Psychoconduction allows the processing of the same patterns by using different areas of the brain. Psychoconduction is implying that many psychological disorders caused by the brain's difficulty in processing the information in certain areas of the brain. The areas responsible for processing of audio, visual, kinesthetic and olfactory information are failing to recognize and properly deal with the complex information of this nature.

The information has established roads for delivery. The kinesthetic information is transferred through the object's movement. The audio information is transferred through sound, and visual information is transferred through images. The simple kinesthetic symbol could be created by any of the movements, mainly of the body or the object. For example, we move with the body from front to back. The simple audio symbol is a simple sound, like a knock. The simple visual information is a dot or dash. Psychoconduction is the way of transporting stimuli through the simple brain cells.

Even damaged brains have no difficulties receiving stimuli provided by these simple symbols. With psychoconduction the brain process the information in the special area allocated for the simple information and moves the processed information to the areas of the complex cells. We achieve the equilibrium when we use different areas of the brain and adequately process the same patterns of the simple symbol through the visual, audio, kinesthetic and olfactory expressions. We are enhancing with the correct and effective information the ineffective or damaged complex cells by using the simple symbols. As a result we simplify and clarify the confusing and even incorrect information from complex stimulus.

Psychoconduction allows references by translating simple symbols in different expressions to ensure the correct translation of the symbol. We use Litvin's Code to create different patterns of simple symbols.

The psychological difficulties begin when the brain does not receive the adequate information to process. It makes conclusions, which are confusing, scarring, and agitating. For example, in scary movies we hear a very intense sound but we do not receive confirming visual information. Sometimes we see the feet of the killer but not his face and then expect that everyone is a killer. Sometimes the scary images are accompanied by the sound, like the telephone ringing, teakettle whistling, etc., which are not related to the plot of the movie. This is how incongruities are created, followed by confusion and fear.

The intense sound prepares us for the alarming situation but visually we do not have enough information to process. We jump to conclusions by using our imagination that the crime will be committed and feel frightened. The movie director does not need to do much work because we process the audio stimuli without any visual stimulus and become very scared. When the information is incomplete or confusing, the responses are fear, anxiety, anger and emotional instability. We also understand that a person, who has difficulty in processing information, avoids the intellectual stimulation in the areas of difficulties. One, who has difficulty with coordination, avoids dancing, and one, who has dyslexia, avoids spelling.

Method

The simple symbol is recognized by the simple brain's cells. We treat subjects with difficulty processing information by translating the sequences of the simple symbol from the one mode of expression to another. We see the improvement in the subject's ability to process complex information and to return to the normal representation of digits and letters.

Psychoconduction consists of four different parts. The first part is the calibration of the brain, with the goal to increase the brain's intellectual capacity. It deals with the training of the brain to achieve congruency in responses to visual, audio, kinesthetic, and olfactory stimuli. The brain is exposed to different pattern of stimuli without explanation of the logical connection in the pattern's design. In the second part, the method is providing changes in the structure of the brain cells and allowing the non-functional cells to be reinforced or replaced. The third part includes the therapy, which reduces the discrepancies in the perceived and real emotional experiences and increases the equilibrium of emotional experiences. The fourth part of psychoconduction, called Litvin's Code, provides unlimited amounts of patterns to balance and restructure the brain cells. Litvin's Code is based on binary arithmetic and is made for easy absorption by the brain and is easily translated to audio, visual, kinesthetic and olfactory stimuli.

Calibration of the brain

By translating the sequences of the simple symbol from one mode of expression to another we create a system of references between different modes expression. This process produces a calibration of the release of the brain chemicals. The complex cells, which are controlling the brain's release of chemicals are damaged and can not regulate the exact amount of chemicals needed to process a stimuli. By translating the sequences of simple symbols from one mode of expression to other, we are balancing the brain. The simple cells are creating the system of references on different modes of expression and are helping to release a balanced amount of brain chemistry.

With psychoconduction we create the congruence between different modes of communication. We provide training for the different parts of the brain to release the congruent amount of chemicals in reaction on the audio, visual, kinesthetic and olfactory stimuli. The training includes the use of patterns, and to translate them to different stimuli. These patterns become more and more complicated. We translate them from one of the modes of expression to the other. The recognition and processing of the same information transmitted in the audio, visual, kinesthetic and olfactory forms provides the calibration of different parts of the brains against each other. The goal of the training is to achieve congruence in the response of different parts of the brain and in the different forms of representation. We are providing the examples of patterns, which are combined in Litvin's code. In our examples Litvin's Code is combinations of positions, which are filled up with symbols or empty. We do not need to know the patterns' meaning and their references. We are just translating the patters from one representation to another.

Example 0.1

•	•	

In example 0.1 the audio representation of the pictures is **knock, knock and double knock**. We are balancing our visual and audio reaction on the symbols knock and double knock. The knock represents a filled up position and double knock represents an empty position. Kinesthetic representation for knock is clamping the right hand and double knock is tightening the left hand into a fist. Clamping the right hand then represents the filled up position and the left hand clamping represents empty position. In our example, kinesthetic representation of the example 0.1 is to clamp the right hand twice then the left once. When we have the board with four sides, we are stepping on one side to show that the first position is filled up. We are moving to another side and step on the board again to show that the second position is filled up. To show that third position is empty we are moving to the next side of the board without stepping on the board.

Example 0.2

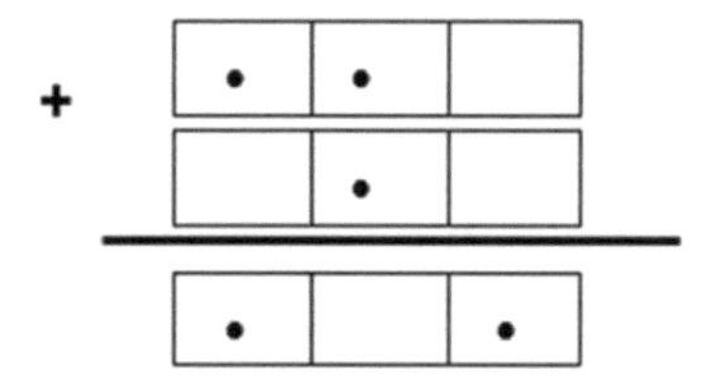

Example 0.2 brings us to a more advanced level of balancing different parts of our brain on the same patterns using different modes of communication. In example 0.2 the audio representation of the pictures is **knock, knock, double knock, tram, double knock, knock, double knock, cling, knock, double**

knock, and knock. We are balancing our visual and audio reaction using the symbols knock, double knock, tram, and cling. The knock represents a filled up position, double knock represents an empty position, tram represent visual sign plus, and cling represents visual sign equal. Kinesthetic representation for knock is clamping the right hand, double knock is tightening the left hand into a fist, tram is extended left hand, and cling is crossed two hands. Clamping the right hand represents the filled up position, the left hand clamping represents empty position, and extending one hand represents sign plus, crossed both hands represent sign equal. Kinesthetic representation for the hand's movement is as follows: clamping the right hand twice, the left hand once, extend the one hand, to clamp the left hand, to clamp the right hand, to clamp the left hand again, to cross both hands, to clamp the right hand, to clamp the left hand, and, eventually, to clamp the right hand.

If we have a board with four sides and we are moving around it, then we are stepping on one side to show that the first position is filled up. We are moving to another side and stepping on the board to show that the second position is filled up too. Eventually, we are moving to the third side of the board without stepping on it to show that the third position is empty. We extend one hand in front to show sign plus. We are moving to the next side without stepping on the board to show that position is empty. We are moving to the next side and stepping on the board to show that position is filled up with a symbol. We are moving to the next side without stepping on the board when the position is empty. We are crossing both hands. We move to another side and stepping on the board because position is filled up. We move to another side of the board without stepping on because the corresponding position is empty. The next position is filled up, and we are moving to another side and stepping on the board.

Figure 0.3

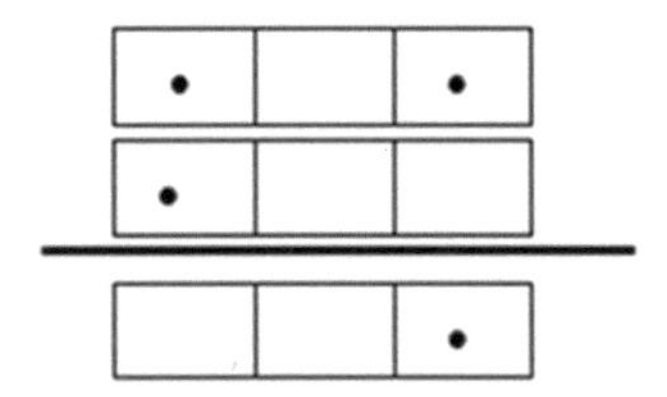

In example 0.3, the audio representation of the pictures is **knock, double knock, knock, double tram, knock, double knock, double knock, cling, double knock, double knock, and knock**. The double tram represents sign minus. The kinesthetic representation of the double tram is both hands extended in front. The kinesthetic representation for hands movement is: right, left, right, both extended in front, right, left, left, both crossed, left, left and right. The kinesthetic representation for the move around the board is:

step on, move without stepping, step on, extend both hands in front, step on, without step, without step, cross both hands, without step, without step, and step on.

Figure 0.4

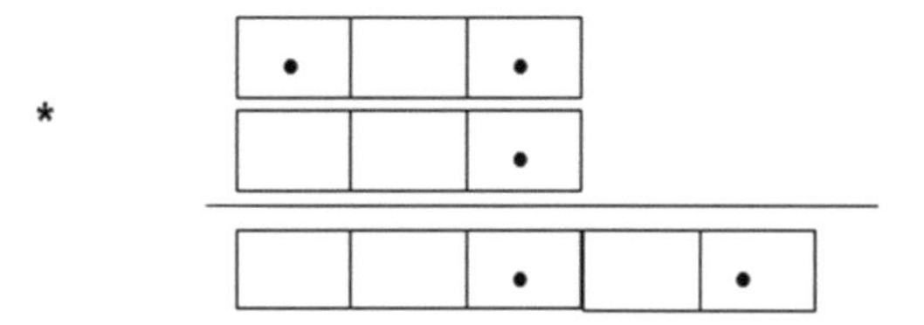

In example 0.4, the audio representation of the pictures is **knock, double knock, knock, blick, double knock, double knock, knock, cling, double knock, double knock, knock, double knock and knock**. The blick represents visual sign to multiply. The kinesthetic representation of the blick is one hand extended up. The kinesthetic representation for hands movement is: right, left, right, one hand extended up, left, left, right, both crossed, left, left, right, left and right. Kinesthetic representation for the move around the board is: step on, move without stepping, step on, extend one hand up, without step o, without step, step on, cross both hands, without step, without step, step on, without step, step on.

Figure 0.5

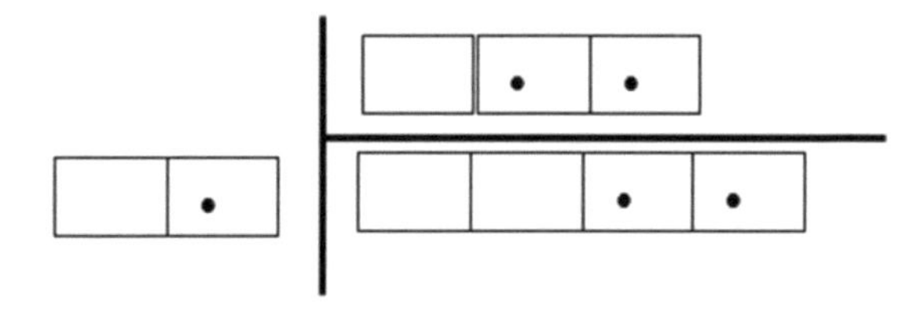

In example 0.5, the audio representation of the pictures is **double knock, double knock, knock, knock, double click, double knock, knock, cling, double knock, knock, and knock**. The double blick represents sign to divide. The kinesthetic representation of the double blick is both hands extended up. The kinesthetic representation for hands movement is: left, left, right, both hands extended up, left, right, both crossed, left, right, and right. Kinesthetic representation for the move around the board is: without step, move without stepping, step on, step on, extending both hands up, without step, step on, cross both hands, without step, step on, and step on.

Figure 0.6

A R M

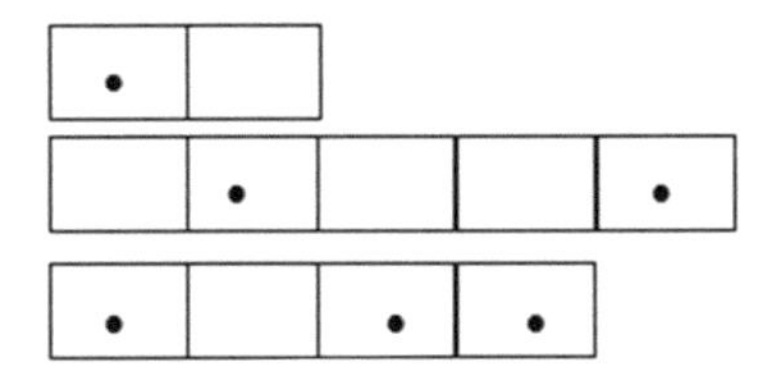

In example 0.6, the audio representation of the pictures is **knock, double knock, chick, double knock, knock, double knock, double knock, knock, chick, knock, double knock, knock, and knock**. The chick represents sign of space between letters. The kinesthetic representation of the chick is one of the hands shaking. The kinesthetic representation for hands movement is: right, left, left, left, one of hands shaking, left, right, left, left, right, one hand shaking, right, left, right, and right. Kinesthetic representation for the move around the board is: step on, without step, without step, without step, shaking one hand, move without stepping, step on, without step, without step, step on, shaking one of the hands, step on, without step, step on, and step on.

Structuring of brain cells

When we describe the structuring of brain cells, psychoconduction can somehow to be compared with the primitive stem cells. We know that the stem cells could be transferred to dysfunctional cells to get them function. We have discovered that, with the help of the simple brain cells, we also can help the complex information, and the simple cells can function as complex cells. As it is evident in the past that the simple brain cells responding to stationary stimuli such as lines, bars, and edges, whereas the complex cell responds to more complex stimuli and more sophisticated forms. The difficulty of the brain to process complex information is mainly attributed to difficulty of complex cells to process complex information and difficulty receiving information because of incomplete or distorted requests. The main goal of the brain's simple cells is to have clear request with the not distorted data to process and then finding of the right address for requested information. Psychoconduction is simplifying the request to process information, and, by modifying the simple cells to process the same information as complex do, is assisting in finding the right address of the complex cells.

By the creation of the new functional cells we can process the complex information. We build the congruence in the response to the audio, visual, kinesthetic and olfactory stimuli. It is the main essence of the psychoconduction. Psychoconduction helps to achieve better usage of simple cells by using simple stimuli in complex patterns to substitute the use of injured or underdeveloped complex brain cells. We use the simple cells, which are responsible for processing the simple stimulus. We enlarge the brain capacity to process the information. We stimulate the brain with the simple symbols and transferring them between the areas of the brain. These transfers are between the areas responsible for different functions. The transfers are: audio to the visual, audio to kinesthetic, visual to audio, visual to kinesthetic, kinesthetic to visual, kinesthetic to audio and etc. We use the area of the brain processing olfactory functions by using simple smell symbols. We also can use tactile stimuli. These simple symbols are the visual, audio, kinesthetic or olfactory stimuli and create different responses in different areas of the brain. They are processed in different parts of the brain, but have the same logical references and are a part of the same system called Litvin's Code.

In psychoconduction we are significantly limiting the use of damaged complex brain cells which are the cause of confusion, anxiety and agitation. The same pattern of the simple symbol translated by the brain to different mode of expression is creating internal communication and enhancing the brain's equilibrium. By using a different part of the brain to process the same pattern, we compensate for inadequate information and help to avoid psychological difficulties. By creating congruent responses, the brain also provides the clarity of the logical patterns of simple symbol and opportunity for all different parts of the brain to enhance performance.

Psychoconduction is a logical system, which allows one to place the simple symbol in the previously assigned positions with the references, which connect all letters and digits in logical order. We could use this system to facilitate processing of information in the different areas of the brain, which are responsible for processing of visual, audio, kinesthetic and olfactory stimuli. We can have unlimited amount of different symbols and patterns.

Psychoconduction is a process in which we are allowed to utilize the complex pattern using simple symbols. The brain is processing only the simple information with simple cells and it is different from the cells which process complex information. With Litvin's Code the complex patterns of information could be processed through the area of the brain by using the channels, which before used only for processing of the simple information. We found out that by transferring information, which contain the same patterns, to the different part of the brain, creates the opportunity for the brain to become more stimulated and to overcome previous difficulty by utilizing the complex cells.

In present time society is more aware that illiteracy or inadequate academic performance is not only caused by social disadvantages. It is also caused, in our opinion, by the neuro-psychological barriers. Those barriers create the difficulties to recognize or learn complex patterns needed to read or to do mathematical calculations etc. The brain has the ability to organize the complex and simple information. It is also important; when we process information received from audio, visual, kinesthetic and olfactory stimuli, we need to have congruency of responses from different areas of the brain to be able to properly process the incoming information. The lack of congruency creates distortions and difficulty to process information. It is also evident that the brain has Difficulty to process information then it needs some augmentation of the processing of the information. The deficiencies are not a barrier but the indication that this information is incomplete and can be enhanced through distinct channels of the brain. If the brain can not process information, it can be because the information would not reach its predisposed destination. It means that the system of addresses, which supervise correct delivery of information, is inadequate. The simple cells have a really simple delivery system and most likely get information delivered better than complex cells.

Psychoconduction provides the channels of the simple information, which are capable to process the complex information. For example, the same way as Braille's alphabet helps people with visual impairments, the psychoconduction helps people with brain deficiencies that have inability to properly process complex information through the complex brain cells. To overcome these difficulties, it is important to be open to the new ways of audio, visual and kinesthetic and olfactory communication. In our discovery instead of the areas of the brain processing complex information, we utilize the parts of the brain, which are responsible for the sequential order and hierarchy of positions. These areas are rarely impaired, and are able to do processing of complicated patterns of information. Through these areas we are able to stimulate the brain and enhance the intellectual capacities of people with the brain's complex cell impairments. Psychoconduction does not replace the complex process of reading or doing mathematical calculations. They only facilitate the process of intellectual stimulation and make it less stressful. Psychoconduction simplifies the recognition of the patterns needed to stimulate the brain and to improve academic performances.

The psychoconduction helps to use the brain's capacity and to enjoy the process of intellectual stimulation. The process of information is starting with stimulation of the brain with simple sounds and then to translate representation of this sound as a written symbols. We can also use the body movement to represent the simple symbols. We create the different channels from the use of audio, visual and kinesthetic faculty to represent simple symbol and to translate it in the different sequences of representation. By creating these sequences we activate the areas of the brain, which are responsible for organizing and keeping the hierarchy of the information. When the brain transfer a simple sound symbol to the same simple written symbol or the same simple kinesthetic symbol, the brain releases the equivalent chemicals while processing the same patterns of the simple stimuli. The brain without hearing or visual impairment is able to distinguish one sound of piano from the double sounds of piano and then to write dash or dot. For hearing and visual impaired brains we use tactile and kinesthetic symbols. The brain is able to associate the two piano sounds of the audio stimulation with dot and dashes of the visual stimuli. To show the presence of the symbol we can use for visual symbol the dot, for audio the knock, and for kinesthetic the movement of the right hand and for olfactory the smell of soap. For the absence of the symbol we use a dash as a visual, for the audio double knock, for kinesthetic the movement of the left hand, and for olfactory the smell of perfume.

One of the most important tasks in psychoconduction is the ability with the same pattern of simple symbols to stimulate the different parts of the brain. It is important, for people with difficulties in processing visual information to process visual symbols as the last and other symbols before to have valid references. In this situation, the first symbol can be the audio or kinesthetic and then they can translate it to the visual symbol. When we transfer the stimulation from the area of the brain responsible for audio information to the area responsible for visual information, we stimulate the brain with the patterns of audio symbol (knock) and then we translate to the same pattern of visual symbol (dot). We use the audio and kinesthetic symbol before we use the visual symbol. We create patterns for visual, audio, kinesthetic and olfactory stimulation.

We are transferring the stimulation from the area of the brain responsible for audio information to the area responsible for kinesthetic information. We stimulate the brain with the patterns of audio symbol (knock), and at the same time the brain responds with the same patterns of kinesthetic symbol (hand movement). Next we stimulate the area of the brain responsible for kinesthetic information with the same patterns to respond with visual information (dots and dashes). We translate the already processed patterns of audio information to the same patterns of kinesthetic information. In the next step we continue with transfer of the audio information to the visual and also from the kinesthetic to the visual. To reinforce the learning of olfactory stimulation we could use the simple smelling symbol (smell of soap) to learn the same patterns.

Our initial assessment clarifies that the brain has difficulty recognizing the patterns of complex symbols and also different area of the brain does not have adequate processing capabilities. With people who have a problem with audio information processing, we stimulate the area of the brain responsible for kinesthetic information and translating it to audio. We begin processing the information by stimulating the area of the brain that shows the most adequate capabilities. For example, if the visual symbol is shown as the best performance in stimulation of the brain, then we process the visual symbol first to create the reference for audio symbol. The stimulus, which has shown the worst performance, we process last. For example, if the audio symbol has shown the worst performance, then we process the audio symbol the last after we have the list of references ready.

To bring people with deficiency in the processing of complex information to the mainstream of general population, we need to bypass the damaged or underdeveloped part of the brain by using simple cells of the brain to bypass damaged complex cells. Our goal is to make this transition as simple and as effective as possible within the limits of individual capacity of the simple brain cells.

In psychoconduction by using Litvin's Code we place the symbol in different pre-learned positions and by doing it we change the meaning of the expression. The simple symbols in the different position contain different information. Litvin's Code is also a method, which is utilizing the brain's communication capacity. The different organic structures of the brain have different physiological functions. The simple cells are the majority in the brain's cell structure and usually they are not used. We start processing information through the simple cells. The simple cells direct the information to the complex cells and if the complex cells receive the right information they may be able to correctly process it.

Psychoconduction protects the complex cells from delivering distorted information and helps cells to correctly process it. Psychoconduction also corrects the information that was initially not received properly.

The brain is using different areas for the complex information like the contours of the words, and for simple symbols, like the dots and dashes and hierarchy of positions. Some of us have difficulty with recognition of complex symbols but are able easily recognize simple symbols.

In some cases we might have difficulty in identifying the numbers but easily recognize the decimal points or the signs of plus and minus. Psychoconduction is utilizing the area of the brain, which is processing the simple symbols. The use of Litvin's Code patterns are similar to the encrypting very sophisticated information, and only the person, who knows the key for decrypting, could understand it. By conditioning the brain to process information with simple patterns we could extend the brain recognition of more complex patterns, which will aid in academic field, social interactions, employment and business.

In the mean time, we do not have any evidence that the brain could not process the same information by utilizing only simple cells. We use the simple patterns as only a way to give the brain an assignment, but the result of the processing of the information is the same as the processing of complex patterns by the complex cells, and the academic training is not jeopardized at all. It is similar to the use of electric jump-start for the car's engine. We want to clear up the information from the distortion. We intensify the learning process, building confidence in us and we achieve success in the academic field by jump-starting our simple cells.

Therapy

In Litvin's Code the emphasis is on the positions to represent digits and letters. For example, usually, to dial the telephone number, we have memorized the number and then dial the numbers on the dial. On the other hand by using Litvin's Code, we remember location of the numbers on the dial rather then memorizing the actual numbers. Little children remember the positions of the numbers on the telephone dial. They also remember which places parents have hidden the toys without remembering the names of the shelves or drawers. The easiness of Litvin's Code is that we do not need the recognition of the actual digits or letters to do mathematical calculation or write a letter. In Litvin's Code instead of digit and letters we just recall the pre-learned position by checking if the position is occupied with a simple symbol or not. The notion of the positions has deep roots in our thinking. It helps us to have order in our thoughts and understand the concept of intensity and quantity. There are several alphabets that have the same symbols for digits and letters (E.G. Chinese, Greek, Russian, Arabic, Hebrew), but Litvin's Code moves further and utilizes the simple symbol to recognize the letters and digits by using the concept of the positions.

The psychoconduction is much wider than Litvin's Code. It is the regulatory process between verbalization and emotion. When the verbal expressions are not congruent with emotional expressions, it attributed to the deficiency of verbal or emotional intellect adequately processing the stimuli. The verbal intellect associated with the richness of the verbal expression. The emotional intellect shows deep feelings of love, compassion, and care. The verbal deficiency is associated with the difficulty to process information verbally. The emotional difficulty is associated with lack of feelings of comprehension, attention, and also with presence of different distortions. The emotional deficiency is also attributed to emotional frigidity, emotional numbness, as well as difficulty to focus and become taken away by different distortions. This approach is based on congruent processing of the verbal and emotional information. This is the self-regulatory process, which allows the person to take conscious control of the expressed emotions. It allows the person to compare the verbal and emotional experiences.

We trainin our modes of communication to be congruent and reduce the overreaction in some mode by opening the dialog with other modes of communication. For example, our brain is taking the reaction of the body sensation and translates it to the verbal expression. By comparing body reaction and verbal expression we could correlate and then recognize overreaction. By using self-regulation, we could reduce the increased heart palpitations and could regulate breathing. Our modes of communication became a congruent system could successfully communicate as a whole system with other people modes of communication, which could be verbal, kinesthetic, and visual.

When a person is experiencing anger and frustration, the body sensations or kinesthetic functions takie over all other functions. It manifests itself in increase heart palpitation, muscles tension and rapid breathing. In a state of anger and frustration, the individual's movement, verbal, and audio communications in this state of mind are down to a very simple level. An individual in this state is using simple movements, has difficulty using the verbal expressions, and difficulty understanding audio information, and he is reacting only on simple audio signals. In this state the complex cells are not working and communication is going on a very simple level. In the affective state of mind, the capacities of different modes of communication are shot down, and the fight or verbal assault is the only available response. When we train the brain to have congruent response to the stimulus, we train the audio, visual and kinesthetic information to respond adequately to the situation. If a person becomes angry and frustrated his verbal and audio modes of communication are adequate to the kinesthetic. Because in the highly emotional situation the complex cells are shut down, the simple cells could work as a complex one. The fighting might be avoided if the person is listening to other suggestions and solutions, and is verbalizing other options, and is using different vocabulary. The training in psychoconduction is helpful in the area of anger management and impulse control. We control different reactivity on the intensity of audio, visual, kinesthetic information and create the equilibrium in processing of information. When we are in the steam room and the unpleasant smell of body odor creates a felling of discomfort and anger, then verbalization of feelings and validation from others makes discomfort less significant. When we observe that the other people feel the same discomfort but tolerate it, then our anger is subdued.

The psychotic symptoms are mainly related to different hallucinations. The audio or visual hallucinations take over the person's life. The stressed or depressed people do not have hallucinations, but have intrusive thoughts, flash backs and nightmares. The hyperactive people have the exaggerated kinesthetic functions. They are not in control of their desire to move fast, appear restless, fidgeting, and producing rapid speech. The inattentive people react on insignificant details, detaching, and daydreaming. After training in psychoconduction the person, who is experiencing hallucinations or hyperactivity, is able to realize that some of his reactions are exaggerated. By providing training in psychoconduction with different patterns of the same symbol, we lower the intensity and exaggeration of response. The stimuli are visual, audio, kinesthetic and olfactory and the congruent responses on those stimuli not support the needs for hyperactivity or hallucinations and also act as the reality testing for hallucinations. Through different modes of communication the brain could get the message that hallucinations are not real, the flash backs and nightmares are about the past, and they do not reflect the safe present. Our reaction on the same precipitated factor is different from one situation to other. On the same stresses we react with accelerated body movement, loud voice, but on similar one we have nightmares and flashbacks. After the psychoconduction therapy the kinesthetic reaction would adjust to visual and audio reaction. The intensity of body reaction will decrease if the audio, visual and olfactory information does not support its intensity.

If kinesthetic response does not support the exaggerated reaction, visual and audio information could be adjusted by kinesthetic response. If we work in a chicken farm and exposed to intensive smell, but visual and audio information does not support any danger, then we are not agitated. The intensive olfactory stimuli do not create agitation because we see the helpless chicken and hear the non-threatening noises. Opposite, in the sauna little smell or body odor creates a different response. In sauna we are not trained to tolerate intense olfactory stimuli. When we train to tolerate one type of intensive stimuli, we are less likely to overreact to another stimulus. When we do not work in a chicken farm and called a mama's boys, we may get angry and insulted. The person, who works on the chicken farm and is exposed to the excessive smell, will not overreact to verbal comment, because he is trained not to overreact. If the intense stimulus is present and reaction of the brain does not support the need for anger and a strong reaction, the reaction on the stimulus is decreased. In other words, in our examples the response on intense olfactory stimulus when is congruent with the verbal response, is not supporting the need for stronger reaction. When one stimulus is intense and the other stimuli are not validated by other stimuli, the need for anger and intense responses are reduced significantly.

Fears and worries work the same way as anger does. If a bus is running late and may not arrive at all, and we pace, then we feel fear, and worry for our circumstances and safety. Conversely, if we visualize two miles of walk and think about nice scenery and an opportunity to have relaxing exercise or listen to soothing music, fears and worries are decreased. Another situation is when our child is late form school; we also may feel worried and fearful as well. Yet, all these situations happen because our audio visual and kinesthetic experiences are not congruent. Fear and worry are usually accompanied by physical reactions such as: increased heart palpitations, shortness of breath and chest heaviness. If visual experience helps us to visualize that our child stayed in school for extra curricular activities, or our audio experience helps us to recall a sports event that our child participated, then our worry and fearfulness are decreased. Instead of getting sick we call the child's coach, or our child's friend. We also experience a change in your physical responses such as lower respiration and regular heart beat. We understand that we need to check lot of things before start to worry.

If we experienced a fall while in the elevator from the tenth floor and miraculously survived due to emergency brakes of the elevator shaft and also sustained limited physical injuries, obviously this experience left a strong psychological trauma. During the fall our thoughts were that we definitely would die, and experienced a shock, where the only recollection we had, was a kinesthetic reaction, also grabbing the pregnant lady in the elevator to clench for our lives. We have flashbacks about the screeching noise of the elevator during the fall and get really distorted. To help us to regain control of our life and get over the shock of our experience we visualized the brakes that slowed down the elevator and saved us. We recalled the noise we were hearing during the fall and realized now that it was noise from the brakes, which saved us. After this work our kinesthetic experience was adjusted by audio and visual experiences and our heart palpitations decreased and respiratory function return to normal.

Psychoconduction is a method of memory enhancement and stimulation of the brain functions.

Psychoconduction is a powerful tool to educate and increase intellectual level. The beneficiaries are people, who suffer from permanent and temporary intellectual losses. Because the injured brain has difficulty recognizing and matching the information, we simplifying this information to the most primitive symbol (dot). In the past there were different approaches to use simple symbol. Morse code is the communication system, which represents letters and digits by using the combination of the simple symbols, like dots and dashes for visual. The combination of knocks and double knocks were used for audio, and flags facing up and down for kinesthetic communication, mainly in communication in the sea of one ship to other. Braille is the communication language for the blind people, consisting of the symbols that blind people could feel on the paper by touch. Sign Language is the communication language for deaf people and is using the position of the fingers to express words. The different alphabets, the sign languages, Braille, Morse code etc. use the complex configurations for each letters and digit without any logical connection between them and without any references.

In the complex configuration the letters or codes do not have a logical connection and each one requires memorization. For example, the alphabets or Morse code not require any logical connection between symbols representing A and B etc., or any digits and, actually, all of them used in Morse code are the random symbols. The random combination of the simple symbols becomes a complex configuration. Litvin's Code does not require memorization of the combination of symbols. It is organizing the simple symbols in logical system by using the hierarchy of positions, which are pre-learned and always contain the same information. The success of the psychoconduction is based on a new way to simplify communication between the different parts of the brain, which are involved in the recognition of the simple symbols and allows the release congruent amount of chemicals in different parts of the brain. A congruent release of chemicals in the brain provides the calibration of the different parts of the brain against each other.

The psychoconduction helps to rehabilitate people with permanent losses and shorten recovery for people with temporary losses of brain's intellectual capacity. The work to overcome illiteracy and difficulty to do calculations is very important and always needs different approaches. Psychoconduction is using different approaches to work around any distractions and overcome the obstacles to understand initial information. In Litvin's Code we do not need to focus on the recognition of the letters. The position and corresponding binary number are shown in the table, below. When we do the different calculations, we are not concerned with recognition of digits. By using the hierarchy of pre-learned position we understand the process of addition, subtraction, multiplication and division and could do successfully any mathematical assignment. Litvin's Code is far away from a calculator. We arrive to the answer by using pre-learned steps in the processing of the information.

Litvin's Code uses a simple symbol, in our example a dot, placed in different positions to represent letters and digits. By moving a symbol to different position we change the representation of letters and digits. In Litvin's Code we focuse on recognizing the simple symbol in different position and is not distracted by the configuration of the letters and digits. Litvin's Code creates the new aptitude to construct the words and sentences and to improve the spelling. By using Litvin's Code the calculations become very simple. We do the multiplication and division without using the table of multiplication. Litvin's Code is not just the combination of the simple symbols, which replace alphabet and digits. This is a method of intellectual enhancement.

Psychoconduction helps to regain loss of intellectual functioning and develop a new way to understand information. By using Litvin's code we exercise our brain to higher ability and avoid distraction related to different physiological impairments. The simple symbol in Litvin's Code contains lots of information about binary numbers, ascending positions. It gives the deep understanding of the factorial functions, but at the same time skillfully simplifies the complex information. Litvin's Code uses binary arithmetic, which is the mathematics where all numbers are represented by combination of a number, which has a base of two to the different power. Litvin's Code allows us to do complicated tasks with the one symbol. The simple symbol is a package with a lot of information, which is easy to understand. When information gets complex again then by using Litvin's Code we could simplify it again. The goals of Litvin's Code is to continuously translate complex information into simple symbol. We do need to know about binary arithmetic which is the basis of Litvin's Code we mainly learn information about position.

Litvin's Code is the important part of psychoconduction. Litvin's Code relies on different pre-learned positions of simple symbols in the same way as a brain relies on different organic structures which have different resistance to electric impulses. The information in a brain is translated when the neurons fire or not fire. In Litvin's code the presence or absence of the simple symbol in different position help to comprehend information. In Litvin's Code we use a symbol (dot) and, when we do not have a signal, we use a dash. We assign a symbol (dot) to represent different binary numbers in the previously assigned positions. We use the dot as a symbol, because it is the simplest one, but it can be any other symbol. Litvin's Code could adjust to any other symbols, body movements, electric and light signals and sounds that are used in the assigned sequences. The kinesthetic expression of Litvin's Code can include the movement of the hands; the left hand represents the presence of a symbol and right hand the absence of it.

Litvin's Code could use the same simple symbol with different kinesthetic movements. Step forward can be a dot (presence of symbol) and backwards is a dash (absence of symbol). If visual expression of Litvin's Code shows the presence of the symbol (dot) in any position, it is the manifestation of the binary number. The absence of the symbol (dot) in any of assigned positions represents only zero or nothing. The combination of the dot and dash give some meaning to this position. To understand the expression we need to add all positions, which contain the symbols and results will be a letter or digit. It could be restated that the presence of the symbol in the position show that the position has a number that gives meaning to this position. If the symbol is not present in this position, this position is not presented in the calculation. The advantage of Litvin's Code is in the simplicity to understand the position and the role of the symbol, and the easiness to recognize, manipulate, and transfer to different parts of the brain the simple symbol, which contains the complex information.

Litvin's Code requires the pre-learned knowledge of the fixed positions, which are easy to learn. Even people with severe retardation or brain damage understand the difference between empty and filled with the symbol positions. Litvin's Code is a combination of the symbols in the fixed positions, which represent the letters and the numbers at the same time. With Litvin's Code you could write a letter and calculate your daily expenses by using the simple symbol in a different position. Litvin's Code can be applied to any existing alphabets. Litvin's Code includes the principle of Binary Arithmetic. Litvin's Code could transmit the written text and digit information by the electric signal, any movement, including the movement of the objects and even of human body in the coded sequence. In the Morse code the letters do not have the logical system provided by Litvin's Code. In the Morse code we just learn the dots and dashes, which need to be memorized without the references to the Binary arithmetic, factorial expressions, and ascending order. The advantage of Litvin's Code compared to the Morse code is that during the processing of information instead of the random combination of symbols, the brain uses the information of pre-learned positions, which are the memory enhancement links and opportunity for brain stimulation. In Litvin's Code, the Binary Arithmetic, factorial expressions, and ascending order provide the logical base to put the letters and numbers in the learning system to allow the brain to utilize it. The Binary Arithmetic uses two as the base to different powers to represent different numbers. Here, the binary numbers represent the digits and letters at the same time.

Claims

Psychoconduction is the new way of producing new functional cells with only psychological approach. With psychoconduction we reduce brain deficiency and helping to reduce many problems, which include dyslexia, attention deficit, anxiety, cognitive problems and memory. The success of the psychoconduction is based on the understanding of the logic behind the neurological functions of the brain and is helping the brain to improve performance. We transfer and translate information through different areas of the brain, which are responsible for audio, visual and kinesthetic information. By using the same patterns, we achieve congruence in processing of the information. The brain bypasses the complex cells, which show an inadequate performance in the collecting and processing of the information.

Psychoconduction is supplying the simple cells with the ability to recognize correct information while utilizing the simplest symbol (dot). The simple cells collect and process the simple information and can be used for doing the complicated work. We can stimulate the brain with kinesthetic symbols and ask the person to translate this information into audio or visual symbols. By recognizing the same patterns of the information, we use the different parts of the brain we overcome the brain's weakness in processing information. We visually stimulate the brain with different patterns of simple symbol then the brain is translating those patterns and is reproducing them in the audio and kinesthetic expression. We assume that after our training the chemicals released in different parts of brain are congruent because we stimulate the different areas of the brain with the same patterns of the simple symbol. The most important benefit of using the psychoconduction is to achieve congruency in the processing of the information and equilibrium of the responses. After the training in psychoconduction the brain seems less overreacting on the complex information and is less neglectful of the simple information, and the brain shows an increase in the appropriate responses.

The psychoconduction helps to enhance intellectual capacities as well as rehabilitate people who have had the loss of intellectual properties. Litvin's Code provide memory enhancement by using links to the binary arithmetic, factorial expressions, and ascending order. In Litvin's Code we start with the easy patterns of simple symbol and then make them more complicated by using the system, which includes audio, visual and kinesthetic simple symbols. Litvin's Code is very

practical and beneficial for people with dyslexia, brain damage, retardation, dementia, autism, attention deficit and many others neurological losses. By transferring the simple symbol from audio to visual and kinesthetic representations in the different sequences and then by creating the meaningful patterns, the psychoconduction helps to overcome learning disabilities and brain malfunction caused by stroke or trauma. Litvin's Code is a very effective method to overcome illiteracy and the difficulty to do mathematical calculations. Litvin's Code deals with confusion and frustration in the learning process and helps to build a strong feeling of confidence and provides tools for successful processing of information.

The notion of the position and hierarchy of the different positions is easy to understand. When the process of recognition and processing of the complex information is impaired or not completely developed, it is easier to use the hierarchy of positions to represent letters and digits than contours letters or digits. We use the healthy area of the brain instead of the damaged, overused or underdeveloped one. The hierarchy of position is simple information and uses simple cells by utilizing Litvin's Code to become more functional assistants for complex cells. The simple cells could use the complex pattern of information and also provide references for easy recognition of information by complex cells. The child has a tendency to remember the location of the toy rather than naming the location. Some of us while driving to a certain familiar location do not utilize the names of the streets, but rely on the position of the streets. The brain remembers the same information in many different ways and stores this information into different complex and simple cells. For the hierarchy of positions and simple symbols we utilize the simple cells and then the complex cells eventually recognize the complex letter and digits. Because the collected information in complex cells can be distorted due to physiological disorders, we correct it with the simple cells. We have easy task: instead of complex information we just recognize the hierarchy of position and to check if position is filled with a symbol or empty.

The psychoconduction helps us to increase the social skills when we have the deficits in adaptive functioning. When we deficient in adaptive functioning and social skills, we are not able to create successful peer relation and are socially isolated. We learn to control our impulses and anger, which interfere with the social interactions. Psychoconduction helps us with the lack of the social and emotional reciprocity and helps to increase our effectiveness in communication and ability to control our emotions.

Psychoconduction helps us, when we have problems with reading and writing, when we cannot make sense of written materials or can not spell. With psychoconduction we could increase the reading achievement and increase the writing skills and spelling as well. Many people have difficulty with coordination and motor controls, which include stereotyped and compulsive behavior. The psychoconduction can increase the motor coordination, breaks the patterns of stereotyped and repetitive motor mannerism. Many people have difficulty with attention and concentration because of hyperactivity, impulsiveness and inattention.

Psychoconduction increase the attention to details, following of instructions, and sustain maximum mental effort within individuals. Many people have difficulty understanding and responding to language. Psychoconduction helps people with the difficulty with receptive and expressive language deficit to acquire new skills. Many people have difficulty organizing activity because of many things on their mind or their mind is too slow.

Psychoconduction helps people organize and plan activities to keep focused on the task. It helps people to stabilize the anxious or dysphoric mood and to receive congruent information from kinesthetic, audio, visual and olfactory stimuli.

Processing of information

Litvin's Code calibrates the brain and allows the congruous release of different chemicals into different areas of the brain. The brain is processing the patterns of simple symbol with audio, visual kinesthetic and olfactory stimuli that have the same meaning. By using Litvin's Code the processing of information which was very difficult and sometimes was impossible before, became a very simple process and takes out ambiguity.

Litvin's Code allows the person to understand the notion of ascending direction and quick recognition of letters and digits. By understanding the relationship between empty and filled position an individual is able to simplify the process of division and multiplication without using the multiplication table. Litvin's Code also provides logical connection between letters by utilizing hierarchy of position and numerical references to the alphabet that makes it easier to connect letters to words and sentences. Because we utilize numeric references, it makes easier for us to remember spelling and recognize words.

The brain not performe adequately due to physiological limitation, the process of the recognition of the complex spatial information is complicated and confusing. When we recognize a letter, first of all, we need to scan this letter up and down and from left to right or vice versa depending on the alphabet. To recognize a letter we should have a good match with the same letter stored in our memory. The identification of this letter is a serious process, because the brain needs to find the approximate address of the matching letters. If we have dyslexia, the brain browses many times through the stored alphabet to find the needed match, but may not find the right letter. When the letter is identified, the brain stores this letter, and needs to identify the next letter to assemble the word. Because of the optical nerve or brain impairment the process of scanning the letters is difficult. If we after the scanning we have difficulty producing the clear image of the letter, then the brain has difficulty correctly matching with the letter in the memory.

The brain continues the matching process and we understand that we cannot produce a right answer. Sometimes, because of our learning difficulties we even refuse to recognize the letter, which is not clear and refuse to say that we do not know the letter. We may name a letter with various degrees of approximation, or names the first available letter of the matching file in our memory. As a result, we recognize the letter "P" instead of "B", and etc. When the brain makes recognition of the second letter, we store it together with the first in the storage area.

When the first letter we recognize wrong, we have difficulty making sense from the combination of the first and second letter. It may be possible that the address of the area, which keeps the first letter, is already lost, and our brain has difficulty finding the first letter. The correct identification of the letters is not a guaranty that the brain could utilize them later. Nevertheless, this process continued until the last letter of the word, and then we need to make a decision how to pronounce these combinations of the letters. In this moment we make a decision to use the approximate contour, and instead of the world "small" we recognize and pronounce it as "little". We recognize contour by the content. It happened because all those contours are stored in the same area. The brain matches them by approximation and picks up instead of right one the next; which is appropriate for this purpose. It is understandable, that it is easier for us to memorize the contour of the word instead of individual letters. This is the more simple and practical way to recognize information when the brain has neurological impairments. In this situation the brain does not need the logical connections of the letters and store them somewhere. It just memorizes the contour of the words. The brain memorizes contours of the words and stores them in appropriate areas but sometime it fails in matching them with the incoming words. This process creates a lot of confusion and anxiety. Please consider that to read or write even simple text we should recognize at least a few thousand words and need to store the same amount of contours. Let us believe that the hardship of recognition of the letters and the ambiguity of the process does not stop us and we make a guess. Intuitively, we use approximation by contours. The approximation helps us to make a guess. The guess could be right or wrong.

We need to recognize the tremendous effort of the brain to bring us to this point. We also need to remember that to recognize the letter the brain needs to transfer it from the right to the left hemisphere. The right hemisphere is recognizing spatial information and left recognizes the words. To make this process less painful and less time consuming, psychoconduction significantly simplified the recognition of the letter and digits, and have enhanced the intellectual capacities of brain. We recognize only the simple symbol without many distractions described above.

Psychoconduction provides mathematical references and uses different areas in the brain to store and retrieve the letters and combine them into words. Because we don't do match through complicated scanning and we can provide numerical reference by checking if the position is filled or empty. We train our brain to overcome weakness in the processing of information. The injured brain cells cannot provide recognition of the letter and digits. We cannot move to the next level of processing the combination of letters and manipulation of the digits. The psychoconduction protects the brain against unnecessary stress of recognition and allows focusing on the process of writing, spelling and calculation, because we have numerical references for recognition. We could tune some brain parts against the other parts of the brain to achieve the congruency in the responses.

The process of reading, writing and calculation is moving the brain from the recognition to the more sophisticated level of the manipulation of the letters and digits. Litvin's Code utilizes the notion of positions, which is completely different from memorizing letters and digits. The symbols are in the different positions change the meaning of letters and digits change. Litvin's Code allows the person to have a stable relationship with the simplest symbol, instead of many symbols representing digits and different letters. We also provide the mathematical references, which correspond to the letters we use. Litvin's Code facilitates the process of calculation, writing and reading.

The patterns used in psychoconduction

In table below we count positions from the left side

	Alphabets							***LITVIN'S CODE*** **Using only dots**				***LITVIN'S CODE*** **Using dots and dashes**		
		1	2	4	8	16	32		1	2	4	8	16	32
1	A	x						Or	x	0				
2	B		x					Or	0	x				
3	C	x	x					Or	x	x				
4	D			x				Or	0	0	x			
5	E	x		x				Or	x	0	x			
6	F		x	x				Or	0	x	x			
7	G	x	x	x				Or	x	x	x			
8	H				x			Or	0	0	0	x		
9	I	x			x			Or	x	0	0	x		
10	J		x		x			Or	0	x	0	x		
11	K	x	x		x			Or	x	x	0	x		
12	L			X	x			Or	0	0	x	x		
13	M	x		X	x			Or	x	0	x	x		
14	N		x	X	x			Or	0	x	x	x		
15	O	x	x	X	x			Or	x	x	x	x		
16	P					x		Or	0	0	0	0	x	
17	Q	x				x		Or	x	0	0	0	x	
18	R		x			x		Or	0	x	0	0	x	
19	S	x	x			x		Or	x	x	0	0	x	
20	T			X		x		Or	0	0	x	0	x	
21	U	x		X		x		Or	x	0	x	0	x	
22	V		x	X		x		Or	0	x	x	0	x	
23	W	x	x	X		x		Or	x	x	x	0	x	

		1	2	4	8	16	32		1	2	4	8	16	32
24	X				x	x		Or	0	0	0	x	x	
25	Y	x			x	x		Or	x	0	0	x	x	
26	Z		x		x	x		Or	0	x	0	x	x	
27		x	x		x	x		Or	x	x	0	x	x	
28				X	x	x		Or	0	0	x	x	x	
29		x		X	x	x		Or	x	0	x	x	x	
30			x	X	x	x		Or	0	x	x	x	x	
31		x	x	X	x	x		Or	x	x	x	x	x	
32							x	Or	0	0	0	0	0	x
33		x					x	Or	x	0	0	0	0	x
34			x				x	Or	0	x	0	0	0	x
35		x	x				x	Or	x	x	0	0	0	x
36				X			x	Or	0	0	x	0	0	x
37		x		X			x	Or	x	0	x	0	0	x
38			x	X			x	Or	0	x	x	0	0	x
39		x	x	X			x	Or	x	x	x	0	0	x
40					x		x	Or	0	0	0	x	0	x

Litvin's Code and Binary Arithmetic's

Litvin's Code just means that instead of numbers and letters we use very simple symbols. With Litvin's Code we could represent any numbers using Binary Arithmetic and as well any letters in any existing alphabet. With Litvin's Code we could create as many letters as needed. The English alphabet has 26 letters, the Russian - 33, the Armenian – 39, etc.

When we do calculations with Litvin's Code, we need to remember that usually all calculations are from left to right. With Litvin's Code we could also do calculations as well from the right to the left. Litvin's Code helps us to use the calculation process in a more adaptable way. We could start calculation with the lowest position to help with quantitative concept and also we could do the same thing from the highest level to the lowest. In the follow up calculations we will provide examples, where the ascending position starts from the lowest level. Litvin's Code, uses the row of ascending binary numbers and could be represented by the row of vectors: f(1), f(2), f(3), f(4), f(5), f(7).....f(n), where n is the position's number. Vectors are the numbers placed in the functional commonality. In our situation all vectors have a base of two.

Positions

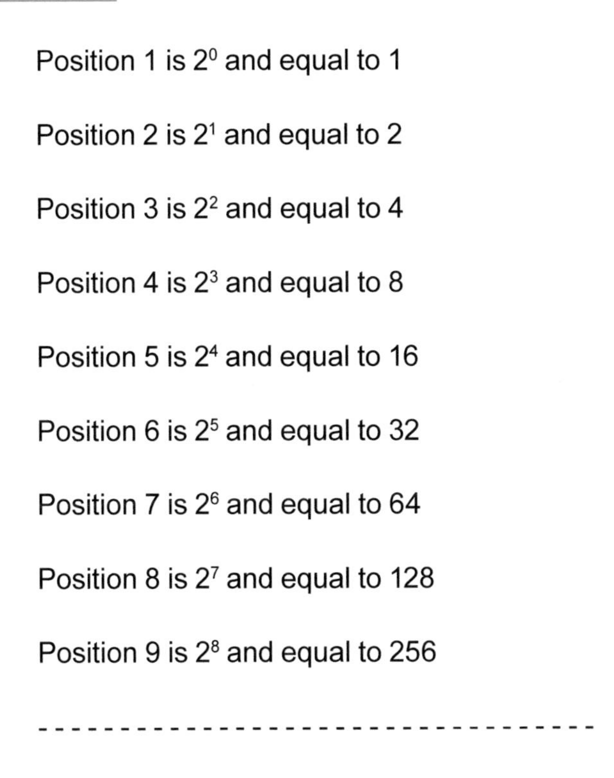

Position 1 is 2^0 and equal to 1

Position 2 is 2^1 and equal to 2

Position 3 is 2^2 and equal to 4

Position 4 is 2^3 and equal to 8

Position 5 is 2^4 and equal to 16

Position 6 is 2^5 and equal to 32

Position 7 is 2^6 and equal to 64

Position 8 is 2^7 and equal to 128

Position 9 is 2^8 and equal to 256

- -

Position n is $2^{(n-1)}$ and equals to $2^{(n-1)}$ times.

The presence of the symbol (dot) is informing us of the presence in this position a symbol, which is equal to the assigned binary number. We keep track of all positions with this symbol. In Litvin's Code, to get the desired number or letter we add the assigned numbers to the positions where the symbol (dot) is present.

We are also using dots and dashes to establish connection between filled and not filled positions and to show the difference between them. The uniqueness of Litvin's Code is in using a dot only to inform that an assigned binary number is present, and the dash to acknowledge that position is empty.

- In ***position one*** we always expect to have binary number 2^0 equal to 1. The number 1 is represented by the presence of the symbol or dot.

In the first position we have "x " The symbol (dot) in the first position is a representation of the number 1 and as well letters A. The absence of this symbol represents number 0.

- In the ***second position*** we expect to have 2^1 equal 2. In Litvin's Code this digit is represented by presence of the symbol (dot) in the second position "0 x ". The first position is empty and the second is filled with a symbol, so this is representation of the number 2 and letter B.

The absence of the symbol (dot) is representing number 0.

- The combination of the ***first and second positions*** filled with the symbol (dot) $2^0 + 2^1 = 3$.

In Litvin's Code it is represented by symbol "x x" or this is the representation of number 3 and letter C.

- In the ***third position*** the presence of the symbol (dot) informs us that we have number $2^2 = 4$.

The first two positions are empty and the third is filled up with symbol, so this is number 4 and letter D and looks as " 0 0 x " .

Example 1.0

When symbol (dot) is in the first and the third positions, and the second position is empty, we are adding numbers in the first and the third positions: $2^0 + 2^2 = 1+ 4 = 5$.

In Litvin's Code it looks like "x 0 x".

This is representation of the number 5 and letter E.

End.

Example 1.1

When the symbol (dot) is in the second and third positions but the first position does not have a symbol (dot) we add only assigned binary numbers in the second and the third position

$2^1 + 2^2 = 2 + 4 = 6$. In Litvin's Code it looks like " 0 x x"

This is representation of the number 6 and letter F.

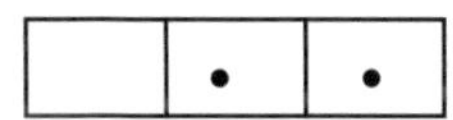

End.

Example 1.2

When the symbol (dot) is in the first, second and third positions, we add corresponding numbers in the first, second and third positions: $2^0 + 2^1 + 2^2 = 1 + 2 + 4 = 7$.

It looks like "x x x"

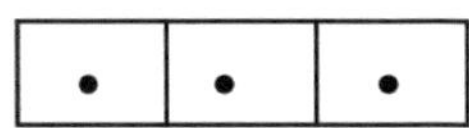

This is representation of the number 7 and letter G.

End.

One Example of Audio Litvin's Code

We introduced the patterns of symbols in the audio part, which we were using in psychoconduction (translation) for the digit and letters. It could be many simple sounds and patterns, which practitioners could use for translation. The combinations of symbols could be introduced in the different directions.

1 A – Knock,

2 B – double knock and knock,

3 C – knock and knock,

4 D – double knock, double knock and knock,

5 E – knock, double knock, double knock and knock,

6 F – double knock, knock and knock,

7 G – knock, knock and knock,

8 H – double knock, double knock, double knock and knock,

9 I – knock, double knock, double knock, and knock,

10J - double knock, knock, double knock, and knock,

11K - knock, knock, double knock, and knock,

12L - double knock, double knock, knock, and knock,

13M - knock, double knock, knock, and knock,

14N - double knock, knock, knock, and knock,

15O - knock, knock, knock, and knock,

16P - double knock, double knock, double knock, double knock, and knock,

17Q - knock, double knock, double knock, double knock, and knock,

18R - double knock, knock, double knock, double knock, and knock,

19S - knock, knock, double knock, double knock, and knock,

20T - double knock, double knock, knock, double knock, and knock,

21 U - knock, dou1111ble knock, knock, double knock, and knock,

22V - double knock, knock, knock, double knock, and knock,

23W - knock, knock, knock, double knock, and knock,

24X - double knock, double knock, double knock, knock, and knock,

25Y - knock, double knock, double knock, knock, and knock,

26Z – double knock, knock, double knock, knock, and knock

27 - double knock, knock, double knock, knock, and knock

28 - double knock, double knock, knock, knock, and knock

+ - ram

- - double tram

/ - blick

* - double click

One Example of Kinesthetic Litvin's Code

We introduced the patterns of symbols in the kinesthetic part, which we were using in translation for the digit and letters. It could be many simple movements and patterns, which practitioners could use for translation. The combinations of symbols could be introduced in the different directions (left to right or reverse right to left).

1 A –clamping the right hand,

2 B –clamping the left hand andclamping the right hand,

3 C –clamping the right hand andclamping the right hand,

4 D –clamping the left hand,clamping the left hand andclamping the right hand,

5 E –clamping the right hand,clamping the left hand,clamping the left hand andclamping the right hand,

6 F –clamping the left hand,clamping the right hand andclamping the right hand,

7 G –clamping the right hand,clamping the right hand andclamping the right hand,

8 H –clamping the left hand,clamping the left hand,clamping the left hand andclamping the right hand,

9 I –clamping the right hand,clamping the left hand,clamping the left hand, andclamping the right hand,

10J -clamping the left hand,clamping the right hand,clamping the left hand, andclamping the right hand,

11K -clamping the right hand,clamping the right hand,clamping the left hand, andclamping the right hand,

12L -clamping the left hand,clamping the left hand,clamping the right hand, andclamping the right hand,

13M -clamping the right hand,clamping the left hand, clamping the right hand, andclamping the right hand,

14N -clamping the left hand,clamping the right hand,clamping the right hand, andclamping the right hand,

15O -clamping the right hand,clamping the right hand, clamping the right hand, andclamping the right hand,

16P -clamping the left hand,clamping the left hand,clamping the left hand,clamping the left hand, andclamping the right hand,

17Q -clamping the right hand,clamping the left hand,clamping the left hand,clamping the left hand, andclamping the right hand,

18R -clamping the left hand,clamping the right hand,clamping the left hand,clamping the left hand, andclamping the right hand,

19S -clamping the right hand,clamping the right hand,clamping the left hand,clamping the left hand, andclamping the right hand,

20T -clamping the left hand,clamping the left hand,clamping the right hand,clamping the left hand, andclamping the right hand,

21 U -clamping the right hand,clamping the left hand,clamping the right hand,clamping the left hand, andclamping the right hand,

22V -clamping the left hand,clamping the right hand,clamping the right hand,clamping the left hand, andclamping the right hand,

23W -clamping the right hand,clamping the right hand,clamping the right hand,clamping the left hand, andclamping the right hand,

24X - clamping the left hand,clamping the left hand,clamping the left hand,clamping the right hand, andclamping the right hand,

25Y -clamping the right hand,clamping the left hand,clamping the left hand,clamping the right hand, andclamping the right hand,

26Z –clamping the left hand,clamping the right hand,clamping the left hand,clamping the right hand, andclamping the right hand

27 - clamping the left hand,clamping the right hand,clamping the left hand,clamping the right hand, andclamping the right hand

28 - clamping the left hand,clamping the left hand,clamping the right hand,clamping the right hand, andclamping the right hand

+ - one hand extended in front

- - both hands extended in front

/ - one hand raised up

* - both hands extended up

DO MATHEMATICS BY USING LITVIN'S CODE

Using Litvin's Code we could do binary mathematics similar to the way Boolean algebra does. Boolean algebra is algebra, which uses the binary numbers. Livin's Code is helpful in arithmetic and spelling. It appears that the damaged brain has difficulty recognizing written numbers but instead will recognize the simple symbol in the assigned position and would be able to do the required calculation. The calculations with Litvin's Code will improve memory and enhance attention and concentration. It is very simple methods and does not require the recognition of decimal numbers.

Many people have difficulty with mathematics because of brain deficiency to recognize and to match written decimal numbers and letters. Litvin's Code provides good opportunity for doing mathematics by using simple symbols instead of decimal numbers. Using Litvin's Code we could do, addition, subtraction, multiplication and division.

ADDITION

To simplify Litvin's Code to represent numbers we use dots, empty spaces. We actually could use any symbols to represent numbers, "X• 0 ", x and zero. To add two or more numbers instead of the digits we use only assigned positions and place over there symbols as specified in Litvin's Code. The following examples will give deeper understanding of the calculations using Litvin's Code.

Rules:

- **When we add two numbers** and the position on the first number is filled up with a symbol and also the corresponding position on the second number is filled up with a symbol, we move the two symbols to the next positions on the answer as one symbol. The next or following position on the answer is corresponding to the decimal number twice bigger than decimal number in the previous position.

- **Two symbols become one symbol** when two symbols form two corresponding positions move forward to the next or following position of the answer.

Example 2.1

We use symbolic (dots) representation of the digits expressed in binary form, where symbols represent the decimal numbers in assigned positions. To add 3 and 2 we will use below only symbols (dots) in the assigned positions. To understand the first number, we need to know that the decimal number one is always assigned to the first position, and the decimal number two is assigned to the second position. We add assigned decimal numbers in the first and second positions, we get decimal number three as a sum, so the first number is three. To understand the answer we know that the first position is one and the third position is equal to decimal number four. When we add one and four than in sun we have five, so answer is equal to five.

•	•		3
+	•		+ 2
•		•	

First number 3 "x x 0"

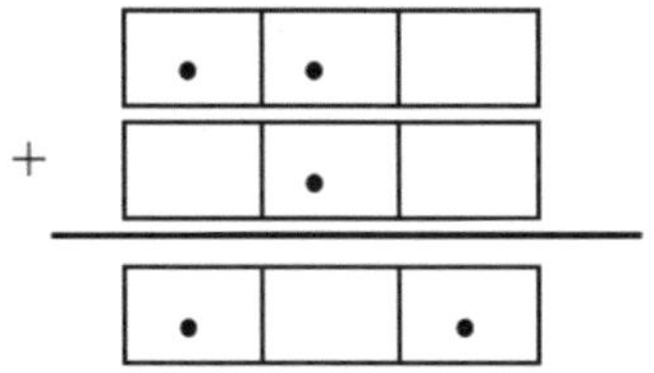

Second number 2 "0 x 0"

The answer is 5 "x 0 x"

In the example 2.1, the **first position** of the first number is filled up with symbol (dot) and, in the mean time, the first position of the second number is empty, and then we simply move the symbol to the first position of the answer. ***On the first number*** two positions are filled with dot and ***on the second number*** only the second position is filled with dot.

Explanation

When on the first number the second position is filled with the dot and the second number's the second position filled with the dot so both second positions are filled up with dot, which making them the parallel positions. Note that next position is always twice bigger than previous. In this situation we move the two symbols from the second position to the next ascending position, the third position, of the answer, where the two symbols are represented by one symbol (dot).

In Summary:

- When we have two symbols in the one position of the answer, then we move one symbol to the next position and leaving empty the second position of the answer.

- When the symbols (dots) are in parallel position, we always move to the next ascending position of the answer.

End.

Example 2.2

In the example 2.2 we have third parallel position filled with symbols (dots) and we move to the next position. If, for example, the next position is already occupied, we move to the next available ascending position. In the first number the positions one and three occupied with symbols, so we add assigned decimal numbers one and four and come up with the number five. Note that the two symbols (dots) in the different positions are corresponding to different decimal numbers. In the second number we have positions two and three occupied with symbol, so adding assigned decimal numbers two and four and come up with number six. In the answer the positions one, two, and four is filled up with symbol. We know, that position one is equal to the decimal number in value of one, position two is equal the decimal value of two, and position four is equal to decimal number of eight, and on the answer in sum we have decimal number of eleven.

First number is 5 “x 0 x”

Second number is also 6 “0 x x“

The answer is 11 “x x 0 x”

•		•

\+

	•	•

•	•		•

5
\+ 6

Explanation

The ***first position*** on the first number has filled up with the symbol. Opposite, the first position on the second number is empty, and we automatically fill up with symbol the first position of the answer***. The second position*** on the first number is empty but the second position on the second number is filled up with a symbol, so we fill up with a symbol on the second position on the answer. ***The third*** parallel position on the first and second numbers is filled up with the symbol and it means that the third position on the answer is left empty and the symbol is moved to the fourth position on the answer.

In summary:

- When the first position on the first number is filled up with the symbol and also the second position on the second number is filled up with the symbol too, then the two positions of the answer are filled up with symbol.

- When the third position on the both numbers are filled up with the symbol and we have two symbols (dots) present in parallel position, we advance one position.

End.

Example 2.3

First number is 3 “x x 0“

Second number is also	3	“x	x	0“	• • □	3
					• • □	+ 3
The answer is	6	“0	x	x”	□ • •	

The first parallel positions of the first and second number are filled up with the symbol. We move the symbol to the second position of the answer.

The second parallel position of the first and second number is filled up with the symbol (dot). So we move the symbol to the next position of the answer.

In summary:

- When we have two symbols (dots) present in parallel position we advance one position.

- The first two symbols on the first parallel position of the two numbers become one symbol on the second position of the answer. The second two symbols on the second parallel position become one symbol third position of the answer. The same principle is applied for parallel position occupied with symbols.

End.

Example 2.4

First number is	3	“x	x	0“		+
Second number is also	3	“x	x	0“		
Third number is	2	“0	x”			
The answer is	8	“0	0	0	x”	

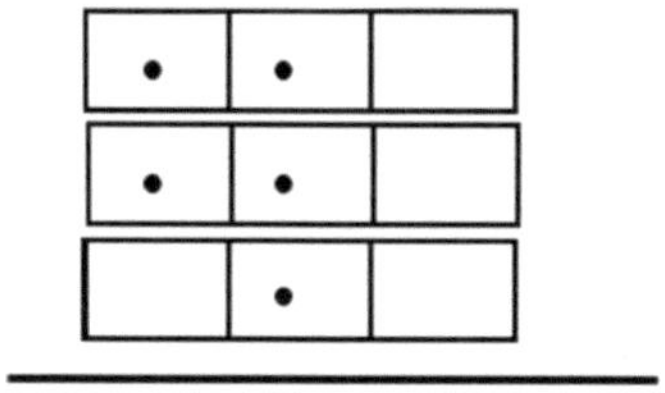

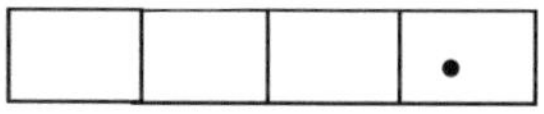

3

+ 3

2

When adding three numbers with parallel positions filled up with symbols (dot) we have a different approach. In the figure 2.4 the first position of the two initial numbers are filled up with symbols (dot). They do not go to the answer. Usually, the two symbols in one parallel position supposed to move as a one symbol to the next position of the answer, but in this case they move to the second advance position of the same number. Now we have four symbols in the second parallel position so we move two position of the answer and fill up the second position with the symbol (dot). In the second parallel position we have **four symbols (dot)**, which is an even number of symbols that's why we move the symbol (dot) to the fourth position of the answer.

The symbols (dot) in the parallel positions represent the different numbers. In this example there are three numbers to add. The two numbers are the same and one is different. In the first position we are left with two symbols, which propel us to move again to the next position and now leaves us only with one symbol (dot), on the second position of the answer.

When we countinue the symbols (dot) in the second positions of three numbers and the answer, the result of this calculation is an even number four. We move the symbol to advance position in the twice less quantity, but the decimal value of the next position is twice bigger. In the second position we have four symbols so we move symbol (dot) on two positions forward of the answer. When we have even number of symbols (dot) in one parallel position, we must move forward until we have one symbol in the answer. We do this by understanding the basic process of move, because in the next position the number of symbols is twice lower but the decimal number is twice bigger. We move two symbols to the next position of the answer and the next position on the answer is filled up only with one symbol (dot). When we have four symbols we move two positions forward in the answer.

Explanation

In the example 2.4, the two first parallel positions of the first and second number are filled up with the symbol, but the first position of the third number is empty. When we have two symbols in one parallel position we move this symbols to the next advance position. Two symbols are represented only by the one symbol when they got transferred to the next advance position, thus leaving the previous position empty. So we move the symbol to the second position. The second position of the first, second and third numbers is filled up with symbol. We also have one symbol moved from the first position. All together we have four symbols and we transfer them to the next level. The four symbols in the second position are represented by two symbols in the third position and then as one symbol to the fourth position. We fill up with symbol the fourth position on the answer. When in different situation, we had in a parallel position an odd number, which is greater then one, then we transfer one symbol to the answer and the rest to the advance position.

In summary:

- When we have two symbols (dots) present in parallel position we advance one position.

- When moving the symbol to the advanced position, which already has three symbols it creates an even number of symbols in this case four (dots).

- When we skip the third position, we advance these symbols (dots) forward to the new position, which is represented by one symbol in the fourth position of the answer.

End.

Example 2.5

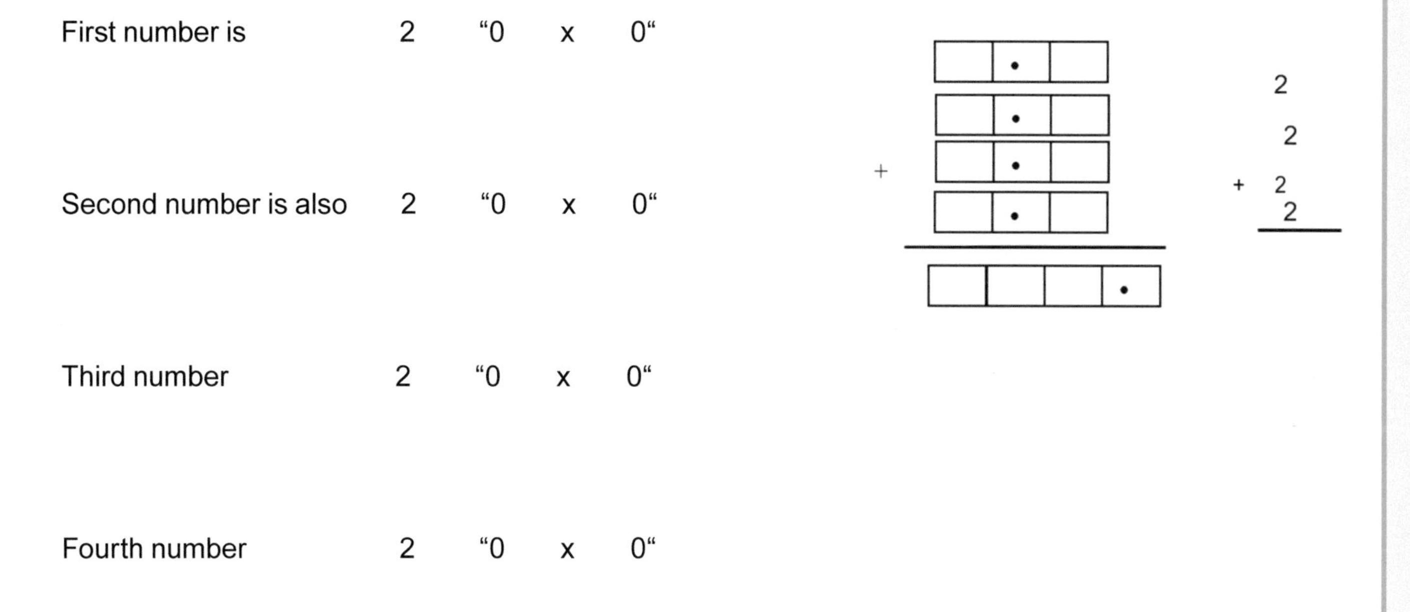

First number is 2 “0 x 0“

Second number is also 2 “0 x 0“

Third number 2 “0 x 0“

Fourth number 2 “0 x 0“

The answer is 8 “0 0 0 x“

In the example 2.5, we add four numbers. The symbols (dot) in the parallel positions correspond to different numbers. The second positions on all four numbers are filled up with symbols. We count the sum of symbols in the second parallel position, and in summary we have four symbols. When we move those symbols to the next position, then the next position represented only by two symbols then we move the symbol further. We move these symbols to the next advance position and continue to move symbols until the position is represented only by one symbol of the answer.

In summary,

- When we have four symbols (dots) presented in the parallel position, we advance two positions on the answer.

- When we move on to the new position, in other words we move the even number of symbols to an advance position until we could represent those symbols with only one symbol on the position of the answer. When we move symbols to the advance position, each advance position has a twice less symbol.

- When the position is correct, it supposed to have only one symbol.

End.

SUBTRUCTION

To subtract one number from another we do not need digits. We could use instead symbols (dots) in the assigned positions.

Rules:

- **When we subtract two numbers** and the same position of the first number is empty but the corresponding position of the second number is filled up with symbol, we borrow from the following ascending positions of the first number a symbol, which becomes two symbols on the prior position.

- When we borrow a symbol from the advanced position, **this symbol becomes two symbols in the previous position** and we can continue borrowing from each higher position to fulfill it with the symbol (dot) the lower position, because we cannot subtract from an empty position in Litvin's Code.

- When we do subtraction we borrow available symbols from the higher position. We need to remember that all symbols borrowed from the higher position are twice larger than one in the prior descending position.

- Once the symbol is borrowed the position is borrowed from becomes empty and two symbols appear in the prior position. If needed we leave one symbol in prior position and taking one symbol to the next descending position which is empty and once again this one symbol becomes two symbols in the next descending position.

- When symbol is borrowed then this one symbol becomes the two symbols on the next descending position.

Example 3.1

In Example 3.1 we could subtract 1 from 5.

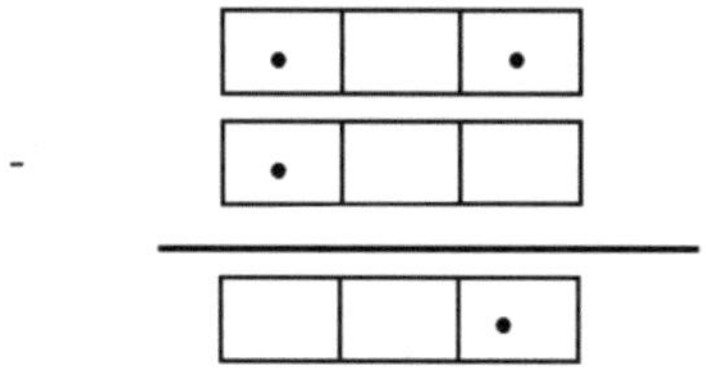

First number is	5	"x	0	x"
Second number is	1	"x"		
Answer is	4	"-	-	x"

•		•
•		
		•

$$\begin{array}{r} 5 \\ -\ 1 \\ \hline \end{array}$$

In example 3.1, we have two numbers with tree position each. In the first number the first and third positions are filled up with symbol. On the second number only third position is filled up with symbol.

In Summary:

- When we have two symbols (dot) in the first parallel positions, position one on the first and on the second numbers, then after the subtraction the first position on the answer becomes empty or zero.

- When the symbol in the third position on the first number is present and the same position on the second number is empty, we move the symbol from third position of the first number to the third position of the answer.

- Generally, when subtract two numbers and one of the positions of the first number has a symbol and the second number does not have a symbol in the same position, we just move the symbol to the same position of the answer.

End.

Example 3.2

First number is	6	"0	x	x"
Second number is	2	"0	x	0"
Answer is	4	"0	0	x"

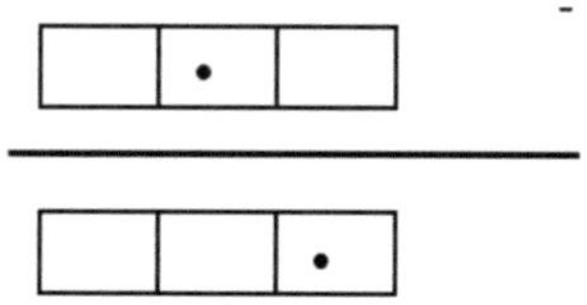

In Example 3.2 we have two numbers and first number with the three position each. In the first number the positions two and three filled up with symbol. On the second number only one symbol is in the second position. The second position on both numbers is filled with symbol and is a parallel position. The parallel position has the corresponding positions filled with symbols (dot).

Explanation

During the subtraction we eliminate the two symbols in the parallel position. As we see below two symbols (dot) in parallel position are replaced in the answer by an empty space or "0". We are moving the single symbol left on the third position of the first number to the third position of the answer.

In Summary:

- During subtraction of the two numbers we see that the first number in the third position has a symbol and the second number in the same position has an empty space then the answer has a symbol in the third position.

- When it is only one symbol in parallel position then it is moved to the answer.

End.

Example 3.3

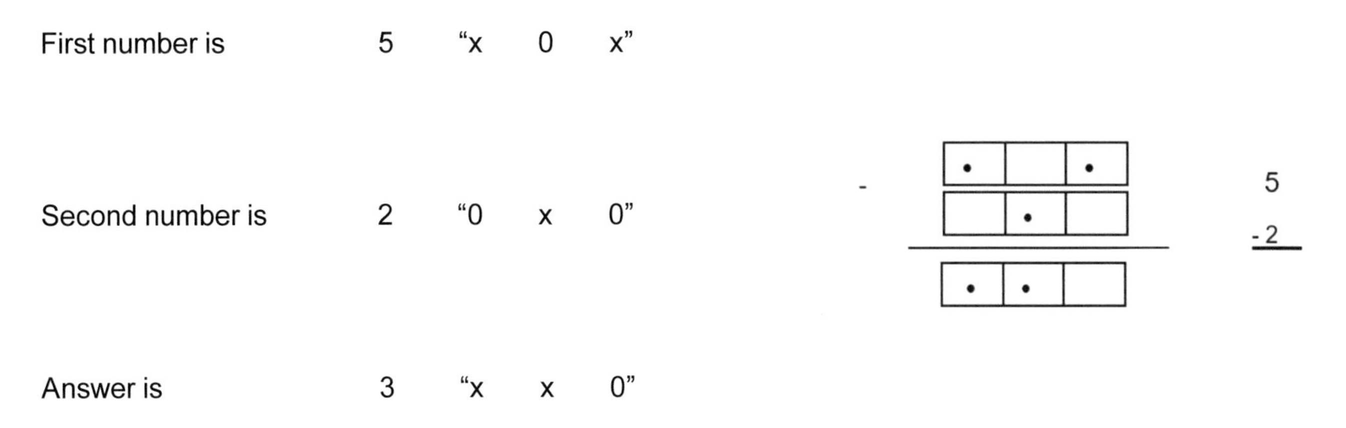

In Example 3.3 we also have two numbers with three positions each. The first number has symbols in first and third positions. The second number has symbol in second position.

In Example 3.3 on the second number in the second position is a symbol, which is different from the example 3.2, where on the second number in the second position is an empty space and symbol is in the third position.

Explanation

In example 3.3 we have borrowed symbol on the first number from the third position to the prior second position. In the third position we see the first number has the symbol (dot), but the second number has an empty space in the same position. We transfer the symbol from the first position of the first number to the answer. Because in Litvin’s Code you cannot subtract from empty space of the first number to the symbol of the second number, we borrow symbol from the third position of the first number. The one symbol from third position becomes two symbols in second position leaving the third position of the first number empty. We subtract one symbol from second position of the first number, because the first number has two symbols in the second position and the second number in the second position has one symbol. Now the answer has the symbol in the first and second position.

In Summary:

- When we borrow the symbol from the third position, we need to remember that binary number in the third position of the first number is twice as big then in prior descending position. In our example the prior descending position is position two in the first number. Now we have two symbols in the second position. The binary number which was expressed with one symbol in the third position now we need to express with two symbols in the second position. Now we subtract two symbols (dot) in the second position of the first number and one symbol from second position of the second number or (2-1) which leaves us with one symbol for the answer, which is placed in the second position of the answer.

- When we subtract only one symbol from the second position on the second number, the symbols from the first and second positions on the first number go to the answer.

End.

Example 3.4

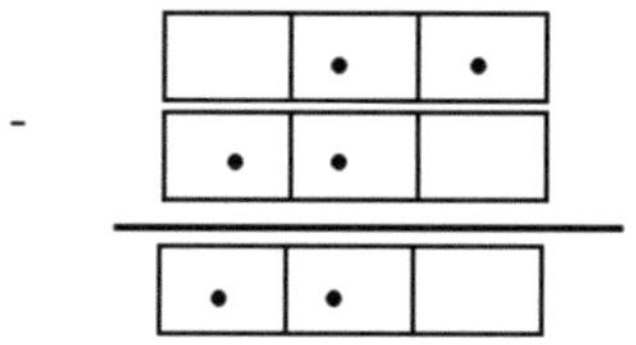

First number is	6	"0	x	x"	
Second number is	3	"x	x	0"	
Answer is	3	"x	x	0"	

		•	•	6
-	•	•		- 3
	•	•		

In example 3.4 we have two number with the three positions each. In the first number the first position is empty, the second and third positions are filled up with symbol (dot). In the second number the first and second positions are filled up with symbols (dot). As mentioned before in Litvin's Code we cannot subtract form the empty position. In this case the first position of the first number is empty so we borrow from the second position of the first number and one symbol becomes two in the prior descending position of the first number, also leaving the second position of the first number empty.

Explanation

After borrowing two symbols from the second position of the first number, we subtract from two symbols in the first position of the first number the one symbol in the first position of the second number. We place one symbol left from subtraction in the first position of the answer. After the first borrowing we have the second position of the first number empty. Now we borrow a symbol from the third position of the first number. One symbol in the third position becomes two symbols in the second position of the first number and third position of the first number is now empty. Now we subtract two symbols in the second position of the first number and one symbol in second position of the second number, and placing symbol to the second position of the answer. In the end we have the first and second position of the answer filled up with symbols (dot).

In Summary:

- During subtraction with first number having empty positions over the second number who has the same position filled up we do the borrowing available from the next advance position of the first number and borrowed position becomes empty.

- In situation when the borrowed position becomes empty and its over the second number where there the position is again filled up, we need to borrow again from the next available position and the symbol from the next position becomes two symbols in prior positions.

Example 3.5

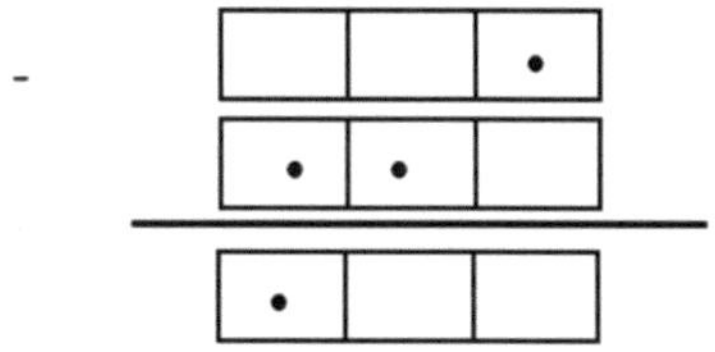

First number is 4 "0 0 x"

Second number is 3 "x x 0"

Answer is 1 "x 0 0"

-			•	4
	•	•		- 3
	•			

In the Example 3.5, we have two numbers with three positions each. The third position on the first number is filled up with a symbol. The first and second positions on the second number are filled up with symbols. If the higher position of the first number (third position) is filled with the symbol and the lowest two positions of the second number (positions one & two) are filled with the symbol, the answer has only one position filled up with the symbol (position one).The two positions of the first number have two empty spaces and the third position has a symbol (dot). Opposite the second number are two symbols (dots) in the parallel position and an empty space. The answer has only a symbol (dot) in the first position but second and third positions have empty spaces.

Explanation

In Example 3.5 we have the third position of the first number occupied with the symbol (dot), but the first two positions are empty. The second number occupied with symbols (dot) the first two positions, but the third position is empty. As we mentioned in previous example we cannot subtract the position of the second number from the position of the first number if the positions are empty in the first number. So we begin the borrowing process of the symbol from the third position of the first number to the second position of the first number. So now the second position of the first number has two symbols and the third position of the first number is empty. We subtract one symbol, which is on the second position of the second number, from the one symbol in the second position of the first number. In result we are left with one symbol in the second position of the first number, continue the same process by borrowing a symbol from the second position of the first number into the first position of the first number and we repeat the same mathematical process. The result is we are left with one symbol in the first position of the answer.

In Summary:

- When the second position on the first number has two symbols, then we borrow a symbol from the next position, which is equal to two symbols in the prior position and the next position becomes empty.

- When we subtract one symbol on the second position of the second number from two symbols on the second position of the first number, the one symbol is left on the second position of the first number. Then we borrow this symbol from the second position of the first number into the first position of the first number.

End.

MULTIPLICATION

When we utilize Litvin's Code for multiplication, table of multiplication is not required. Rather we use simple steps. The second multiplicand is used as an indicator for moving of the first multiplicand and indicates how many steps the multiplication process has. At the end we sum up the steps to receive the answer. In the first step we move the first multiplicand on the number of position prior to the first occupied position of the second multiplicand. We want to mention specifically, if the first position of the second multiplicand is filled up with a symbol, then the result of the first step is equal to the first number. As mentioned before the first position indicates binary number 2^0 which is equal to 1. So any number multiplied by 1 is equal to the first multiplicand. On the other hand if the first position of the second multiplicand is empty, but the other position is filled with a symbol (dot) we move the first number on the number of empty position prior to one filled up with symbol on the second number. In the next step we allocate the next occupied position of the second number. When we do multiplication, using Litvin's Code we move the first number on the number of position prior to any symbol of the second number.

Once we are complete to move the first number according to the filled up positions of the second multiplicand we summarize steps using Litvin's Code. We add the results of all steps and arriving to the answer. We want to emphasize the need to understand the difference between the powers of binary number in different positions, and the sequential number of position. For example if the power of the binary number is 2 then the position number n is 3. It happened because the first power of binary number is 0 but is located in position equal to 1. The power is less by one number than the position. The power of the binary number goes as 0, 1, 2, 3, 4, …, n-1. Corresponding positions go as 1, 2, 3, 4, 5, …, n. when the power of binary number is equal to 0, it corresponds to the position equal to 1. In general when the power of each position equals to (n –1), the corresponding position is equal to n.

Rules:

When we multiply two numbers we do it in several steps, which are equal to the number of positions filled up with the symbols on the second number.

When we multiply two numbers, we move the first number to the ascending direction on the number of positions filled with a symbol on the second number. In each step move the first number on the number of empty the position prior one to the position filled with symbol on the second number.

When the first position of the second number is filled up with the symbol then in this step, we do not need to move the first number and the result of this step is equal to the first number. Note, if the second number has the first position filled up with the symbol then it is the same situation when the first position of the second number is equal to the one, then the result of this step is equal to the first number.

When the second number has two or more positions filled up with the symbols so at the end we add the results of each step and after the addition we arrive to the final answer of the multiplication.

Example 4.1

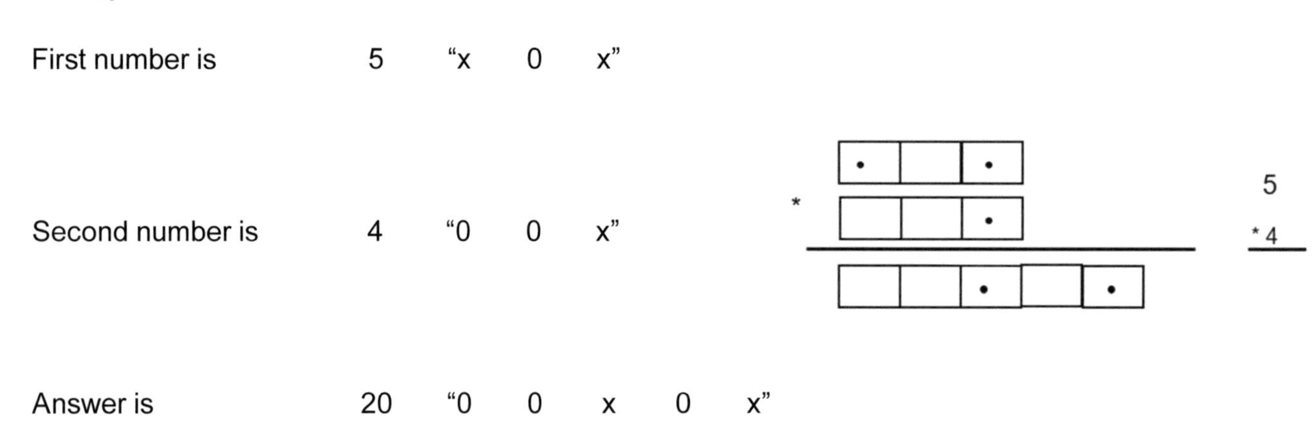

In the Example 4.1 the two numbers have three position each. To have decimal equivalent of the number from the binary representation we sum together the quantity of positions with symbols.

Each position contains 2 in power of (n-1), where n is a position on the number. If the second number it has only the third position filled up with a symbol, then the second number equal to 2 to the power of 2. To make this multiplication, we move the first numbers to the ascending direction on 2 position, which is equal to (n-1), power of filled position on the second number. In example 4.1 the (n-1), position 3 minus one is equal to 2. The second number is equal to 4, 2 in power of 2.

Explanation

In the Example 4.1, the second multiplicand is an even number and has only one symbol in the n=3 position. On the answer we move the first multiplicand on (n-1) positions in the ascending direction. The second number has the third position filled with a symbol. The power of the binary number in this position is also 2 and the number is four. We move the first number on two positions to the ascending direction to get the result of this step. This method of calculation required moving symbols on specified position corresponding to the power of the binary number in this position or of the number of positions, which are prior to the position filled up with the symbol. We move the binary equivalent of the number 5 on 2 positions to the right and the answer is binary equivalent of decimal twenty.

In Summary:

- When the second number has two empty positions prior to the position filled with symbol, we move the first number on two positions to the ascending direction.

End.

Example 4.2

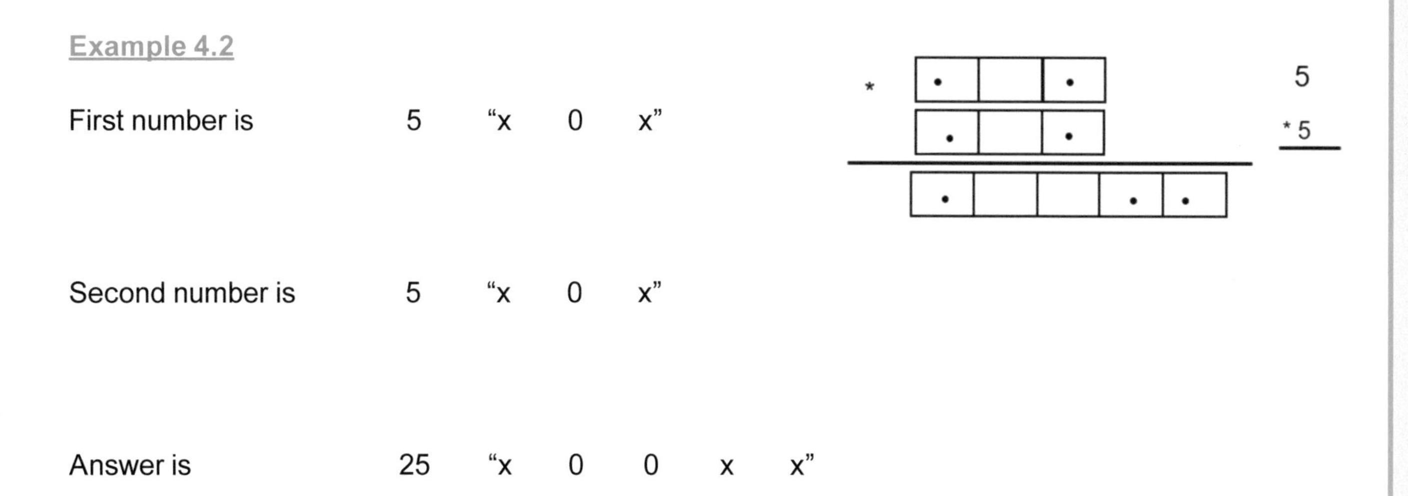

First number is 5 “x 0 x”

Second number is 5 “x 0 x”

Answer is 25 “x 0 0 x x”

In the Example 4.2, we have two numbers with the three positions each. The first and the second numbers are the odd decimal numbers and are also the equal numbers, and are represented in Litvin’s Code by the two symbols in different positions. We move the first number according to filled up with the symbol position of in the second number. We have the two position of the second number filled up with the symbol, so we move the first number twice. The first position of the second number is filled up with the symbol and since its decimal value is one we do not move the first number. If the second symbol of the second multiplicand is in the ***n*** position, we move the first multiplicand on the (***n***-1) positions in the ascending direction. The result of the first step is the number equal to the first number. The result of the second step is the first number moved on two positions to the right. At the end we add the results of the moves, and arrive to the answer of the multiplication.

Explanation

First step for example 4.2,

On the second number the first position is filled up with the symbol and is equal to one, so we just leave the first number as it is. If we multiply any number by 1, we have the same number in the result of this step. When we use the formula for moving (***n***-1) position, then we subtract 1 from 1, and as a result we have 0. We should move the first number on 0 position, which means that we do not move the first number at all.

Result for the first step is -5 "x 0 x"

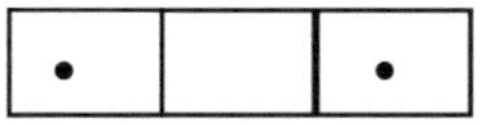

•		•

Second step for example 4.2.

On the second number the next filled up with symbol position is the third one and is corresponding to the decimal number four. The third position is represented by 2 to the power of 2 and is equal to 4. As we discussed before, in multiplication we are moving the first number on the (***n***-1) positions, which are two positions in the ascending direction. (***N***-1) formula is equal to the power of the binary number in this position, which is equal to 2.

The power of the binary number in this position is the sequence number of this position minus one. To make things simple, in the result of the second step the second number is moved two positions to the ascending direction,

Result for step two is 20 "0 0x 0 x"

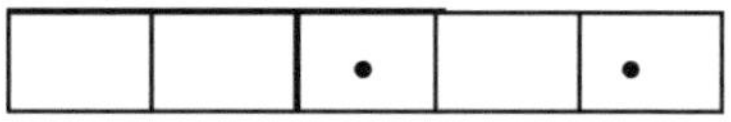

		•		•

Third step for example 4.2,

We add results of the first and second steps, and this will be the answer of the multiplication, which is equal to the number 25 and is represented in Litvin's Code by five positions, of which three fill up with symbol.

First step is 5 "x 0 x"

Second step is 20 "0 0 x 0 x" +

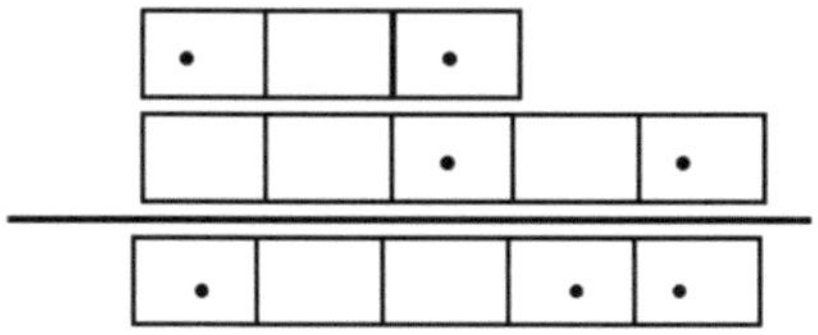

Answer is 25 "x 0 0 x x"

- When any number multiplied by one then the result is the same number. In this case, since the power of binary number is 0 so the number is equal to one. Any number in the power of 0 is equal to 1.

- When the first position on the second number is filled with symbol, then in the first step we do not move the first number and answer for the step one is equal to the first number.

- When we do the second step, we move the first number two positions to the ascending direction and at the end we add the results of two steps.

- When we use this method of calculation described in Litvin's Code, we move numbers on specified positions, which correspond to the power of the binary number in this position. The power of binary number in the third position is equal to 2 and we moved the first number on two positions.

End.

Example 4.3

First number is	3	"x	x	0"	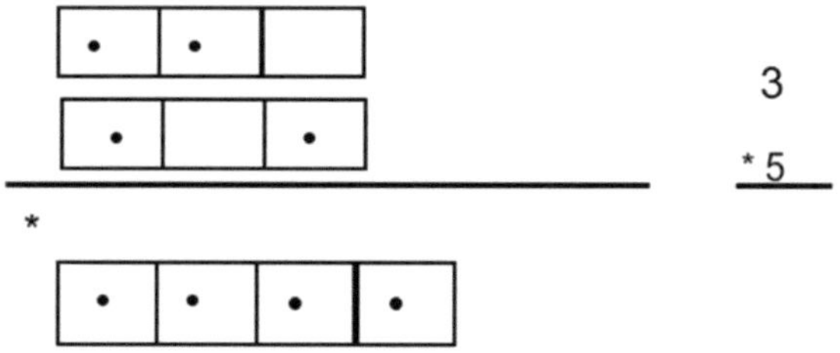
Second number is	5	"x	0	x"	
Answer is	15	"x	x	x	x"

•	•		
•		•	
•	•	•	•

3
* 5

In the Example 4.3, we have two numbers with three positions each. if the first and second numbers are the equivalent of odd decimal numbers. The first number is having the symbols in the first and second positions. The second number has first and third positions occupied with the symbols. By looking on the positions filled with the symbols on the second number, we move the first number to the ascending direction. The symbols in the second number are in positions 1 and 3, which represent binary number in the power of 0 and 2. To receive the answer we need to move the first number twice, first time on 0 positions and second time on 2 positions, and then to add the results of the steps.

Explanation

Step one for the Example 4.3 the second number has the first position filled with symbol, then the power of the binary number in this position is 0. When power of binary number equal to 0 and any to the power of zero is equal to 1. We do not move the first number and the result of the first step will be equal to the first number.

Step two for the Example 4.3. The second number has the third position filled with a symbol, then the power of the binary number in this position is 2. When power of binary number in this position is equal to 2, we move the first number on two positions, and the result of the second step will be the first number moved two positions in the ascending direction.

Result for step two is 12 “0 0 x x”

Third step for Example 4.3. In this step we add the results of the first and the second steps and arrive to the result of the multiplication.

The first number is 3 “x x”

The second number is 5 “x 0 x”

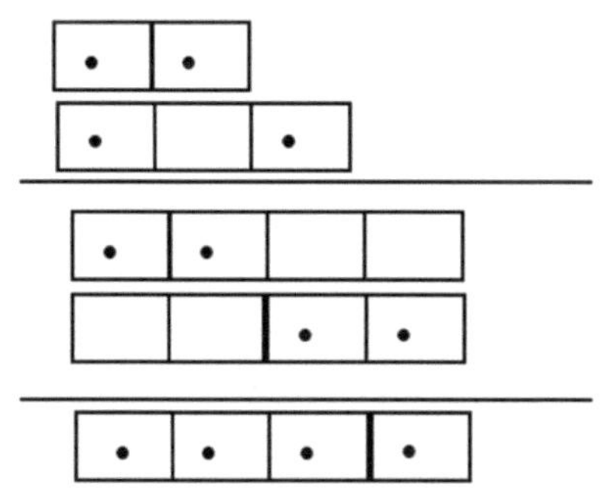

Result of step one is 3 “x x”

Result of step two is 12 “0 0 x x”

Answer is 15 “x x x x”

In Summary:

- When we do calculations and the first position of the second number is filled with the symbol, then the result of the first step is equal to the first number.

- When we do the second step, we move the first number on two positions to the ascending direction and then add the results of two steps.

- All the time, when we do multiplication by using Litvin's Code, we just move symbols to the ascending direction.

End.

Example 4.4

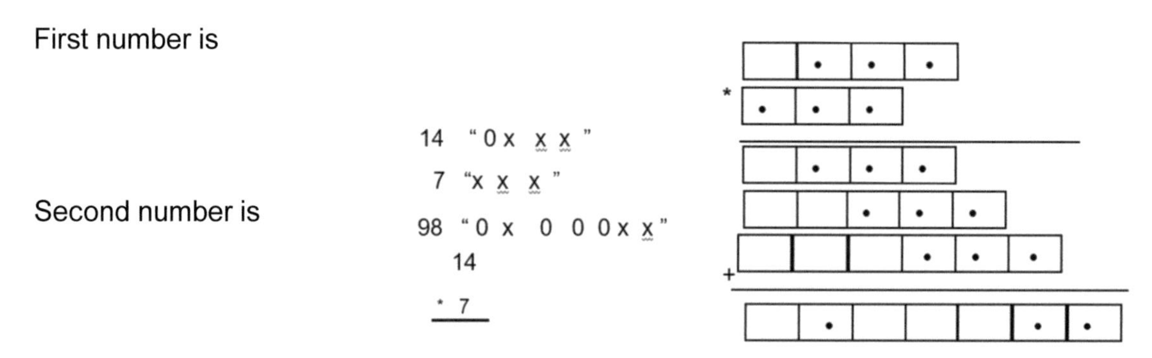

Answer is

In example 4.4 we do the multiplication with two numbers with four positions on the first number and three positions on the second number. The second number has all positions occupied with symbols. First of all, we understand that in this example, positions 1, 2, and 3 are filled up with symbols and the powers of binary numbers (***n***-1), which are equal to ***0, 1, and 2***. The filled up positions on the second number represent digits 1, 2, and 4 and the sum is 7.

In the first step we do not move the first multiplicand on any position because the power of binary number in the first position of the second multiplicand is 0. If we move the number on 0 positions then we have the result equal to the first number. In the second step we move the first multiplicand on the one position to the ascending direction because the power of the binary number in the second position of the second multiplicand is equal to 1. In the third step we move the first multiplicand on the two positions to the ascending direction, because the power of the binary number in the third position of the second multiplicand equal 2.

Explanation

***First Step* for Example 4.2 –**

On the second multiplicand the first position is equal to 1,

When (n-1)=0 and n=1, where (n-1) is the power, or $2^{n-1}=2^0=1$.

In the first position we always have a power of 0. Any numbers in power of 0 is 1. We do not move the first multiplicand, because if we multiply by 1 the answer is the same. It means we do not do anything.

The result of step 1 is equal to the first multiplicand or 14. " 0 x x x"

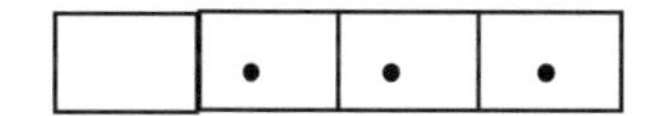

***Second step*.**

The second occupied position on the second multiplicand is n=2 and the power of binary number is (n-1)=1. The power of binary number in the second position of the second multiplicand is equal to 1. We need to move the first multiplicand by 1 position to the ascending direction. We move the first multiplicand by one position and the result of Step 2 is 28.

" 0 0 x x x"

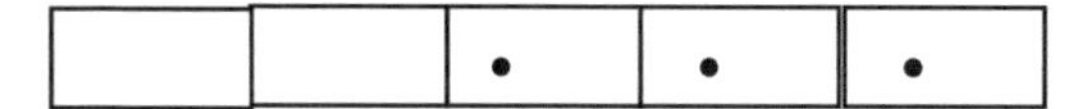

Third step

The third occupied position on the second multiplicand is 3. We need to move the first multiplicand on (n-1) positions. The (n-1) is corresponding to the power of binary number in this position and is equal to 2; (n-1) = 2, n=3

We moved the first multiplicand on the two positions to the ascending direction and result of this step is 56. " 0 0 0 x x x"

The Intermediate steps are presented below. We add all binary numbers in filled with symbol the positions of the answer.

First step 14 “0 x x x”

Second step 28 “0 0 x x x”

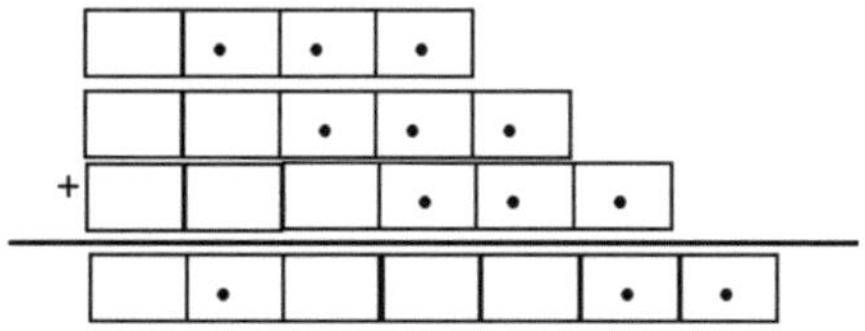

14
+ 28
56
98

Third step 56 “0 0 0 x x x”

Answer 98 “0 x 0 0 0 x x”

In Summary:

- When we do multiplication with Litvin’s Code, we move symbols to the ascending direction. When we do the first step and the first position on the second number is filled with the symbol, then the result for the first step is equal to the first number, which is 14.

When we do the second step, we move the first number on one position to the ascending direction and then the result of the second step is 28.

When we do the third step, we are moving the first number on two positions to the ascending direction and then the result of the second step is 56, and then we add all steps and have an answer of 98.

End

DIVISION

Part I

Litvin's Code is different from division in the standard decimal numbers, where we try to find the answer by multiplying the divisor by an approximate number to get closer to the dividend. In the decimal division after we multiply the approximate number by the divisor, we arrive to an approximate answer, which should be less than dividend. We subtract the approximate number from the dividend. We continue in the next step the process of multiplication until we get the results of subtraction which yield to zero.

In Litvin's code we don't need to multiply approximate number on divisor to get approximate answer. In division by Litvin's Code we right away subtract the divisor from the dividend until the result yields to zero. The division by Litvin's Code is a process of subtracting two numbers, which are represented by symbols in different positions. The division of the two numbers required, that the first number was bigger or equal to the second number, because we are not using fractions in present format in the follow up examples.

If the corresponding positions in the beginning of both dividend and divisor are empty then they are automatically eliminated.

The way of division is the subtraction of two numbers until we reach 0. We usually have several intermediate results of the subtractions until we reach 0. We do division through Litvin's Code by using subtraction and subtract the divisor from the dividend. Our goal is to get the first position empty on the answer as a result of subtraction. It means that subtraction was successful. When the subtraction in the step is successful, then on the answer of the division we fill with a symbol (dot) the corresponding to the step position.

When we see that the first position of the dividend is empty and the corresponding position of the divisor is filled up with symbol, we move the corresponding empty position to answer of the division, and in the mean time eliminating it on the dividend.

When after subtraction we have empty first position on the result of the subtraction, we eliminate the first empty position from the result of subtraction, and fill up with symbol (dot) the corresponding position on the answer of the division. In another example, if after subtraction we have several positions empty on the result of subtraction then we have a different situation. First of all, we fill up with a symbol (dot) the first empty corresponding position on the answer of the division and eliminate the first position on the result of subtraction. When more positions in front on the result of the subtraction are empty, we move the corresponding position without symbol (dot) to corresponding position on the answer of the division and eliminating those positions from the result of the subtraction.

When subtraction is not successful, and the intermediate result has the first position still filled with a symbol (dot), we keep the first position on the result of subtraction and continue subtract divisor in the next step. We also keep empty the corresponding position on the answer of the division. To be more precise, when after subtraction the first position of the result of intermediate step is filled up with a symbol, then we keep the corresponding position on the answer of the division empty. It could be that the divisor is too big and we need to continue subtracting this number in the next steps until the first position on the result of subtraction become empty. Until the first position result of subtraction is empty, we continuously keep empty corresponding positions of the answer until the result of the subtraction has the first position filled up with a symbol.

Rules of the Division:

- When in the beginning both numbers have equal number of empty positions, then we eliminate them from both numbers.

- When during the division the subtraction is producing the empty position on the beginning of the result of the subtraction, we eliminate the empty position from the result of subtraction for this step. We place the symbol to the corresponding positions on the answer of this division.

- When the subtraction does not produce empty space on the beginning of the result for the step, then we leave empty the corresponding position on the answer for this division, and continue to subtract divisor.

- When the beginning of the first number is an empty position and the second number does not have empty positions, we automatically transfer the empty position from the first number to the corresponding position to the answer of the division.

Example 5.1

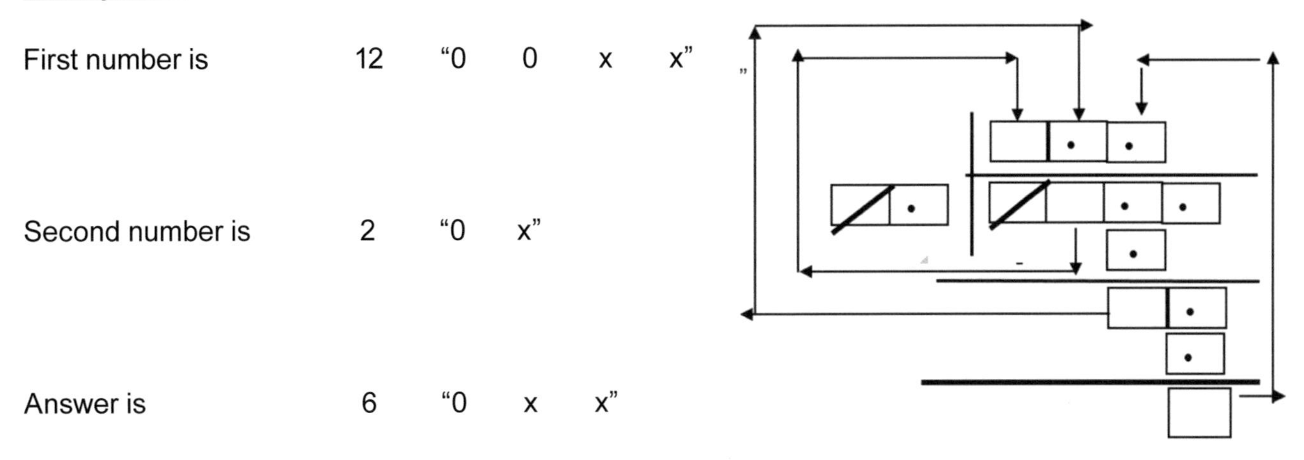

In example 5.1 we divide two numbers, where dividend has four positions and divider has two positions. The positions three and four on the dividend are filled up with symbols. The divider has second position filled up with a symbol. The dividend has two empty positions in front, and the divisor has one empty position. Before the subtraction we are eliminating the equally empty corresponding position in the beginning of both numbers. The equally empty positions are in the first positions of both numbers and we eliminate them. The dividend had four boxes and divisor had two boxes. After elimination of the corresponding empty positions, we have three boxes in dividend and one box in divisor. After the first step dividend has an empty position in front but divisor does not, and we move the empty position to the answer.

Explanation

Both numbers have the empty space on the first position. We compare this to the digit “0” of the regular decimal division. In the regular division, when we eliminate 0 at the end of the decimal number, the number becomes ten times smaller. When we eliminate the first position from both numbers, then both numbers become twice lower.

New first number is 6 “0 x x”

New second number is 1 “x”

We know, by dividing any number into 1, we will have the same number we started with. In this example we just want to demonstrate below how Litvin's Code applies to this division. We also eliminate the empty position from the first number, and bring the first empty position to the answer. When the dividend has the first empty position and the first position of the divisor is filled up with a symbol (dot) we move the first empty position from dividend to the answer of the division and in mean time we eliminate the empty position on the dividend.

New first number is	3	"x	x"	• •
The second number is	1	"x"		•

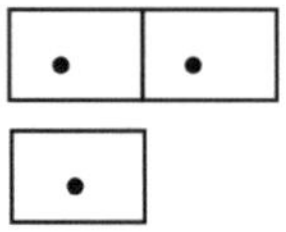

Now we have two steps for subtraction. ***In the first step,*** after erasing the corresponding empty positions, we subtract the new divisor from the new first number and as a result of subtraction there is an empty space. When we get an empty space as a result of subtraction this indicates that first number was bigger than divisor, and the subtraction was successful. When subtraction is successful, we fill up with a symbol (dot) the corresponding position on the answer of the division. ***In step two*** we also subtract the divisor by the new first number and in the result of subtraction is an empty space. As we mentioned before, when we have an empty space as a result of the subtraction, we fill up with a symbol (dot) the corresponding position on the answer of the division.

In Summary:

- When we divide 12 by 2 by using Litvin's Code, we have on the beginning of both numbers one empty position and we eliminate this position from both numbers and the numbers became 6 and 1.

- Since number 6 has an empty first position and the divisor has no empty spaces, then we move the empty space in the dividend to the answer. When we do division using using Litvin's Code and the first number equal 6, we move the first empty position to the answer of the division and the numbers become 3 and 1.

- When in division by using Litvin's Code, we subtract 1 from 3, and we have an empty position of the result of subtraction. We fill up with a symbol the next corresponding position on the answer of the division. The numbers became 1 and 1.

- When by using Litvin's Code we subtract 1 from 1 and we get 0 or empty position on the result of the subtraction, we fill up with a symbol the next position on the answer of the division.

End.

Example 5.2

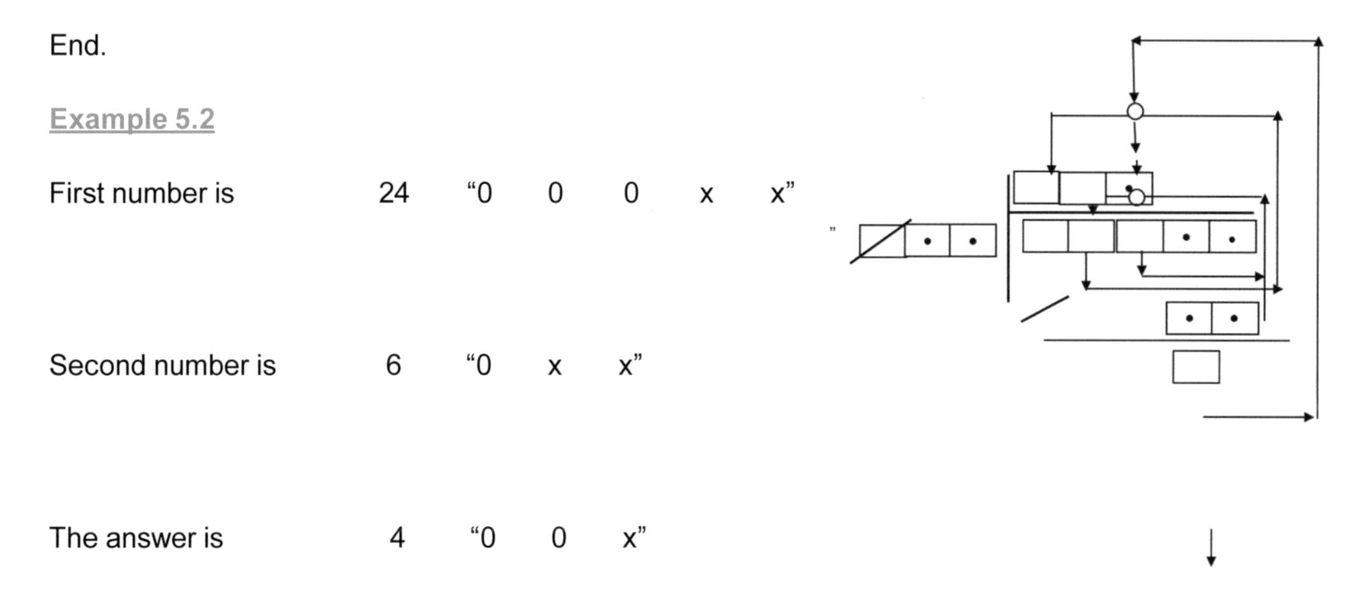
First number is 24 "0 0 0 x x"

Second number is 6 "0 x x"

The answer is 4 "0 0 x"

In example 5.2, there is the division of two numbers. The first number has five boxes and the second number has three boxes. The dividend has positions four and five filled up with symbols. The divisor has positions two and three filled up with symbols. The dividend has three empty spaces, and divisor has one. In division by using Litvin"s Code before subtraction we eliminate or transferring the empty positions.

Explanation

The both numbers have the empty positions in the front. We eliminate the corresponding empty positions in the front of both numbers. After the eliminating of the first corresponding empty positions, the first number left four boxes and the second number two boxes.

The second number does not have any empty position in the front, but the first number still has in front the two empty positions, which are automatically moved into the answer of the division. We eliminate the empty position from the first number. Now we have in the first number only two boxes remind. The remained two boxes on first number are equal to the decimal number 3. The second number is equal to 3. Now, if we subtract 3 out 3, we will have zero or an empty position as a result of the subtraction. As we have discussed about the successful subtraction, the next position is filled up with a symbol on the answer of the division.

In Summary:

- When we divide 24 by 6, we eliminate one empty position from both numbers and the numbers become 12 and 3.

- When we divide 12 by 3, we transfer two empty positions of the number 12 to the answer of the division the numbers will be 3 and 3.

- When we subtract 3 from 3, the result for this step becomes an empty position. We fill the next position on the answer of the division with a symbol.

End.

Example 5.3

First number is 65 "x 0 0 0 0 x x"

Second number is 5 "x 0 x"

The answer is 13 "x 0 x x"

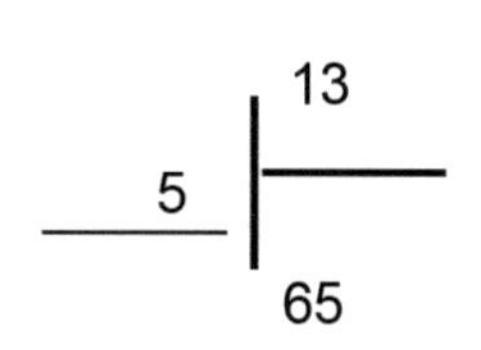

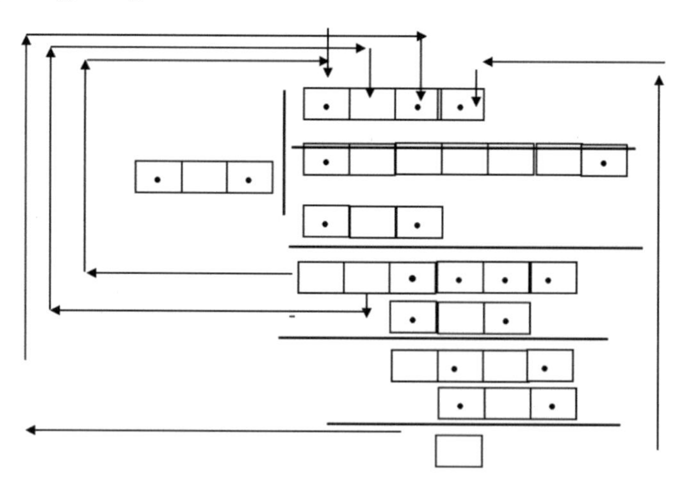

In the Example 5.3 we are dividing two numbers with seven positions on the first number and three positions on the second number. On dividend the positions one and seven are filled up with symbols. On divisor positions one and three filled up with symbols. The five positions of the first number in the middle are empty, and the divisor has an empty position in the middle.

The way to divide numbers in Litvin's Code is to do subtraction step by step until 0. The division could be done as subtractions. In step one after subtraction we have six boxes as a result of subtraction. The first two boxes on the result of the subtraction are empty. If, after subtraction, the first position on the result of the subtraction is empty, we erase this position and fill with a symbol in the corresponding position on the answer of the division. Now we left with five boxes on the result of subtraction. We still have an empty box in front of the result of the subtraction. We transfer the next empty position from the result of the subtraction to the answer of the division as an empty position. We left with four positions on the result of the subtraction. In other words, if after the first subtraction the first position of the result of subtraction is empty, we erase the empty position and fill with a symbol the corresponding position of the answer of the division.

Explanation

In the example 5.3, in the first step we subtract the second number from the first. We borrow the last symbol from the last position of the first number to the prior positions. The symbol from the last position does not appear anymore in the calculation because the last number is twice as big than prior, it is the same as having two symbols in the prior position. We leave one symbol in this position, but the other symbol we borrow again to the prior position. Again, we have the same situation that this symbol is twice as big as the regular one and is appears as two symbols on the prior position. We leave one symbol (dot) in this position and move one symbol to the descending position.

We move the next symbol to the prior position. In other words the position seven has one symbol, which we borrowed, and then positions four, five and six have one symbol, but position three has two symbols. In our example the third position on the divisor has a symbol two. In the meantime, on the third position of the dividend we have two borrowed symbols from the next position. We subtract one symbol from the first number and then move the second symbol to the result of the subtraction. In step two the new first number, which is the result of the subtraction, becomes sixty. We see that the two first positions on the result of the subtraction are empty. If the first position of the results of subtraction is empty, it means that the subtraction was successful. We eliminate the one empty position from the result of the subtraction and move the symbol to the first position of the answer. After we eliminate the second empty position from the result of the subtraction, and place the empty space on a corresponding position on the answer of the division. The new first number became four times smaller. We see that number 60 has two empty positions in the front.

The new first number is 60 “0 0 x x x x”

The second number is 5 “x 0 x”

In step three, we see that after eliminating the first two positions the number becomes 15. We subtract from the new first number the second number. We eliminate the empty space on the result of the subtraction. We move the symbol (dot) to the corresponding position on the answer of the division. When we subtract 5 from 15 the result is 10.

The new first number is 15 “x x x x”

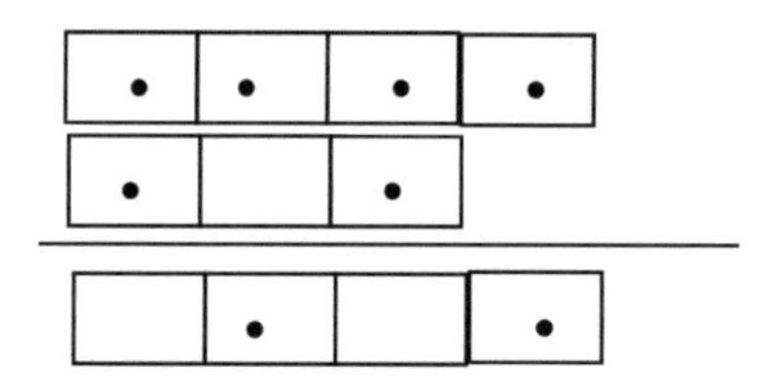

The second number is 5 “x 0 x”

The result for step is 10 “0 x 0 x”

In step four we repeat the same sequence. We have one empty position in front of the results of the subtraction. We transfer this position to the answer of the division, which is filled up with a symbol (dot). The result of the subtraction was successful.

The new first number is 10 "0 x 0 x" -

The second number is 5 "x 0 x"

As a result of eliminating the first position of number ten, we have five as our first number.

In another words, during the subtraction we eliminate the empty position from the result of the subtraction. We move the symbol to the corresponding position on the answer of the division.

The new first number is 5 "x 0 x"

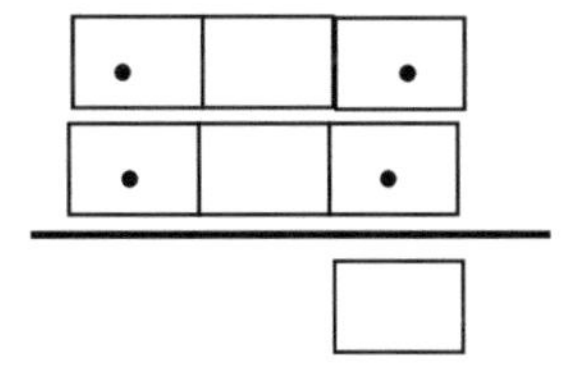

The second number is 5 "x 0 x"

The result for this step is 0 "0"

Once again we fill with a symbol (dot) the corresponding position on the answer of the division.

The final answer is 13 "x 0 x x"

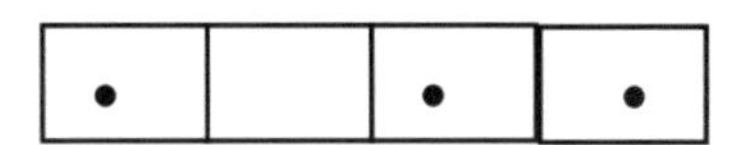

In Summary:

- When we do division, we subtract two numbers.

- When we do division and the beginning positions (first, second, ext.) of the first number are empty, we transfer the symbol to the corresponding positions on the answer of the division. After the subtraction, the first position on the result of subtraction is empty, then we erase this position and fill up with the symbol the corresponding position on the answer of the division, and the second position we just transfer as empty.

- When we divide 65 into 5, in the first step we subtract 5 from 65 and the answer is 60. The result of the subtraction for the first step is the number sixty, and is represented by Litvin's Code. The first position of the first number is empty. We eliminate it from the result of subtraction and fill with a symbol (dot), the corresponding position on the answer of the division. The next position of the result of subtraction has the empty space too. We eliminate the next empty position from the result of the subtraction. We transfer the second empty position to the answer of the division as empty. After the transfer the number 60 became 15.

- When we subtract 5 from 15, the result of subtraction for this step is 10 and has the empty space in the first position. We eliminatr the first position from the result of subtraction for this step. We simultaneously fill up with a symbol (dot) the corresponding position on the answer of the division. In the meantime, after the transfer the 10 became 5.

- When we subtract 5 from 5 the result of the subtraction is 0 or empty space. We fill up with a symbol the corresponding position on the answer of the division. The answer of the division is 13.

End.

Example 5.4

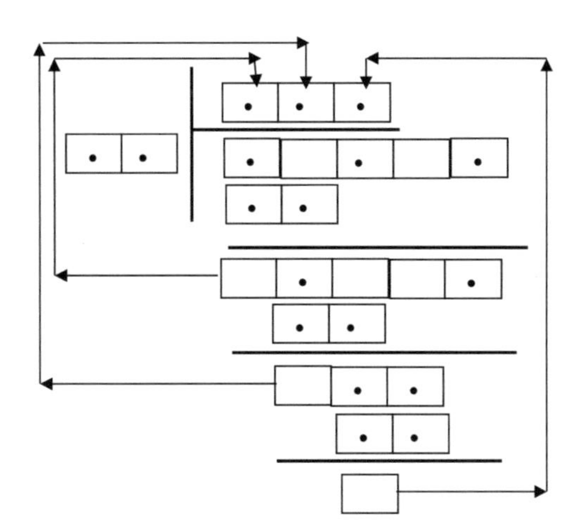

Dividend is 21 “x 0 x 0 x”

Divisor is 3 “x”

The final answer for it is 7 “x x x”

In example 5.4 we divide the two numbers with five positions on the first number and two positions on the second number. The dividend has positions one three and five filled with the symbols. The divisor has positions one and two filled with symbols. The positions two and four on dividend are empty. This division we do step by step. It requires three steps to complete this calculation and get the final answer:

Explanation

First step for Example 5.4

Subtract from dividend 21 “x 0 x 0 x” -

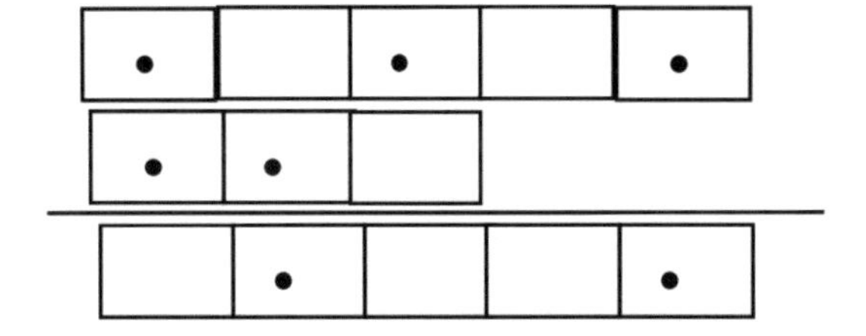

Divisor 3 “x x”

First result 18 “0 x 0 0 x”

Because the first position of the result of subtraction is empty, we fill with a symbol the first position of the answer of the division. In other words, we erase the first position from the result of subtraction and fill with a symbol for the answer of the division.

Result of subtraction

After the erasing of the first position, the result is 9 “x 0 0 x”

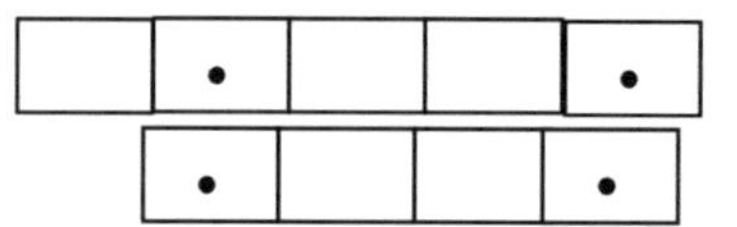

When we are erasing the empty position of the result of

subtraction, we fill with a symbol the corresponding position on the answer of

the division.

Answer for the results of division after the subtraction for step one is 1 “x 0”

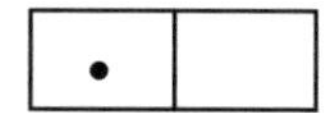

Second step

Subtract from the new first number 9 “x 0 0 x”

Divisor 3 “x x”

Second result 6 “0 x x”

After the subtraction 3 from 9 the answer is 6. The first position on the answer of the subtraction is empty. We eliminate the first position of the result of the subtraction and filling the answer of the division with a symbol (dot) in corresponding position. After eliminating the empty position the number 6 become number 3. If the first position is empty on the result of the subtraction, then the second position of the answer of the division is filled up with a symbol. The first position of the second result of subtraction is erased.

Erase the first position and the result is 3 “x x”

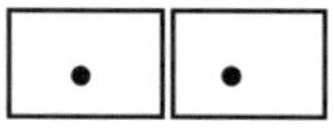

When we erase the empty position of the result of

subtraction, we fill the results of the division with a symbol.

The answer for the results of division after the subtraction for step two is 3 “x x”

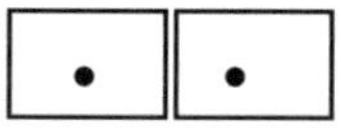

Third step

Subtract from the new first number 3 “x x”

Divisor 3 “x x”

The result of third subtraction is 0 “ –”

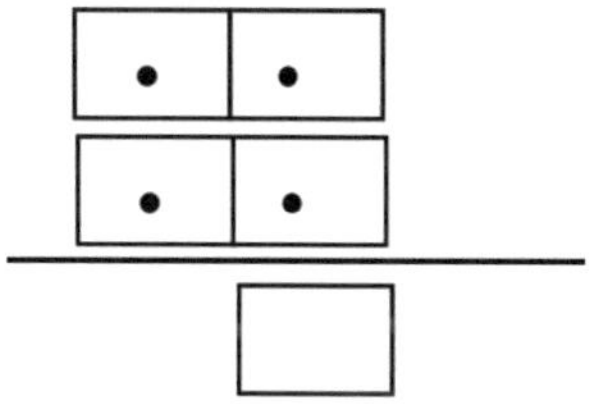

When the result of the subtraction is zero, we are filling up with a symbol (dot)

correspond to the position of the answer of the division.

Answer for results of the division is 7 “x x x”

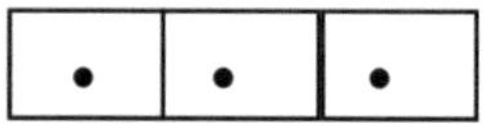

When we look at the answer of the division, we could see that three subtractions were successful, and we have three positions filled up with a symbol (dot) of the result of the division.

In Summary:

- Whenever we get the first empty position in the result of the subtraction, then the corresponding position on the answer of the division is filled up with a symbol. The first position of the result of the step is erased.

- When there are no more steps for subtraction and the result of subtraction is 0, then it is the last step and we have arrived at the final answer.

- When we divide 21 by 3, we first subtract 3 from 21 and result is 18. In the number 18 expressed in Litvin’s Code, the first position is empty. We eliminate this position from the result of the subtraction and fill up with a symbol (dot) the corresponding position of the answer. After eliminating the first position of the result of subtraction, which is number 18, it becomes 9.

- When we are subtracting 3 from 9, the result is 6, and the first position of the number 6, which is expressed in Litvin's Code, is empty. We eliminate this position from the result of the subtraction and we fill up with a symbol on the second position of the answer of the division. The number 6 became 3. We subtract 3 from 3. The result of subtraction is an empty space or 0. We are filling up with a symbol on the third position of the answer of the division. The answer of the division is 7.

End.

Example 5.5

Dividend is 65 "x 0 0 0 0 0 x"

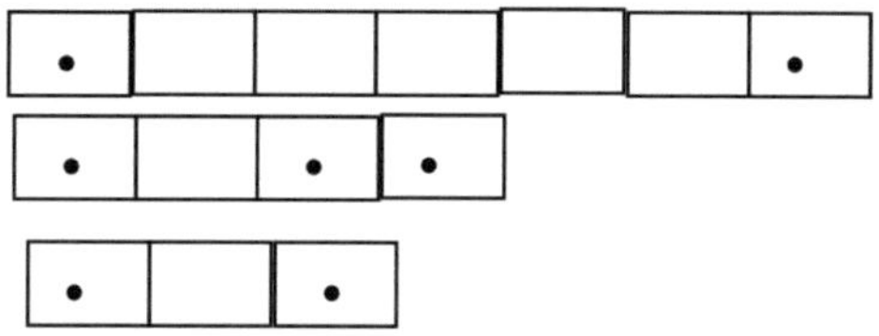

•						•

Divisor is 13 "x 0 x x"

•		•	•

•		•

The final answer is 5 "x 0 x"

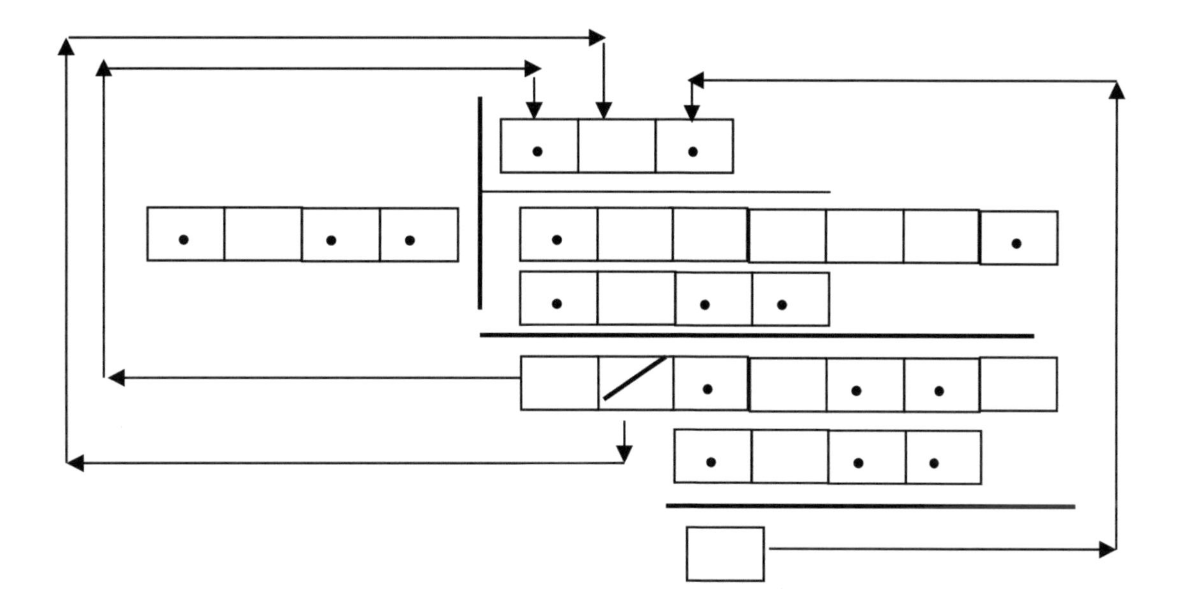

In example 5.5 we divide two numbers. Dividend is equal to 65. It means that the first and last position in this number is occupied with symbols. The last position is the 7th position. Between first and last position we have five empty positions. To do division it means to subtract from dividend the divisor until we get the result of zero. When we have number 65 as a dividend and 13 as a divisor we borrow a symbol from the last position to the prior positions. This division requires two steps of subtraction to complete this calculation and to get the final answer.

Explanation

First step for example 5.5

Subtract from dividend 65 "x 0 0 0 0 0 x"

Divisor

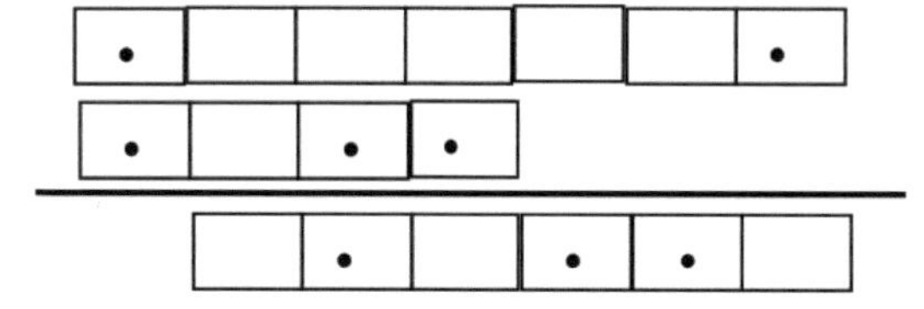

First result of subtraction is 52 "0 x 0 0 x x"

When we subtract 13 from 65, the result of the subtraction is 52. We see that the first result on the subtraction number 52 and first position is empty. It means the subtraction was successful. We eliminate the first position of number 52 and number 52 become number 26. We also see that number 26 has an empty position in the front. We move this empty position to the corresponding position of the answer of the division.

After the move of the first position the number 52

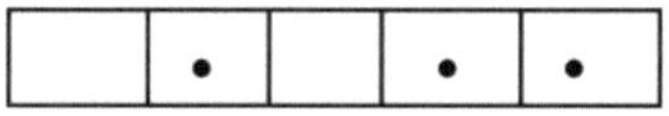

Becomes half and the result is 26 “ 0 x 0 x x”

After the step the answer of division has one filled up position and one empty.

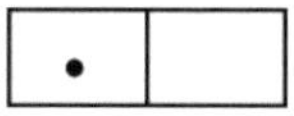

In our first subtraction in this division the result is 52. The number 52 has two empty positions in front. The first empty position indicates successful subtraction, which transferred the first position of the results of the subtraction as filled up with a symbol (dot) to the corresponding position on the answer of the division. When we erase the first position in number 52, it yields to 26. The number 26 also has the first position empty. This empty position is transferred to the corresponding position of the answer of the division as empty. Now, the answer of the division has one filled up position and one empty. After erasing the first position from 26 the result is 13.

Second step for Example 5.5

Subtract from third step result 13 “x 0 x x”

Divisor 13 “x 0 x x”

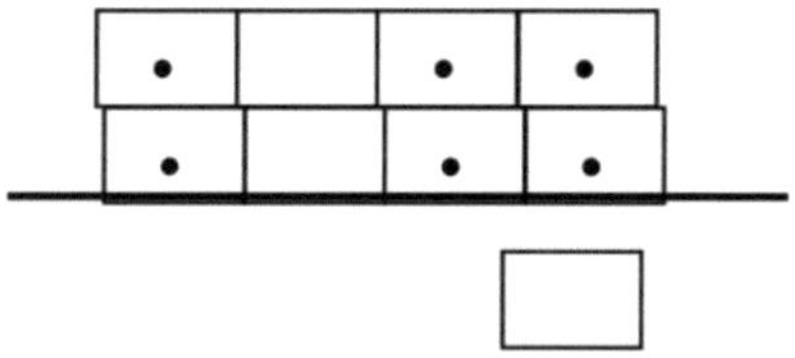

Second step result 0 “0”

Now we subtract the divisor (13) from the result (13). The result of the second step is an empty space after the subtraction. When we have an empty space as a result of the subtraction for a final result, we fill with a symbol (dot) corresponding position on the answer of the division.

Final answer is	5	"x	0	x"	•		•

The final answer shows that two steps had successful subtraction and two corresponding positions on the answer of the division are filled up with a symbol (dot). We see that in the middle of the final answer of the division is an empty position. It means that after the first step, as a result of subtraction we had two empty positions, one of which was transferred as a filled corresponding position in the answer of the division with a symbol (dot). The other, which was left, will transfer as an empty position to the answer of the division.

In Summary:

- When we are dividing 65 by 13, we are subtracting from the first number equal to 65 from the second number equal to13.

- When we have the result of subtraction equal to 52, expressed in Litvin's Code, the two front positions are empty.

- When we got the result of 52, we erased the first empty position and the number became 26. This fills up the first position of the answer of the division with a symbol (dot).

- When we see the first position on the number 26 is empty, which is not the result of subtraction, we transfer it as empty to the corresponding position on the answer of the division. The result after the transfer becomes 13.

- When we subtract 13 from 13 the answer is 0. When we have the result of 0 for the subtraction, then it was successful, and then we fill up with a symbol (dot) on the third position of the answer.

Example 5.6

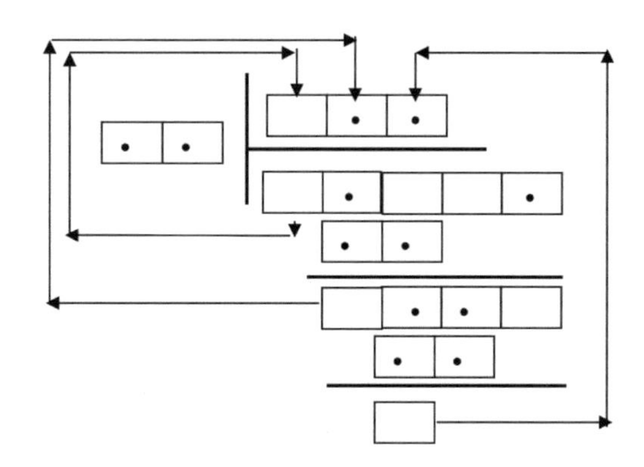

Dividend is	18	"0	x	0	0	x"
Divisor is	3	"x	x"			
The final answer is	6	"0	x	x"		

In example 5.6 we are dividing two numbers. Dividend is equal to 18. The first position on the number 18 is empty. The second position is filled up with a symbol (dot) and the last position is the 5th position that is also filled up with a symbol (dot). Between the second and last position we have two empty positions. To do division it means to subtract the divisor from the dividend until we get the result of zero. The first position of the number 18 is empty and it is not the result of subtraction, then we transfer this position as an empty position to the answer of the division. The number 18 after the transfer of the first position become 9. When we have number 9 as a dividend and 3 as a divisor we borrow a symbol from the last position to the prior positions. This division required three steps to achieve this calculation or final answer:

Explanation

First step for Example 5.6

Because the first position on the dividend is empty and the first position of the divisor is filled up with a symbol, we do not do any subtractions. The first position of the dividend is empty then the first position of the dividend is erased from the dividend.

Erase first position of dividend

18 “0 x 0 0 x”

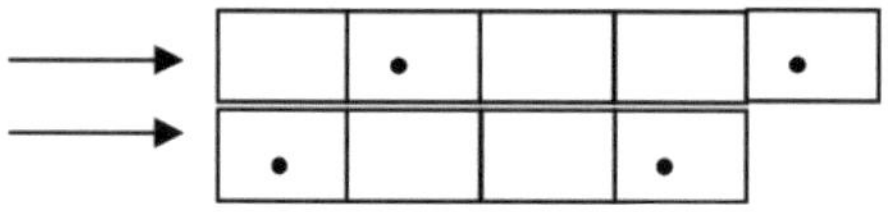

Result is 9 “x 0 0 x”

For the first step we erase the first empty position of the dividend and then

Result for step one is 9 “x 0 0 x”

Answer for step one is 0 “0”

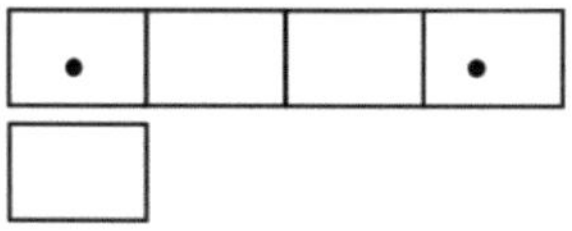

Second step for Example 5.6

When we subtract 9 from 3 the answer for the second step is 6. When we see number 6 using Litvin's Code, we understand that the first position of this number is empty. The subtractions were successful. It means that the answer for the division will have the second position filled up with a symbol (dot). This is a reverse operation. We erase the first position of the result of subtraction and fill up with a symbol (dot) the corresponding position on the answer to the division.

Subtract from the new first result 9 “x 0 0 x”

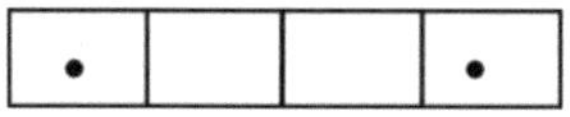

Divisor 3 “x x”

Result is 6 “0 x x”

erase the first position and the result for the second step is 3 “x x”

Answer for step two is 1 “x”

Third step for Example 5.6

When we subtract from the second step result 3 “x x”

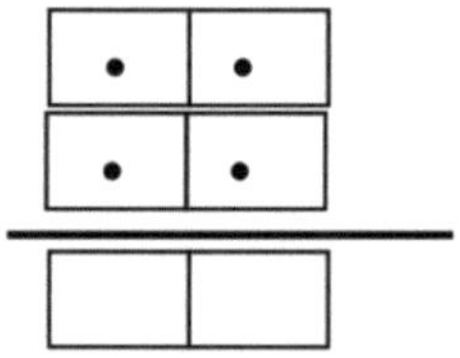

Divisor 3 “x x”

Third result 0 “0”

Answer for the step three 6 “0 x x”

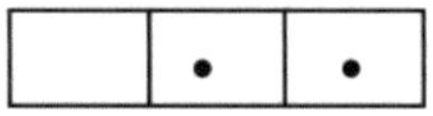

In Summary:

- When we divide 18 by 3, we see that number 18 has the first position empty but the number 3 has the first position filled up with a symbol. We then eliminate the first position from the number 18 and we move the first empty position to the answer of the division. The number 18 became number 9. We subtract 3 from 9 and the answer is 6. The number 6 has the first position – empty. It means that subtraction was successful and we need to eliminate this position.

- When we eliminate the empty position from the result of the subtraction and fill up the second position on the answer with a symbol, the number six became three. We are subtracting 3 from 3. The result of subtraction is an empty space or 0. We are filling up with the third position of the answer with a symbol. The result of division is 6.

End.

Example 5.7

Dividend is	24	"0	0	0	x	x"
Divisor is	6	"0	x	x"		
The final answer is	4	"0	0	x"		

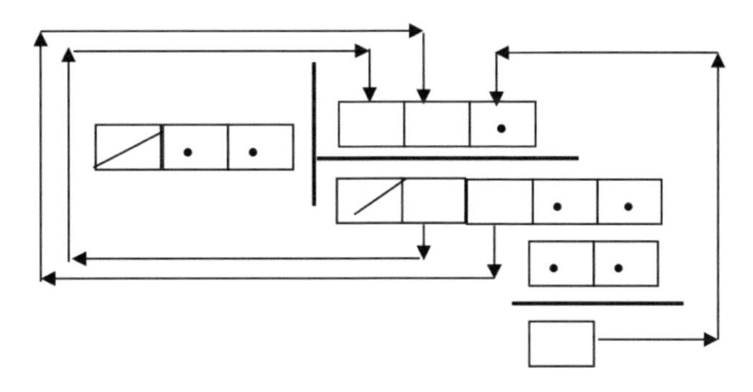

In example 5.7 we divide two numbers. The dividend is equal to 24. The first three positions on the number 24 are empty. The last two positions are filled up with a symbol (dot) and the last positions are the 4th & 5th position. The divisor is equal to 6 and has one empty position in front. Positions two and three are filled up with a symbol (dot). When the first position on the dividend and divisor is empty, we eliminate them without adding any positions to the answer of the division. Now the divisor has only two positions filled with a symbol (dot) and those positions are one and two which are equivalent to number three. After eliminating the first position of the dividend, on the dividend we still have two empty positions in the front. Now we must move these two empty positions from the dividend to the answer of the division. After the elimination of the first position in the dividend, the number is equivalent to 12. When we eliminate the second position, the number becomes six. Then we eliminate the third position by transferring to the answer of the division as an empty position as well. After eliminating these three positions on the dividend we are left with two filled up positions which are equal to the number 3. To do division it means to subtract the divisor from the dividend until we get the result of zero. This division requires two steps to achieve this calculation or final answer.

Explanation

First step for Example 5.7

Erase first, second and third positions on the dividend and first position of the divisor.

Dividend is 24 "0 0 0 x x"

Divisor is 3 "x x"

3 "x x"

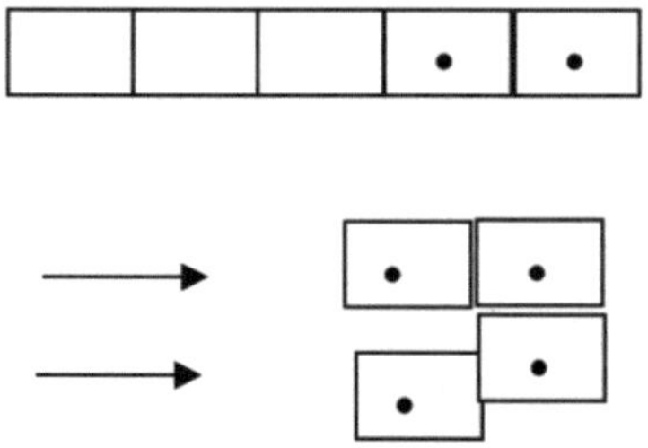

Because the first, second and third positions on the dividend are empty and first position of the divisor is empty too, we do not do any subtractions. We eliminate the empty positions on the dividend and divisor. We transfer the second and third empty positions of the dividend to the answer of division as empty. The first, second, third positions of the dividend are erased from the dividend.

Second step for Example 5.7

When we subtract

Dividend 3 "x x"

Divisor 3 "x x"

The result of subtraction is 0 "0 0"

Answer for the division 4 "0 0 x"

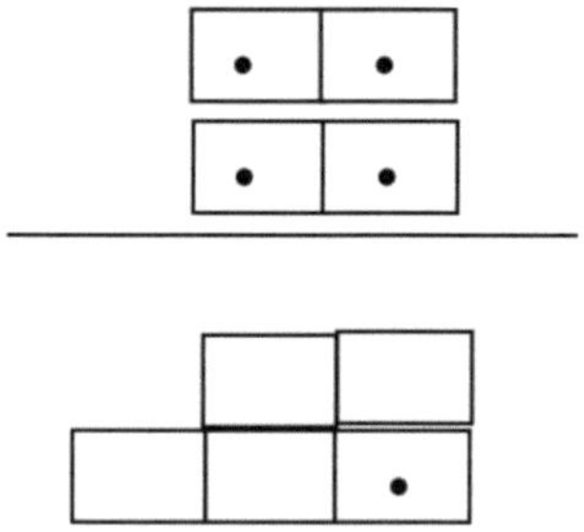

End

Because the first position of the result of the subtraction in step two is empty then the third position in the answer of the division will be filled up with a symbol (dot). Because after the subtraction the last step's result is empty, then the answer of the division in the third position is filled up with a symbol (dot).

In Summary:

- When we divide 24 by 6, we see that the first position on the number 24 and the number 6 is empty. In Litvin's Code we automatically eliminate the first empty positions of the dividend and divisor, and we are left with two empty positions on the dividend and the divisor has no empty positions left to eliminate. The number 24 and 6 became 12 and 3. We see that number 12 has the two empty positions in front but number 3 has not.

- When we eliminate those empty positions from the dividend and are transfer two empty positions to the answer, the number 12 becomes 3. We subtract 3 from 3, which is equal to zero. Division was successful and result zero is transferred to the answer of the division as a symbol (dot). The result of the division is equal to 4.

End.

LETTERS

The letters in Litvin's Code is represented by positions in the alphabet and are equal to the sequence number of positions. By memorizing the position of the letter we do not need memorize the letter itself.

We have several references, which include the sequence number of the position in the alphabet and the letter is equal to the position in the alphabet represented in Litvin's Code. In Litvin's Code we represent the letter as a binary number, which is equal to the position of the letter in the alphabet. The letter "A" is in first position and equal representation of one in binary form in Litvin's Code, and the letter "B" is equal to representation of two in binary form in Litvin's Code etc.

In our brain the access to complex cells cells may not be available then we have difficulty processing complex information. In the brain we have simple cells which are used to process simple information and never before used to process complex information. The brain can have no clear access to the complex cells due to physiological problems. It causes the loss of address, and the complex cells do not receive the right information from the stimulus. We automatically have difficulty processing complex information, and this limits us in learning new things, creates difficulty to write and read and ultimately inability to communicate properly. In the meantime the huge amount of simple cells in the brain is not used and the access to them is very easy. In psychoconduction we use the simple cells to process complex information. The patterns, which are used in Litvin's Code, become more and more complex and simple cells get trained more and more, and actually become conduits of complex information. The simple cells become sophisticated and do the same job, whatever the complex cells must perform; which means simple cells can be substitute for complex cells. For people, who had difficulty know how to read and write psychoconduction provides a unique opportunity to acquire these skills.

In the beginning psychoconduction used very simple patterns and then patterns became more complex and provided the transition to the mainstream alphabets. The psychoconduction does not have any limitations and boundaries of any existing alphabets. Psychoconduction adapts easil because it does not use any letters, but only positioning of the letter in the alphabet. The simple cells contain the stable address and easily could be found to process information from the stimuli. And now example letter "A" is somehow contained in the complex cell and the brain can not provide the right address and as a result the person has difficulty reading. With psychoconduction the simple cells function as complex cells and have correct address of information and can easily process information from the stimuli.

Example 6.1

This word in example 6.1 does not show any letters only shows the position of the letter in the English alphabet. To remember the letter we need to utilize complex brain cells. To deal with the position of the letter translated in Litvin's Code, we use the simple brain cells.

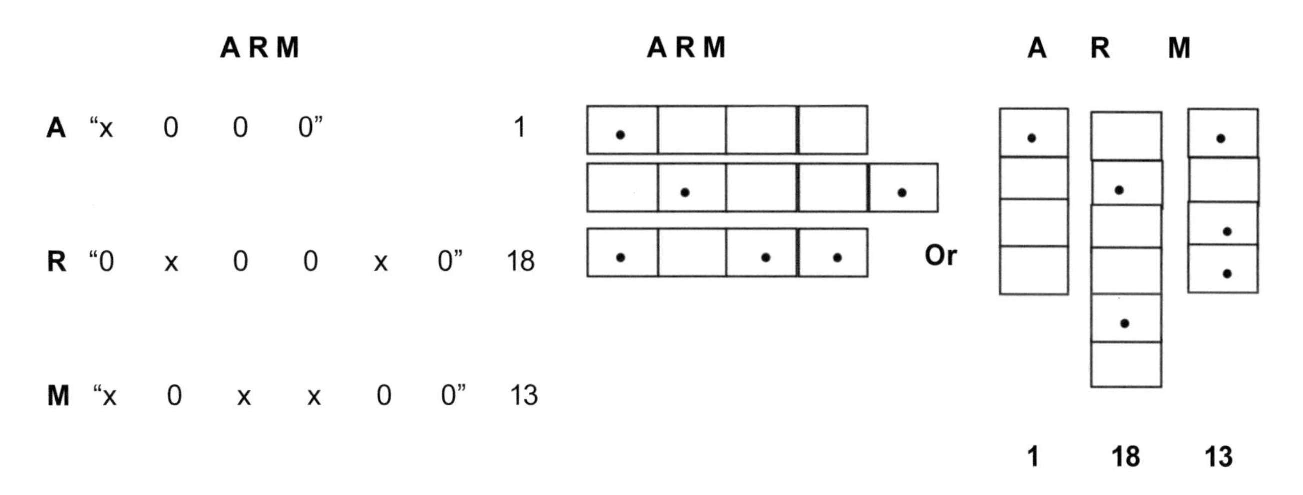

The first position in the English alphabet has occupied by the letter "A". In the example the first letter is "A" and in Litvin's Code is equal to 1. The letter "A" is the first letter of the alphabet.

The second letter is "R" and in Litvin's Code is equal to 18. The letter "R" is the 18-th letter in the alphabet. The letter "M" is equal to number 13 and is in the 13th position in the alphabet.

In Litvin's Code the letters are equivalent to the binary digits and also are represented by the position of the letter in the alphabet.

Example 6.2

CAMEL

C	“x	x	0	0	0”	3
A	“x	0	0	0	0”	1
M	“x	0	x	x	0”	13
E	“x	0	x	0	0”	5
L	“0	0	0	x	x”	12

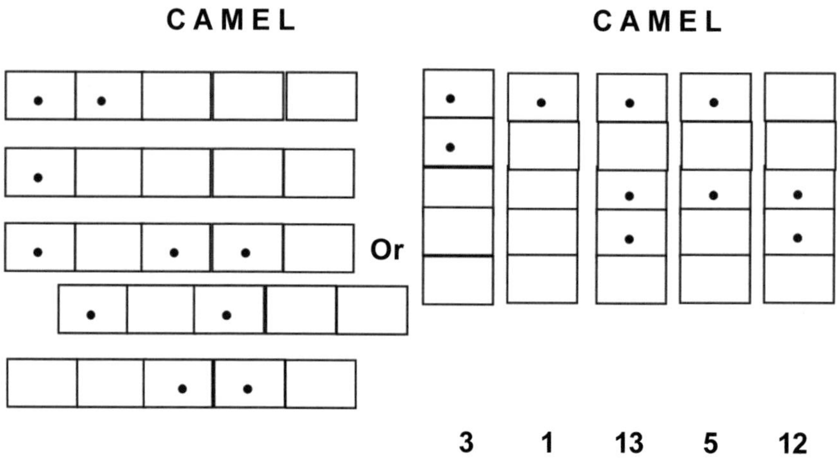

CAMEL

Or

CAMEL

3 1 13 5 12

The first letter in this example is “C” and is equivalent to three, which is the position in the alphabet for this letter in English alphabet. The second letter is “A” and has the first position in the alphabet. The letter ”M” has 13th position and is equal to number 13 represented in Litvin’s Code. The letter “E” equals 5 and the letter “L” is equal to 12. All letters are equal to their position’s numbers. In psychoconduction we could represent the letter with audio, visual, kinesthetic and olfactory stimuli. By processing the letters in the different areas of the brain with psychoconduction, the brain is achieving the tune up and the calibration as well. We transfer the same letters to different areas of the brain using visual, audio, kinesthetic and olfactory stimuli. Because we use the same patterns of information but the different stimuli, the amount of chemicals released during the processing of this information in different areas of the brain is congruent. To make calibration more effective we use the simple symbols (dot, knock, and step). For the visual information we use the dot to show presence of the symbol, and dash to show absence of the symbol. When we use audio information, one knock represents presence of the symbol and double knock represents absence of the symbol.

We use some sound indicators to inform that the letter is completed and then the same for the words and sentences. When we use kinesthetic symbol we step or clenching our fist, or doing other body movement. For smell we use different odors, one for the presence of a symbol and another for the absence. In psychoconduction we transmit the words in visual, audio, kinesthetic and olfactory mode using the same patterns. In the usual classroom the symbols representing visual and sound letters and digits are completely different from binary representation in Litvin's Code and are processed in the complex cells. In our work we provide the gradual transition from Litvin's Code to the mainstream letters and digits by utilizing simple cells to assist comprehension and to provide the correct address to process stimuli in the proper way. The psychoconduction set up communication paths through the brain that allow different parts of the brain to clarify the complex patterns of simple symbols and to be confident that the patterns are understood and the symbol is processed in the suitable way.

Example 6.3

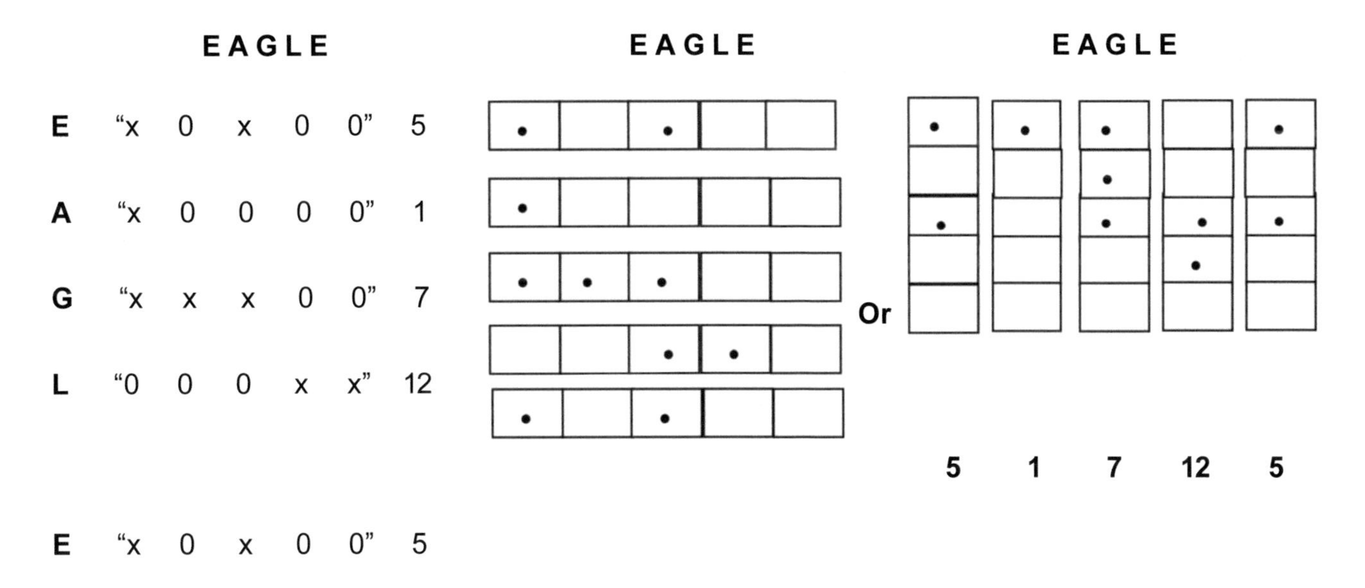

The first letter E has the fifth place in the English alphabet. The letter A is occupying the first position in the alphabet and represented by the number one. Letter G is occupying the seventh position and is represented by number seven in the English alphabet. Letter L is represented by the number 12 and letter E is represented by number 5.

Example 6.4

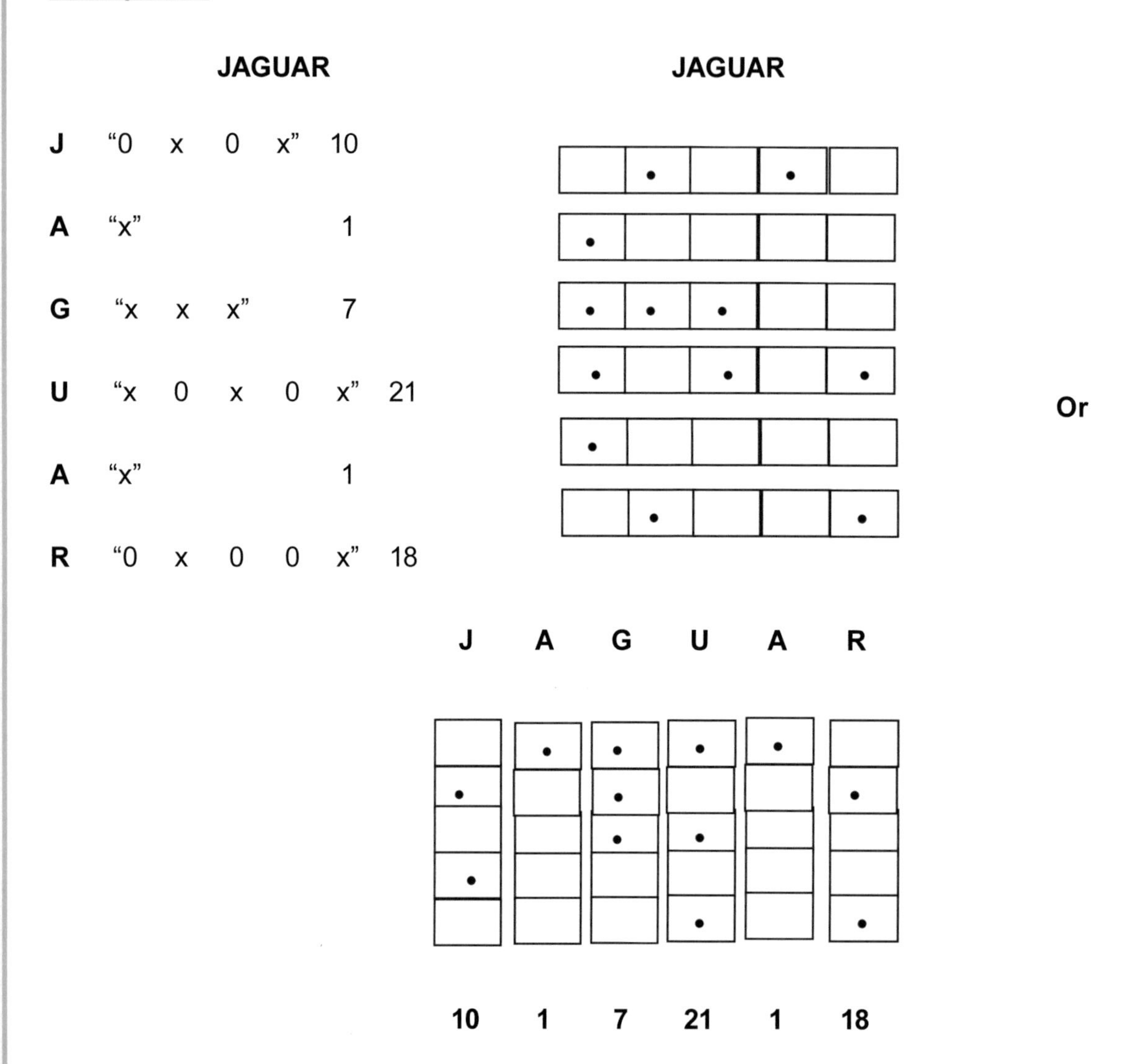

The first letter J is in the tenth position in the English alphabet and is represented by the number ten in Litvin’s Code. The letter A occupies the first position and is represented by number one in the English alphabet. The letter G has the seventh position, letter U is in the twenty first position of alphabet, and numbers 7 and 21 represents them. In Litvin’s Code the letters, “A” and “R”, are represented by numbers 1 and 18. The psychoconduction is using special algorithm, which is Litvin’s Code. If we do not yet know the meaning of the Code, we still transfer the patterns of simple symbol from visual to kinesthetic or audio or olfactory stimuli. By doing it we calibrate the different areas of the brain to release the congruent amount of the neurotransmitters. The congruency of releasing chemicals in different parts of the brain is one of the goals of psychoconduction. It creates self-regulation of the brain and triggers the appropriate response to the stimuli.

SENTENCES

To write the sentences we use Litvin's Code. With Litvin's code we also could translate letters to the audio, kinesthetic and olfactory representation. By using simple symbols in different positions we could make this sentence sound. We use knock to signify the presence of the symbol and double knock for absence of the symbol. In kinesthetic representation of simple symbol the clenching of the right hand is the sign that symbol is present, and clenching of the left hand is absence of the symbol.

In our work we use the aerobic board and when the symbol is present, we step on the board, and when the symbol is absent, we just move to the other side without stepping on the board.

I AM LEARNING LITVIN'S CODE

This sentence has five words. We learn by clenching our hand and stepping on the aerobic board. We are using knocks and a double knock to sound it out.

Word 1

The sentence begins with the word of one letter "I", which has the ninth position in the English alphabet.**I**

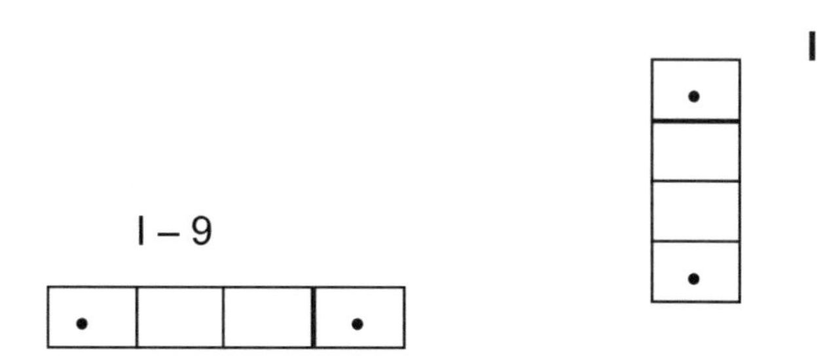

Word 2

The second word is "am" and has two letters. The first letter "a" has the first position and letter "m" is in the thirteenth position. The sequential position in alphabet is represented by Litvin's code.

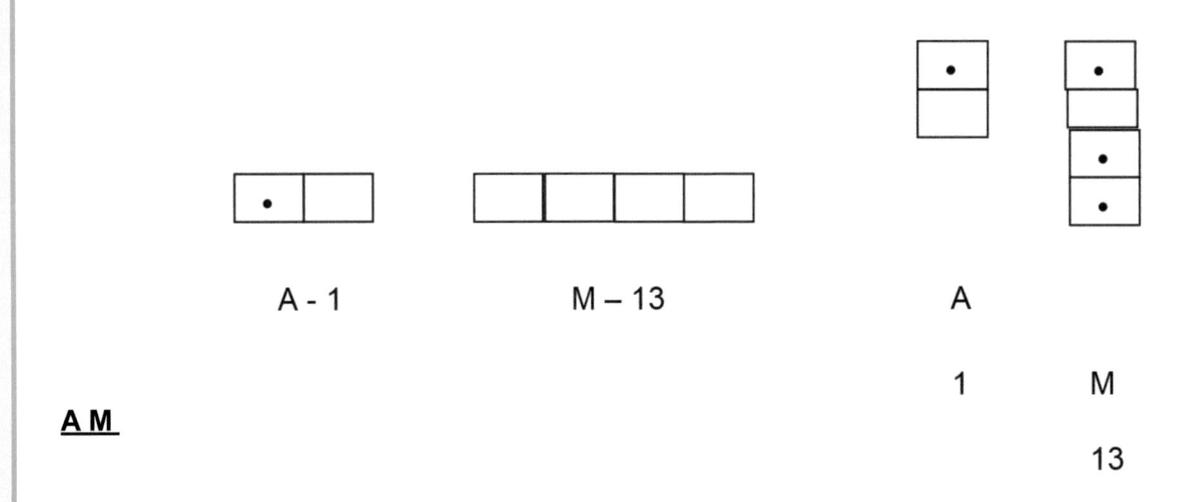

A M

Word 3

The third word is "learning", which has eight letters. The first letter is "l", which is in the tenth position in the alphabet. The second letter is "e" and occupies the fifth position of the alphabet. The third letter is "a" and we discussed this letter in the second word. The fourth letter is "r" and occupies the eighteenth position in the alphabet. The fifth and seventh letter is the same and is "n", which is in the fourteenth position in the alphabet. The sixth letter is "i" and we discussed it in the first letter. The last and the eighth letter in the word is "g" and has the seventh position in the alphabet.

LEARNING

L - 10 E - 5 A - 1 R - 18 N - 14

I - 9 N - 14 G - 7

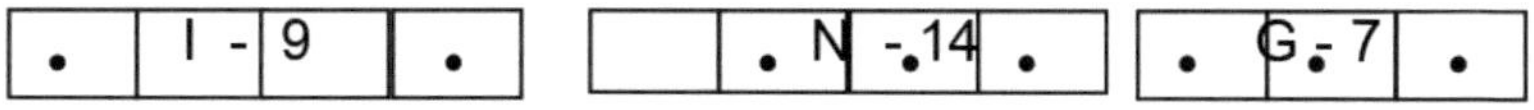

L E A R N I N G

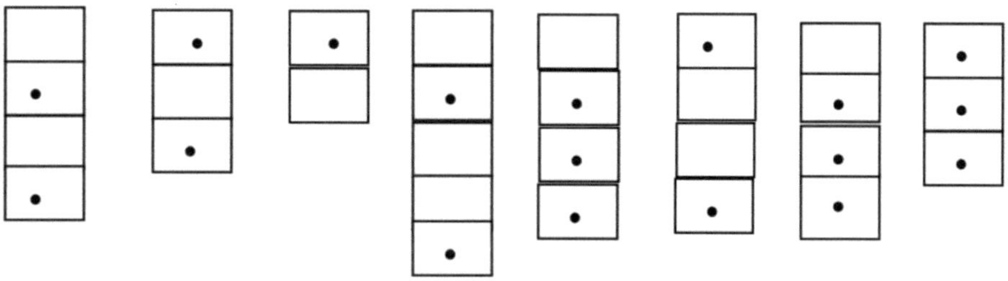

1

5

7

10 14 9 14

18

Word 4

The fourth word starts with the same letter "L" as the third word. The second and fifth letter in the third word is "I' and we discussed this letter before. The third letter is "t" and has the twentieth position in alphabet. The fourth letter is "v" and has the twenty second position in the alphabet. The sixth letter is "n" and we discussed it in the second word. The last and seventh letter for this word is "s" and is in the nineteenth position in alphabet.

LITVIN'S

L - 12 I - 9 T - 20 V - 22

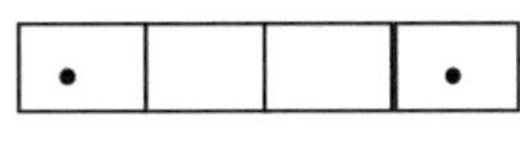

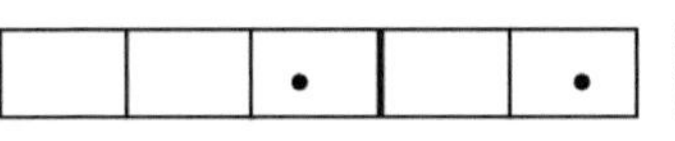

I - 9 N - 14 19 - S

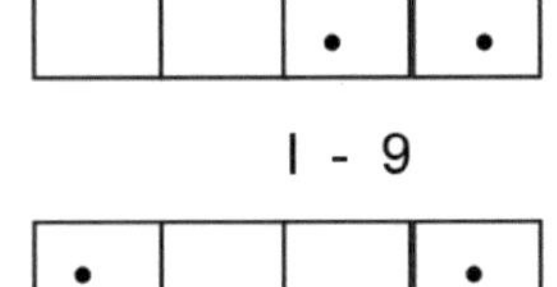

L I T V I N 'S

12 9 20 22 9 14 19

Word 5

The fourth word starts with the same letter “c” and has the third position in the alphabet. The second word is “o’ and is in the fifteenth position in the alphabet. The third letter is “d” and is in the fourth position in the alphabet. The fourth letter is “e” and we discussed it in the third word.

CODE

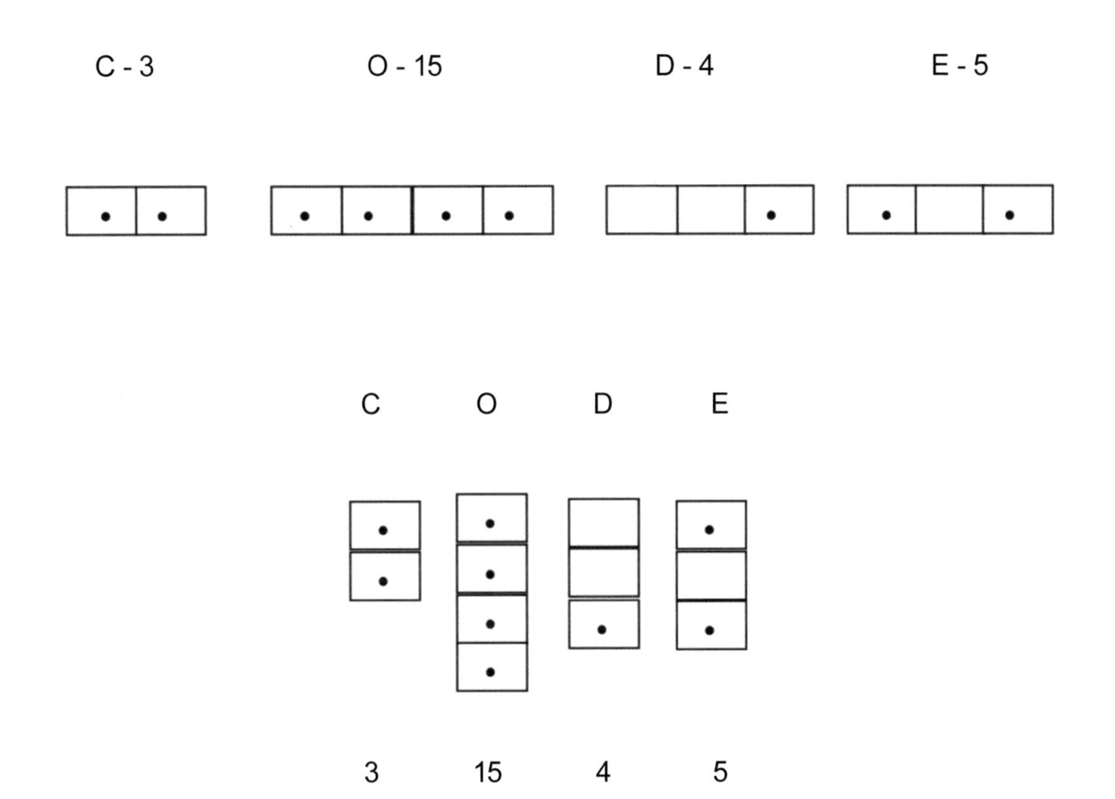

End of sentence.

This sentence is easily translated in kinesthetic stimuli, when clenched right hand represents presence of stimuli in the box and clenched left hands showing that absence of stimuli. The kinesthetic stimuli could be easily translated in many different ways from visual and audio stimuli. One example of kinesthetic translation is: when the person is moving around the aerobic board and steps on the boards when symbol is present, and just moves to the other side of the board when symbol is absent.

DIVISION (Reversed)

PART II

In Part II, the count of the positions is starting from the right side. We provide an example of division to show that the same result could be achieve when the count of positions starts from the right side.

Positions (Reversed)

Reverse order (Reversed)

The count starts from the right side and the order of position

increases to the left. The first position is the last on the right side.

Position 1 is 2^0 and equal to 1 (Reversed)

Position 2 is 2^1 and equal to 2 (Reversed)

Position 3 is 2^2 and equal to 4 (Reversed)

Position 4 is 2^3 and equal to 8 (Reversed)

Position 5 is 2^4 and equal to 16 (Reversed)

Position 6 is 2^5 and equal to 32 (Reversed)

Position 7 is 2^6 and equal to 64 (Reversed)

Position 8 is 2^7 and equal to 128 (Reversed)

Position 9 is 2^8 and equal to 256 (Reversed)

- -

Position n is $2^{(n-1)}$ and equals to $2^{(n-1)}$ times.

The reverse patterns used in psychoconduction

Decimal numbers-Alphabets-LITVIN'S CODE-LITVIN'S CODE

Using only crosses
Cross means Symbol present

Using crosses and zeroes
Cross means Symbol present
Zero means symbol Absent

		32	16	8	4	2	1		32	16	8	4	2	1
1	A						x	Or					0	x
2	B					x		Or					x	0
3	C					x	x	Or					x	x
4	D				X			Or				x	0	0
5	E				X		x	Or				x	0	x
6	F				X	x		Or				x	x	0
7	G				X	x	x	Or				x	x	x
8	H			X				Or			x	0	0	0
9	I			X			x	Or			x	0	0	x
10	J			X		x		Or			x	0	x	0
11	K			X		x	x	Or			x	0	x	x
12	L			X	X			Or			x	x	0	0
13	M			X	X		x	Or			x	x	0	x
14	N			X	X	x		Or			x	x	x	0
15	O			X	X	x	x	Or			x	x	x	x
16	P		x					Or		x	0	0	0	0
17	Q		x				x	Or		x	0	0	0	x
18	R		x			x		Or		x	0	0	x	0
19	S		x			x	x	Or		x	0	0	x	x
20	T		x		X			Or		x	0	x	0	0
21	U		x		X		x	Or		x	0	x	0	x
22	V		x		X	x		Or		x	0	x	x	0

		32	16	8	4	2	1		32	16	8	4	2	1
23	W		x		X	x	x	Or		x	0	x	x	x
24	X		x	X				Or		x	x	0	0	0
25	Y		x	X			x	Or		x	x	0	0	x
26	Z		x	X		x		Or		x	x	0	x	0
27			x	X		x	x	Or		x	x	0	x	x
28			x	X	X			Or		x	x	x	0	0
29			x	X	X		x	Or		x	x	x	0	x
30			x	X	X	x		Or		x	x	x	x	0
31			x	X	X	x	x	Or		x	x	x	x	x
32		X						Or	x	0	0	0	0	0
33		X					x	Or	x	0	0	0	0	x
34		X				x		Or	x	0	0	0	x	0
35		X				x	x	Or	x	0	0	0	x	x
36		X			X			Or	x	0	0	x	0	0
37		X			X		x	Or	x	0	0	x	0	x
38		X			X	x		Or	x	0	0	x	x	0
39		X			X	x	x	Or	x	0	0	x	x	x
40		X		X				Or	x	0	x	0	0	0

Example 5.1R (reverse)

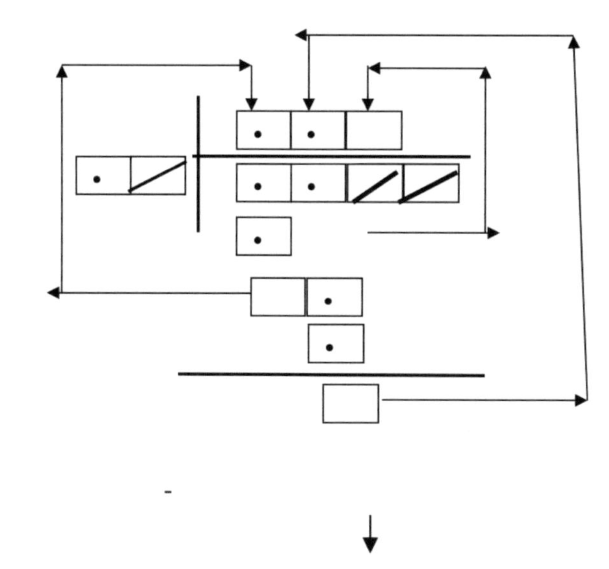

First number is 12 “x x 0 0”

Second number is 2 “x 0”

Answer is 6 “x x 0”

6
2 | 12

Both numbers have the empty space on the last position. We compare this to the digit “0” of the regular division. In the regular division, when we eliminate 0 at the end of the number, the number becomes ten times smaller. In binary arithmetic when we eliminate the last position from both numbers and then both numbers become twice lower.

New first number is 6 “x x o”

New second number is 1 “x”

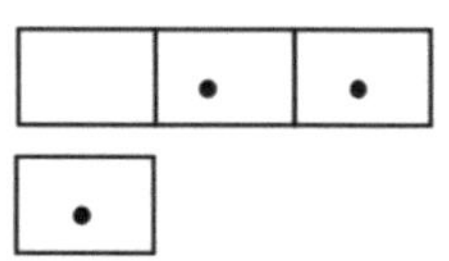

We know, by dividing any number into 1, we will have the same number we started with as the answer. In this example we just want to demonstrate below how Litvin’s Code applies to this division.

New first number is 3 “x x”

The second number is 1 “x”

Now we have two steps for subtraction. In the first step we subtract the divisor by the new first number and as a result of subtraction there is an empty space. When we get an empty space as a result of subtraction we fill up the corresponding position with a symbol (dot) on the answer of the division. In step two, we also subtract the divisor by the new first number and in the result of subtraction is an empty space. As we mentioned before when we have an empty space as a result of the subtraction we fill up the corresponding position with a symbol (dot) on the answer of the division. We also eliminate the empty position from the first number, and bringing the last empty position to the answer. When the dividend has the last empty position and the last position of the divisor is filled up with a symbol (dot) we are moving the last empty position from dividend to the answer of the division and eliminating the empty position on the dividend.

In Summary:

- When we divide 12 by 2 we have on the beginning of both numbers one empty position and we eliminate this position from both numbers and the numbers became 6 and 1.

- When in division we subtract 1 from 3 we have an empty position of the result of subtraction and fill up with a symbol the next corresponding position on the answer of the division. The numbers became 1 and 1. Since number 6 has an empty last position and the divisor has no empty spaces then we move the empty space in the dividend to the answer. When we do division using number 6, we are moving the last empty position to the answer of the division and the numbers become 3 and 1.

Example 5.2R (reverse)

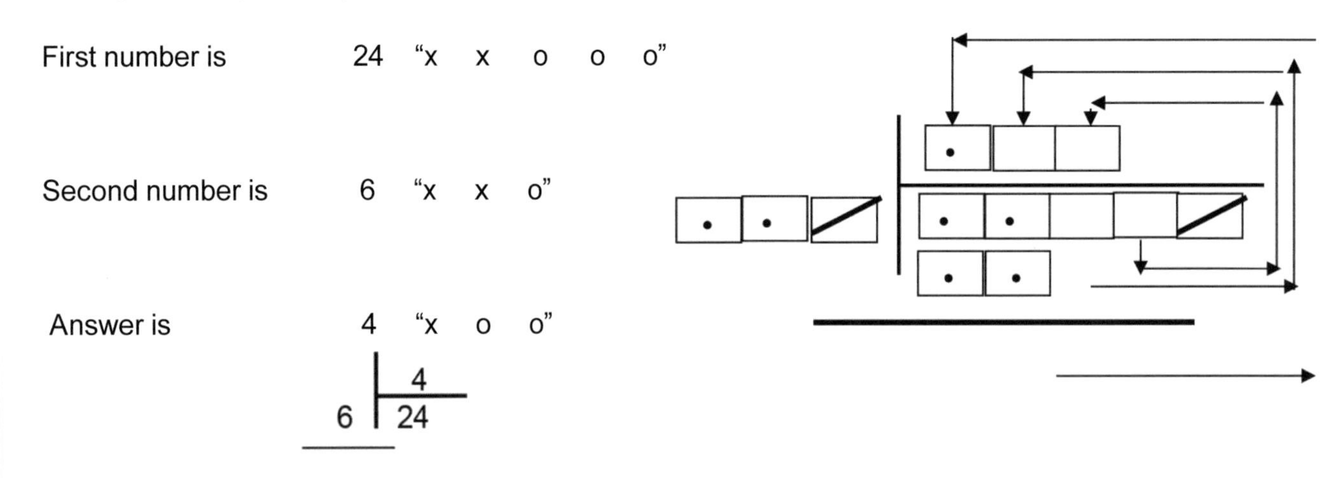

Example 5.3R (reverse)

First number is 65 "x o o o o o x"

Second number is 5 "x o x"

Answer is 13 "x x x 0 x"

$$5 \overline{)65}^{\,13}$$

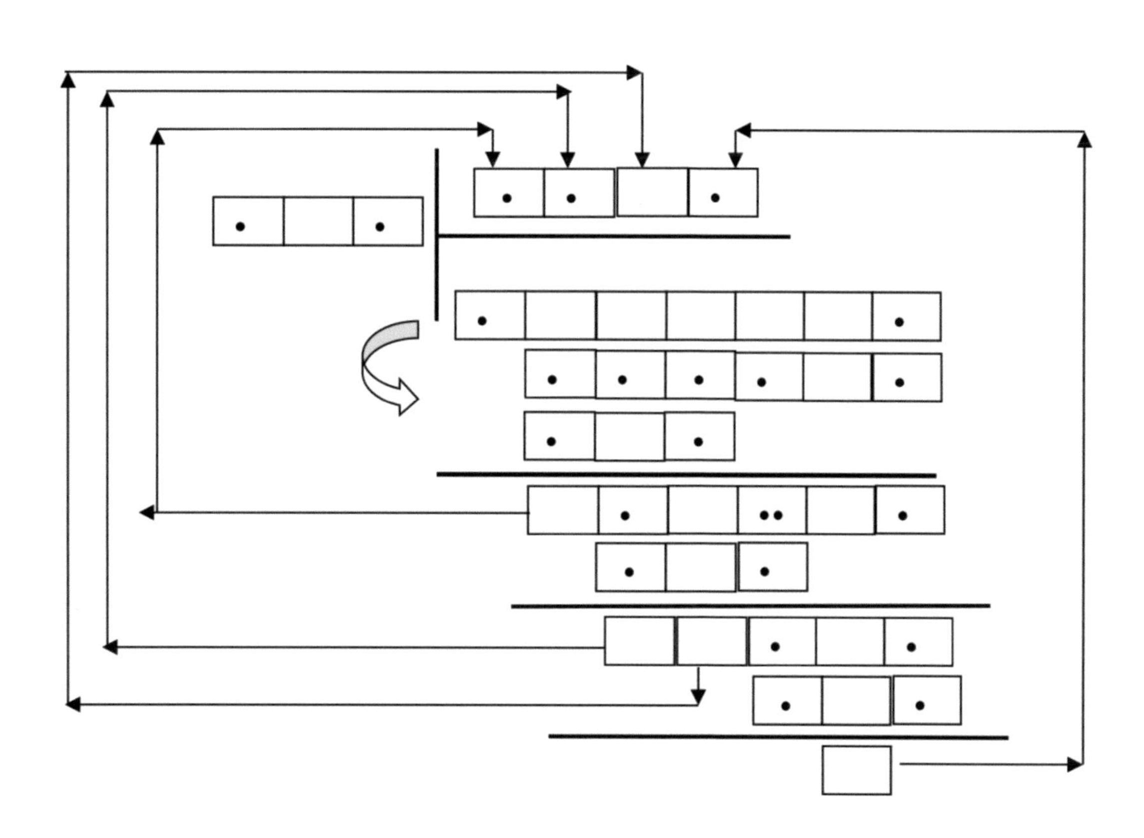

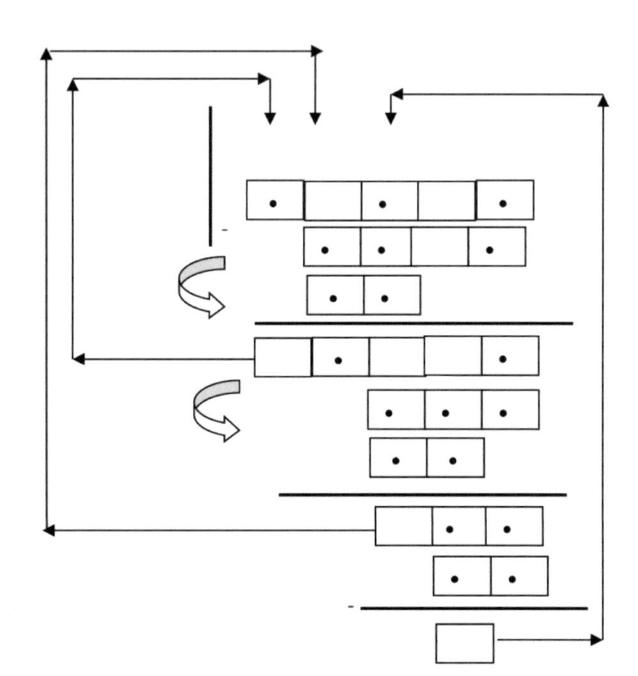

Example 5.4R (reverse)

First number is 21 “x o x o x”

Second number is 3 “x x”

Answer is 7 “x x x”

$$3 \overline{)21}\; 7$$

0

Example 5.5R (reverse)

First number is 65 “x o o o o o x”

Second number is 13 “x x o x”

Answer is 5 “x o x”

$$13 \overline{)65}\; 5$$

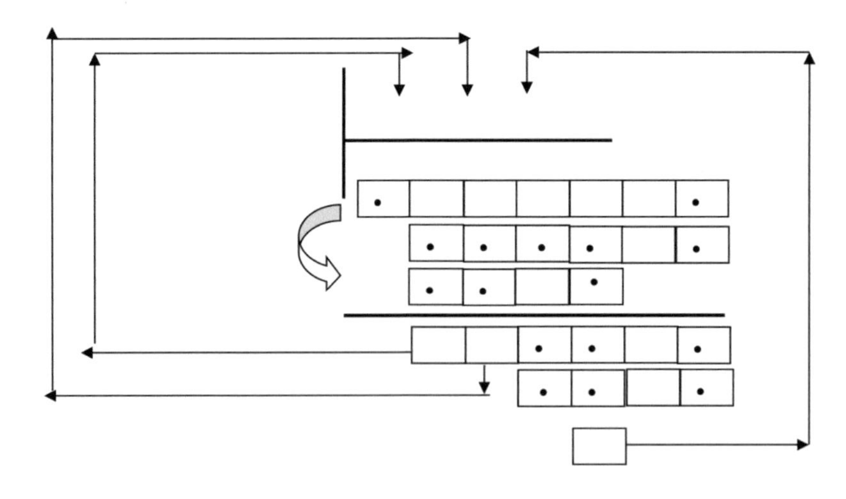

Example 5.6R (reverse)

First number is 18 "x o o x o"

Second number is 3 "x x"

Answer is 6 "x x o"

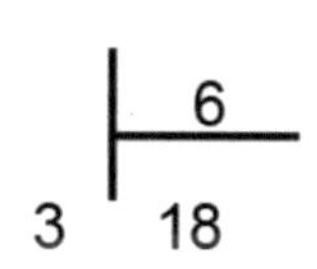

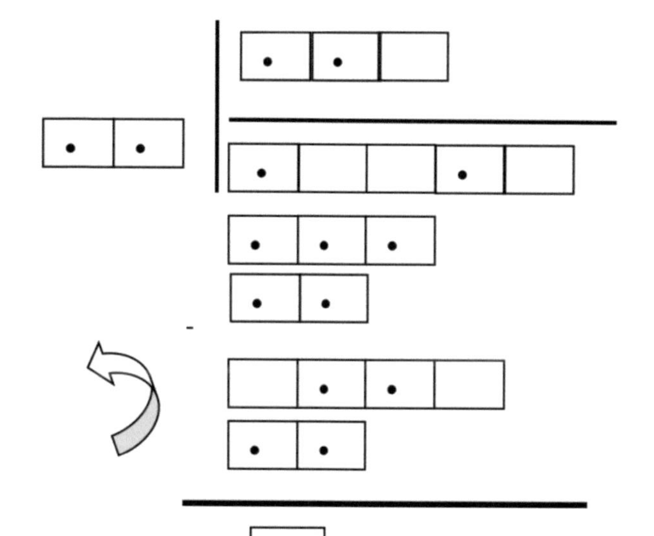

Example 5.7R (reverse)

First number is 24 "x x o o o"

Second number is 6 "x x o"

Answer is 4 "x o o"

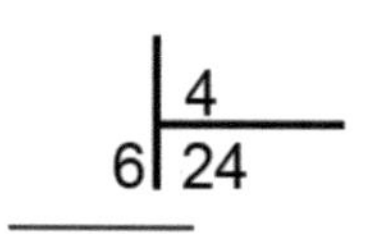

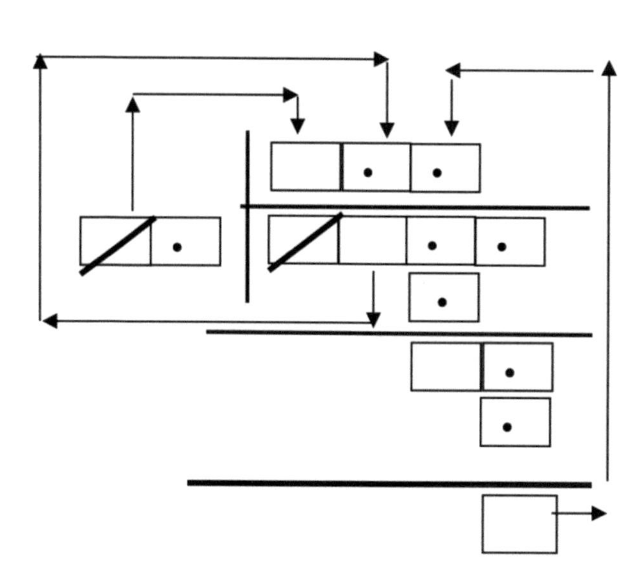

Disclosure

Psychoconduction creates an opportunity for the brain to use the simple cells to function as complex cells without any physiological and pharmacological intervention and is only using the psychological training. Psychoconduction provides training of simple cells by using the different patterns of simple symbols, which become more complicated with advance training. It also facilitates the congruent release of chemicals in different parts of the brain in the reaction to audio, video, kinesthetic, and olfactory stimuli. The training includes the processing of simple symbol using the same patterns of the simple symbol but as different stimuli to process in different areas of the brain. Some preliminary trails have shown the alarming incongruence in the release of the chemicals in different parts of the brain on the responses to the same patterns of audio, visual, kinesthetic or olfactory stimuli.

Many of our responses are either excessive or inadequate and they can be the common cause of mental impairments. The psychological therapy based on the principle of psychoconduction is addressing many psychological problems. Psychoconduction is presently only one existing method, which restructures the brain's function and creates new functions for brain cells by using psychological interventions. By using psychoconduction we calibrate and regulating appropriate release of chemicals in different areas of the brain without using any medications.

The psychoconduction helps if we have deficits and impairments in present adaptive functioning, social interactions and deficit in social and emotional reciprocity, helps to increase the effectiveness in communication and increases an ability to control our emotions. With psychoconduction we can increase the reading achievement and writing skills and spelling as well. The psychoconduction can significantly increase our motor coordination and break the patterns of stereotyped and repetitive motor mannerisms. Psychoconduction increases the attention to details, following of instructions and sustains maximum mental effort. Psychoconduction helps people with the difficulty with receptive and expressive language deficit to acquire new skills. Psychoconduction helps people organizing and planning activities to keep focused on the task. Psychoconduction helps people with anxious or dysphoric mood to stabilize, and helps to receive congruent information from kinesthetic, audio, visual and olfactory stimuli.

Dyslexia

In *dyslexia*, when the information could not get to the purposed complex cells for processing, the simple cells could help to find the right address. It is similar to the Post Office processing of mail, when the address on a letter is wrong, but the mailman recognizes the familiar name and is able to deliver it to the right address. When in the brain the address of information is lost because of difficulties with complex cells, we stimulate a brain with a simple symbol, and getting the needed information to the right place through the simple cells. When a brain gets familiar with the complex sequences of the simple symbols it establish the correct address and deliver them to the right place for processing. The correct address was found because a brain was able to recognize the sequences of the simple symbols. We create the references by translating the sequences from a one mode of expression to another.

After the brain has the established references and correct purposed address, the next step is the translation of the information back to the complex symbols the decimal digits and letters. We bypass the damaged complex cells, which contain the wrong addresses and with the help of the simple cells deliver the information to the right place.

Advantages of psychoconduction

- The psychoconduction allow the brain to process the audio, visual, kinesthetic information as one harmonious system. It provides congruency and equilibrium by transferring information to the different parts of the brain. We use the brain's auditory centers to process the auditory stimulus and etc. We can change the modes of expression by using translation. We begin with visual or audio patterns of simple symbol and then translate symbol from visual or audio to different expressions of the same patterns. The result of the training is the congruency in the processing of the patterns of the simple symbol and equilibrium achieved in the responses.

- Psychoconduction relieves stress of the overreaction to the stimuli and increasing the chances of correct estimation of the information. It gives us a chance to respond in proper way to the stimulus. It allows the brain to choose the most appropriate reaction on processed information. The training in psychoconduction helps us to understand that the reactions are not always physical and the verbal responses can be as effective as physical.

- The psychoconduction is translating the patterns of the simple symbol in a different way to express the same patterns and triggers the congruent release of the chemicals in different parts of the brain. The calculations and linguistics using Litvin's Code is the perfect way of increasing our academic performance, when we have difficulty recognizing digits and letters. It makes the calculation a fun process. Litvin's Code is a tool to increase memory and intellect. When we suffer from brain injuries, by using Litvin's Code we can achieve the higher level of functioning, which can be possible for our injury.

- The psychoconduction is a comprehensive system and unite the audio, visual, kinesthetic and olfactory information by using the same patterns of simple symbol. In psychoconduction the brain is calibrated against different areas. With psychoconduction we train the brain to provide appropriate responses to the stimuli. When the different parts of the brain process the same pattern, the brain produces different types but congruent amounts of chemicals.

- Psychoconduction improves attention; concentration, and planning ability by emphasizing the importance of the positions and recognition of the simple symbol.

- Psychoconduction effectively helps us when we have emotional instability, attention deficit or have difficulty recognizing the shape and meaning of the letters and digits, and could not easily write the letters and digits. It is a powerful tool to train the brain to do the intellectual work without being limited by injury or defects.

- Psychoconduction uses references, which are transferable to audio, visual, kinesthetic and olfactory information. To each letter in Litvin's Code is an assigned binary number. It is important to remember that the binary numbers are also used for the references to each letter and can assist us in moments of confusion. The references help us when we suffer from the brain impairments to find the letters, because they are always in an ascending order.

- With psychoconduction it is easy to memorize information and understand the relationship between letters and digits. It enhances the intellectual capacities and increases memory. It provides a chance to use the brain in the most efficient way without distractions of physical limitation, and helps to focus on intellectual process.

The evaluation of a brain's difficulty to process information.

The people with damaged or wrongly directed cells on one of the modes of expression the first stimulate the functional cells to build the base of references, and understand the hierarchy of symbols. We translate the stimuli to the damaged mode. When the visual complex cells damaged but kinesthetic and audio complex cells work, we process the information by using translation of kinesthetic sequences of simple movements to the audio symbols (knock, double knock) and then to the kinesthetic (clench right hand, clench left hand) and then to the visual sequences with the crosses and zeros. We stimulate the most functional parts of the brain by developing the references for data by using appropriate simple stimuli and then we translate the sequences by using the simple stimuli, which correspond to the mode of expression of the less functional part of the brain. We use the patterns, which get more complicated, for translation of the sequences of the simple symbol to stimulate the brain's cells. We use the simple cells to help or replace the complex cells.

We indirectly assess the damages to the complex cells by severity of the problem, measured by the neuro-psychological testing. The many neuro-psychological tests evaluate the brain activities, and indirectly deal with damaged complex cells, which might be completely or partially damaged, or just incorrectly addressed. There are many tests with different approaches in measuring the brain impairments. These tests provide the standardized testing results in standard deviations from the mean, which is the normal brain activity. The result of those tests cannot say approximately how many cells are damaged. They only can tell us how severe are the impairments in a brain functioning, which is the same as saying that complex cells are completely or partially damaged, or are incorrectly addressed. For the completely damaged cells the degree of severity of the damages is too high and the complex cells are not functioning as they supposed. The neuro-psychological testing assess and provide the standardize measurements of the degree of severity of damaged cells.

<u>The completely and severely damaged, partially damaged, or incorrectly addressed complex cell are assessed by secondary effect with standardized neuro-psychological tests and progress is measured by the same tests.</u>

We assess damages to the complex cells by the secondary effect of the damages. There are many different neuro-psychological tests evaluating the brain impairments, which in a certain degree are the effects of the damaged complex cells. They have the standardized appraisal of damages. The subject in the visual test was provided with a picture of a simple figure, which later became more complex. We asked him to copy this figure and to recognize a figure from others similar in shape. A subject with completely and severely damaged complex cells has difficulty to make a copy of the provided figure, even it is a simple one. Our subject cannot mark out the provided figure from similar others. A subject without damaged vision and blindness show the severe impairments by the criterias of the particular test. We assumed that he has the severely damaged or completely damaged complex cells, which process information on visual mode. If a subject shows moderate impairments on neuro-psychological tests then we presume that he has partially damaged complex cells. If a subject demonstrate the reversibility of the figure or guess and has difficulty recognizing the figure, then, if the vision is not damaged, we believe that the visual complex cells are incorrectly addressed. There are many neuro-psychological and educational tests that by answers can identify the partially damaged, severely damaged and incorrectly addressed visual cells. The sounds in audio tests, when deafness is not a problem, require of the subject repeating or recognizing a sound, and according to severity of impairment we assess the damages, which are complete damages, or partial, or incorrectly addressed cells. In the kinesthetic test, when the other medical problems beside the brain impairments are ruled out, the object is a movement, which subjects need to copy or recognize, and again by the test result we assess the severity of damages and incorrectly addressed complex cells.

The difficulty to perform in a standardized educational test of spelling writing, reading and mathematics is the secondary effect incorrect functioning of complex cells and the progress of the treatment is measured by the same tests.

The first educational test will identify the area of impairment of ability. After training in psychoconduction we repeat only the part of the educational test, which deal with a particular impairment. The treatment of translation deals with impairments, which are the digits, letters and spelling. The secondary effect of incorrectly functioning complex cells is a deficit in attention and is identified by standardized attention test.

In psychologically disturbed subjects are a deficit in attention, memory, and perceptual abnormality, which included audio, visual, and kinesthetic, and is the secondary effect of the non- functioning properly complex cells.

The standardized test will identify different diagnoses of psychological disturbances and after the ten sessions the self report of subjects will show improvement in perception.

The correction of a brain's impairment.

The correction of the complex cells is achieved by stimulating the simple cells with simple stimuli, which could be visual, audio, kinesthetic, olfactory and tactile and then translating the stimuli from one way of expression to others. The curative process is in translation from one mode of expression to another. We have several steps in our treatment, which could be easily be repeated by anyone. In the treatment, the main thing is the encouragement of translation of the patterns from one mode of expression to another. The complicity of the patterns is dependent on the impairment and the age of the person. For little children all work could be done in step 1.

Step 1 is an introduction of different simple symbols and examples on more functioning modes of expressions to build the reference base, to translate them to less functioning mode of expression and to prepare for introduction of more complicated patterns of simple symbols. We did not want to limit creativity of practitioners to use their own symbols.

Step 2 is the connection of combination of symbols on different modes of expression to the numbers. The symbols could be introduced in assenting or dissenting order and could be introduced from right to left or opposite. The purpose is correctly translating the symbols to different modes of expression.

Step 3 is introduction of the symbols for addition and the reverse addition, when the patterns of simple symbols are in opposite directions. The purpose is the translation of the symbols from one mode of expression to another is connecting them to the decimal numbers.

Step 4 is introduction of the symbols of the subtraction and reverse subtraction, which is in different directions, only to translate simple symbols from one mode of expression to another.

Step 5 is introducing the symbols for the multiplication and reverse multiplication for purpose of translating.

Step 6 is introducing the symbols for the division and the reverse division only for the purpose of translation.

Step 7 is introducing the symbols for letters and reverse letters for purpose of translation.

Step 8 is introducing the symbols for the words for translation.

Step 9 is introducing the symbols to the sentences for translation.

Step 10 is connection symbols on different modes of expression to the decimal mathematic and letters of the alphabet, words and sentences.

When we have complicatedness in visual perception, we need to use our not damaged kinesthetic and audio ability to build the base of references and to translate the kinesthetic and audio stimuli to visual. When our cells damage or wrongly directed and we have difficulty in audio or kinesthetic perception, to regain our abilities we use translation from the more functional mode of expression to a less functional one. Note, we can use any of symbols accepted by the simple cells will work. The simple visual stimulus can be a dot, x, 0 and many other symbols, which could be accepted by the simple visual cell. For the audio stimuli we use a simple sound (knock), for kinesthetic we are using the simple movement (clenching of hand); for tactile we touch parts of the body, which can be touched on the left or right hands, or shoulders, and for olfactory stimulation we use the simple odors (coffee, soaps, different perfumes).

We translate patterns of the simple stimuli from one mode of expression to another. When we have damaged or wrongly directed cells on one mode of the expression, we use the more functional mode of expression, where we feel comfortable and then translating the patterns to the impaired mode of expressions. When we suffer from *dyslexia*, we begin stimulation process with kinesthetic stimulation and building the base of references to process complex information by patterns of simple symbols. Later, the information from kinesthetic modes we transfer to audio and visual modes.

The person with *dyspraxia* identifies the patterns in kinesthetic mode, building a reference base then translates the patterns to the audio and visual modes. The person with the *dysphasia* starts treatment with his most functional mode by identifying and translating the patterns of corresponding simple stimuli to build the table of references then translates the patterns to various modes and back. The patient with *ADD* starts stimulation with more functional modes of expression, builds references, and translates to the weakest one with the help of the acquired before the data base of the collected references. With the patient with *ADHD*, *Dementia* we starts identification of patterns by gentle touch of the left and right hands or shoulders, or fingers, etc. in the provided sequence. In any stimulation we stimulate the most functional parts of our brain by using the corresponding simple stimuli to reach those parts. Later, by developing the references data the patients translate the patterns to the less functional part of the brain.

5a. Because we could not say exactly how many simple cells the brain needs to stimulate and how many complex cells need to be replaced, only by increase in performance and by the results of testing we see that the process is working. When the complex cells are not able to process information, we introduce the stimulation of the brain by simple symbols in the patterns, which became more complex, and as a result we see that the brain regains the ability of processing the complex information.

When before the training the complex cells were not able to process the information, we know that patterns get processed by the help of the simple cells. We can see by the speed of improvement that the complex cells work again. When the complex cells bring their share to the process of the information, then the process is becoming much faster, because one complex cell could process as much information as many of the simple cells.

By translation of patterns we empower the simple cells to facilitate or replace a complex cell and to help the brain to process complex information, and also to guide the stimuli to find the right address.

5b. The goal is to process the complex information with the simple symbols with the patterns, which contain the complex information. In many instances the address where information is processed, is lost or unclear and our brain cannot process information correctly. It happens because the address is too complex and the cells which are responsible for the correct address could be completely damaged, partially damaged or keeping incorrect addresses.

The simple symbol can easy find the area, where information can be processed, because the addresses are easy to find. The different simple shapes or shades, or simple sound or simple movement is easier to process that the letters, paintings, music and dances. The dyslexic patient always recognizes a shape of a dot, a patient with dysphasia understands a knock, the patient with dyspraxia is clenching his hands, and a patient with dementia understands the touching of the hand or shoulder. The visual simple stimuli is processed in visual part of the brain by simple cells, and the simple audio, kinesthetic, tactile and olfactory stimuli are processed by simple cells in the corresponding parts of the brain. The training is the stimulation of the simple cells with the coded simple symbols, which contain the information, which is getting increased in complicity. By using this approach we use the simple cells to process the complex information and stimulate, if possible, the complex cells. By translation of the patterns of the simple symbols, which are recognized by simple cells, we help the damaged complex cells to regain the lost functioning. During the process of translation we create the base of references from the most functional mode of expression to a damaged one. If the lost function is not possible to regain by the complex cells, then the simple cells replace the complex cells.

We use for the kinesthetic representation and translation of the patterns of simple symbols the clenching of the right and left hand. For the audio representation and translation we use knock and double knock. For the visual representation and translation we use the combination of presence of symbol and the absence of symbol. All symbols can be introduced from the left to right sight and opposite. The symbols could be introduced from up to down, like in Chinese, and opposite. We have all personal freedom to introduce the patterns of simple symbols. We use the patterns of stimulation, which we call Litvin's Code, and translate them between different modes of expressions.

Litvin's Code is the patterns of simple symbols, which could be translated from one mode expression to another, kinesthetic, visual, audio, tactile, and olfactory to balance the brain processing of information. We create the complex patterns by using the simple symbols. For a simple symbol in visual mode we can use many symbols, which include empty space, zero, dot, cross, column, etc. For simple symbol in kinesthetic mode we could use clenching of the right hand, clenching of the left hand,clamping the left hand,clamping the right hand, extending the right hand, extending the left hand, stepping on the board from the side, passing the side of the board, etc. For a simple symbol in audio mode we could use knock, double knock, etc. For a tactile mode the simple symbol could be touch of the left shoulder, touch of the right shoulder, etc. For olfactory mode the simple symbols can be the smell of soup, smell of coffee, etc.

For the patterns of translation we use the complex sequences of the simple symbols, which represent digit and letters, do calculations, including addition, subtraction, division, multiplication, writing words and sentences.

The goal is to encourage patients to translate the patterns between different modes of expression in organize manors and to create references, to provide smooth transition to regular numbers and letters, which, for our patients, are decimal numbers and Latin letters. The all different methods work in the encouragement of translation of the patterns, which are similar to Litvin's code, between different modes of expressions.

Printed in the United States
By Bookmasters